Achim Müller
Andreas Dress
Fritz Vögtle (Eds.)

From Simplicity to Complexity in Chemistry – and Beyond

Part I

Achim Müller
Andreas Dress
Fritz Vögtle
(Eds.)

From Simplicity to Complexity in Chemistry – and Beyond

Part I

Produced by Lengericher Handelsdruckerei, Lengerich

ISBN 978-3-642-49370-6   ISBN 978-3-642-49368-3 (eBook)
DOI 10.1007/978-3-642-49368-3

Preface

An impression of the ubiquity of chemistry becomes manifest by means of a critical phenomenological examination of the material and biological world, or rather, of inanimate and animate appearances in *Nature*. The obvious paradox is that, on the one hand, we try to understand *Nature* using our present modern methods of chemical research and, on the other hand, would like to profit and learn from her secret intriguing pathways in "being and becoming". We strive to acquire possession of her (intrinsic) chemical procedures, try to mimic and decipher them in order to optimize *our* methods of inquiry so as, subsequently, to facilitate our ultimate goal, the creation of more and even more appropriate materials. (But we should constantly bear in mind that we, ourselves, form a part of this *"Nature"*.)

All biological processes are essentially concerned with complex systems and structures. The problem of decoding *Nature's* different pathways from simple to more complex systems, demands the collaboration of so many different research fields, chemistry, physics, biochemistry, biophysics, genetics and materials science, just to mention a few. The challenge of attempting to solve the problems concerning more complex structures lies, in the first instance, in the fascinating field of supramolecular chemistry (as opposed to single molecular chemistry) and demands the interdisciplinary exploration of current developments in the whole area of natural science. Through the astounding progress made in this field, the synthetic chemist, nowadays, is not forced only to fiddle around with functional groups based on molecules with orders of magnitude smaller than proteins. He is in a position to set his own goal and create large molecular assemblies without going through the difficult task of forming covalent bonds one by one.

The aim of this book is to stimulate even more the interest in multi-, inter-, and transdisciplinary aspects of chemistry and science in general and to provide the reader with up-to-date information (reviews) on some of the most exciting relevant topics of the present day. They range from inorganic, bioinorganic, bioorganic, and supramolecular chemistry (including crystal growth, metal clusters, and topology) to catalysis, biochemistry and even to some relevant philosophical aspects.

Some specific mechanisms of *large-scale molecular interactions* and their consequences for (supra)molecular geometry and architecture are treated by *C. A. Hunter*, by *G. R. Newkome* et al., by *M. T. Pope*, by *M. Lahav* et al. and by *J. F. Stoddart* et al.. *C. A. Hunter* describes the design of synthetic receptor molecules for quinone as guest and reports on the template synthesis of new catenated structures based on specific hydrogen bonding patterns and the orthogonal arrangement of the reaction partners. *G. R. Newkome* refers to molecules of the "dendrimer" type not only by reaching out to high generations but also by incorporating functions allowing guest nestling and chemical reactions inside their hollow space, whereas *M. T. Pope* deals with self-assembly reactions leading to high-nuclearity polyoxometallates starting from simple entities. *M. Lahav's* studies on crystallite formation and crystal growth lead likewise to an insight with applications, particularly emphasizing enantiomer recognition at the crystal interface, enantiomer separation, and the process of shaping selected crystal surfaces by addition of tailored additives. How to synthesize topologically new architectures and the construction of molecular devices therewith are discussed by *J. F. Stoddart* (oligo-catenanes and -rotaxanes).

The enigmatic emergent magnetic and electronic properties of mesoscopic materials and their amazing technological potentiality are analyzed by *G. Schmid*, by *L. J. de Jongh*, and by *D. Gatteschi* et al..

Three case-studies concerning the use of large-scale molecular interaction in the regulation of fundamental biochemical processes are presented by *H. Ringsdorf* et al., by *W. Saenger* et al., and by *C. K. Biebricher*. *H. Ringsdorf* studies methods to control – via structure and composition of designed lipid molecules – the recognition dynamics and catalytic reactions at interfaces of synthetic and of natural membranes. *W. Saenger* reports on the specific recognition, binding, and bending of DNA sequences by two distinct proteins, and *C. K. Biebricher* presents experimental and theoretical studies on how such interactions and the resulting reaction kinetics can bring about molecular evolution through selective amplification and repression.

On a more theoretical level, *E. Szathmáry* discusses chemical pathways which – starting with a primordial RNA-soup and combining chemical recognition and self-assembly processes in primordial molecular metabolism – could have led to the origin of the genetic code, and *H. Herzel* et al. present cunning statistical methods to detect long-range correlation in the biopolymers used presently to encode basic biological information – an indispensable prerequisite for understanding the molecular organization of life processes.

Finally, all these different topics are complemented by three articles (two in German with English abstracts, one in English) which discuss some related philosophical or general features. *A. Müller* and *H. Hörz* report on philosophical aspects of chemistry *(Its essence: universality and continuity of change)* emphasizing the importance of mutual dialogue between philosophers and chemists, and *N. Luhmann* describes his sociological *system theory* suggesting that system theory, in general, may be a method of bridging the gap between the natural sciences and the humanities, a viewpoint that was criticized in Snow's famous book *"Two Cultures"*. Obviously, it is not necessary to reduce cultural evolution to biological, chemical or physical evolution in order to apply system theory interdisciplinarily. Contrary to any reductionistic kind of naturalism and physicalism the complex system approach bears relationships with regard to the characteristic intentional features of human society. In order to characterize our world more fully it is necessary to consider also its sociological aspects. Concepts of social sciences have often been influenced by theories of natural sciences (c.f. T. Hobbes's famous book *"Leviathan"*, which describes the state as a machine and the citizen as a cog-wheel). The fundamental question here should be: How can a knowledge of *simple systems*, acquired by the instruments of natural-science research, be transferred to (more) *complex systems* and be utilized to improve our understanding, even of society? (In this context we have to realize that the behaviour of a *system* is characterized (programmed) by the *system's laws* and the boundary conditions). In Chapter 1, which considers the range of molecular systems to more complex ones, some general aspects relevant to that problem and to the contents of this book have been outlined by *A. Müller* and *K. Mainzer*. In particular, they discuss conservative systems with the creation of molecular complexity by stepwise self-assembly, complexity far from equilibrium, a system of high complexity like human society, and refer to the mathematical and methodological background of these models.

If we want to leave the prevailing dangerous one-way street specialization of today, we have to take general principles of interdisciplinary character into consideration and try to activate scientists from different fields to cooperate. *The Center for Interdisciplinary Research (ZiF), University of Bielefeld*, which supports our relevant work, has addressed itself to the task of stimulating and promoting collaborations of this kind.

Most contributions are written versions of lectures that were presented at a workshop *"From Simple Material (Chemical) Systems to Complex Ones"* held at the ZiF. The editors wish to thank the participants (see photo) for their efforts in providing manuscripts of their presentations and the ZiF authorities for their financial support and the use of facilities for this workshop. They are extremely grateful to Dipl.-Chem. Jens Osterodt for formatting and Dr. Angelika Schulz from the Vieweg-Verlag, Wiesbaden, for acting as a lector.

A. Müller
A. Dress
F. Vögtle

1: Bernt Krebs, 2: Günter Schmid, 3: Christopher Hunter, 4: Philip C. H. Mitchell, 5: Peter Janich, 6: Christof K. Biebricher, 7: George R. Newkome, 8: John Fraser Stoddart, 9: Ernst Ruch, 10: Meir Lahav, 11: Jean Pierre Launay, 12: Fritz Vögtle, 13: Jean-Marie Lehn, 14: Andreas Dress, 15: Helmut Ringsdorf, 16: Eberhard Neumann, 17: Achim Müller, 18: Eörs Szathmáry, 19: Michael T. Pope, 20: Dante Gatteschi, 21: Olivier Kahn, 22: Larissa Dloczik (not shown above: Werner Ebeling, L. Jos de Jongh, Niklas Luhmann, Wolfram Saenger)

Contents

Organization, Information and Codes in Biological Systems

Organization of Chemical Systems

From Molecules to Materials

Some Reflections on Simple Systems and More Complex Ones

1 From Molecular *Systems* to More Complex Ones Some Preliminary Ideas

Achim Müller and Klaus Mainzer

*"I can hardly doubt
that when we have some control of the arrangement of things on a small scale
we will get an enormously greater range of possible properties
that substances can have."*
R. P. Feynman [1]

1.1 General Aspects

Complexity is a modern science subject, sometimes quite difficult to define exactly or to detail accurately its boundaries. Over the last few decades, astonishing progress has been made in this field and, finally, an at least relatively unified formulation concerning dissipative systems has gained acceptance.

We recognize complex processes in the evolution of life and in human society. We accept physico-chemical and algorithmic complexity and the existence of archetypes in dissipative systems. But we have to realize that a deeper understanding of many processes – in particular those taking place in living organisms – demands also an insight into the field of "molecular complexity" and, hence, that of equilibrium or near-equilibrium systems, the precise definition of which has still to be given.

One of the most obvious and likewise most intriguing basic facts to be considered is the overwhelming variety of structures that – due to *combinatorial explosion* – can be formally built from only a (very) limited number of simple "building blocks" according to a restricted number of straight-forward "matching rules". On the one hand, combinatorial theory is well-equipped and pleasant to live with. Correspondingly, it is possible to some extent to explore, handle, and use combinatorial explosion on the theoretical and practical-experimental level. On the other hand, the reductionist approach in the natural sciences has for a long time focused rather on separating matter into its elementary building blocks than on studying systematically the phenomena resulting from the cooperative behaviour of these blocks when put together to form higher-order structures, a method chemists may have to get accustomed to in the future in order to understand complex structures [2].

Independent progress in many different fields – from algorithmic theory in mathematics and computer science via physics and chemistry to materials science and the biosciences – has made it possible and, hence, compels us to try to bridge the gap between the micro- and the macro-level from a structural (as opposed to a purely statistical, averaging) point of view and to address questions such as:

1. *What exactly is coded* in the ingredients of matter (elementary particles, atoms, simple molecular building blocks) with respect to the emergence of complex systems and complex behaviour? The question could, indeed, be based on the assumption that a *"creatio ex nihilo"* is not possible!

2. During the course of evolution, how and why did *Nature* form just those *complicated* and – in most cases – *optimally functioning, perfectionated molecular systems* we are familiar with? Are they (or at least some of them) appropriate models for the design of *molecular materials* exhibiting all sorts of properties and serving many specific needs?

3. While, on the one hand, a simple reductionist description of complex systems in terms of less complex ones is not always meaningful, how significant are, on the other hand, phenomena (properties) related to rather simple material systems within the context of *creating complex* (e.g., biological) *systems from simpler ones*?

4. In particular: Is it possible to find *relations which exist between supramolecular entities*, synthesized by chemists and formed by conservative self-organization or self-assembly processes, *and the most simple biological entities*? And how can we elucidate such relations and handle their consequences?

In any case, a precondition for any attempt to answer these questions is a sufficient understanding of the "Molecular World", including its propensities or potentialities [2, 3] (see also Chapter 15).

1.2   Complexity in *Systems* far from Equilibrium

The theory of nonlinear complex systems [4] has become a successful and widely used tool for studying problems in the natural sciences – from laser physics, quantum chaos, and meteorology to molecular modeling in chemistry and computer simulations of cell growth in biology. In recent years, these tools have been used also – at least in the form of "scientific metaphors" – to elucidate social, ecological, and political problems of mankind or aspects of the "working" of the human mind.

What is the secret behind the success of these sophisticated applications? The theory of nonlinear complex systems is not a special branch of physics, although some of its mathematical principles were discovered and first successfully applied within the context of problems posed by physics. Thus, it is not a kind of traditional "physicalism" which models the dynamics of lasers, ecological populations, or our social systems by means of similarly structured laws. Rather, nonlinear systems theory offers a useful and far-reaching justification for simple phenomenological models specifying only a few relevant parameters relating to the emergence of macroscopic phenomena via the nonlinear interactions of microscopic elements in complex systems.

The behaviour of single elements in large composite systems (atoms, molecules, etc.) with huge degrees of freedom can neither be forecast nor traced back. Therefore, in statistical mechanics, the deterministic description of single elements at the microscopic level is replaced by describing the evolution of probabilistic distributions. At critical threshold values, phase transitions are analyzed in terms of appropriate macrovariables – or "order parameters" – in combination with terms describing rapidly fluctuating random forces due to the influence of additional microvariables.

By now, it is generally accepted that this scenario, worked out originally for systems in thermal equilibrium, can also be used to describe the emergence of order in open dissipative systems far from thermal equilibrium (Landau, Prigogine, Thom, Haken, etc. [4]; for some details see Supplement 1.5). Dissipative self-organization means basically that the phase transition lies far from thermal equilibrium. Macroscopic patterns arise in that case according to, say, Haken's "slaving principle" from the nonlinear interactions of microscopic elements when the interaction of the dissipative ("open") system with its environment reaches some critical value, e.g., in the case of the Bénard convection. In a qualitative way, we may say that old structures become unstable and, finally, break down in response to a change of the control parameters, while new structures are achieved. In a more mathematical way, the macroscopic view of a complex system is described by the evolution equation of a global state vector where each component depends on space and time and where the components may mean the velocity components of a fluid, its temperature field, etc. At critical threshold values, formerly stable modes become unstable, while newly established modes are winning the competition in a situation of high fluctuations and become stable. These modes correspond to the order parameters which describe the collective behaviour of macroscopic systems.

Yet, we have to distinguish between phase transitions of open systems with the emergence of order *far from* thermal equilibrium and phase transitions of closed systems with the emergence of structure *in* thermal equilibrium. Phase transitions in thermal equilibrium are sometimes called "conservative" self-organization or self-assembly (self-aggregation) processes creating ordered structures mostly, but not necessarily, with low energy. Most of the contributions to this book deal with such structures.

In the case of a special type of self-assembly process, a kind of slaving principle can also be observed: A template forces chemical fragments ("slaves"), like those described in Section 1.3, to link in a manner determined by the conductor (template) [5], whereby a well-defined order/structure is obtained. Of particular interest is the formation of a template from the fragments themselves [6].

1.3   Taking Complexity of Conservative *Systems* into Account and a Model *System* Demonstrating the Creation of Molecular Complexity by Stepwise Self-Assembly

A further reason for studying the emergence of structures in conservative systems can be given as follows: The theory of nonlinear complex systems offers a basic framework for gaining insight into the field of nonequilibrium complex systems but, in general, it does not adequately cover the requirements necessary for the adventurous challenge of understanding their specific architectures and, thus, must be supported by additional experimental and theoretical work. An examination of biological processes, for example those of a morphogenetic or, in particular, of an epigenetic nature, leads to the conclusion that here the complexity of molecular structures is deeply involved, and only through an incorporation of the instruments and devices of the relevant chemistry is it possible to uncover their secrets [*].

---

[*] Interesting in this context is that nonlinear *chemical*, dissipative mechanisms (distinguished from those of a *physical* origin) have been proposed as providing a possible underlying process for some aspects of biological self-organization and morphogenesis. Nonlinearities during the formation of microtubular solutions are reported to result in a chemical instability and bifurcation between pathways leading to macroscopically self-organized states of different morphology (Tabony, J. *Science*, **1994**, *264*, 245).

Complex molecular structures exhibit multi-functionality and are characterized by a correspondingly complex behaviour which does not necessarily comply with the most simple principles of mono-causality nor with those of a simple straight-forward cause-effect relationship. The field of genetics offers an appropriate example: One gene or gene product is often related not only to one, but to different characteristic phenotype patterns (as the corresponding gene product (protein) has often to fulfil several functions), a fact that is manifested even in (the genetics of) rather simple procaryotes.

Several nondissipative systems, which according to the definition of W. Ostwald are metastable, show complex behaviour. For example, due to their complex flexibility, *proteins* (or large biomolecules) are capable of adapting themselves not only to varying conditions but also to different functions demanded by their environment; the characteristics of *noncrystalline solids* (like glasses), as well as of *crystals grown under nonequilibrium conditions* (like snow crystals) are determined by their case history; *spin-glasses* exhibit complex magnetic behaviour [7]; *surfaces of solids* with their inhomogeneities or disorders * can, in principle, be used for storing information; *giant molecules (clusters)* may exhibit fluctuations of a structural type **. Within the novel field of supramolecular magnetochemistry [8], we can also anticipate *complex behaviour*, a fact which will require attention in the future when a unified and interdisciplinarily accepted definition of complex behaviour of conservative systems is to be formed.

But is elucidating complexity *as a whole* an unsolvable, inextricable problem, leading to some type of a *circulus vitiosus?* Or is it possible to create a theory, unifying the theories from all fields that would explain different types of self-organization processes and complexity in general? The key to disentangle these problems lies in the elucidation of the relation between conservative and dissipative systems, which in turn is only possible

---

*   Defects, in general – not only those related to the surface – affect the physical and chemical (e.g., catalytical) properties of a solid and play a role in its history. They form the basis of its possible complex behaviour.

**   Fluctuation – static or nonstatic, equilibrium or nonequilibrium – usually means the deviation of some quantity from its mean or most probable value. (They played a key role in the evolution.) Most of the quantities that might be interesting for study exhibit fluctuations, at least on a microscopic level. Fluctuations of macroscopic quantities manifest themselves in several ways. They may limit the precision of measurements of the mean value of the quantity, or vice versa, the identification of the fluctuation may be limited by the precision of the measurement. They are the cause of some familiar features of our surroundings, or they may cause spectacular effects, such as the critical opalescence and they play a key role in the nucleation phase of crystal growth (see Supplement 1.5). Fluctuations or their basic principles which are relevant for chemistry have never been discussed on a general basis, though they are very common – for example in the form of some characteristic properties of the very large metal clusters and colloids, for instance those discussed in this book (see also Sawada, S.; Sugano, S. *Structural Fluctuation of Au_{55} and Au_{147} Clusters*, in: Jena, P.; Khanna, S. N.; Rao, B. K. (eds.), *Physics and Chemistry of Finite Systems: From Clusters to Crystals*, Vol. 1, Kluwer, Dordrecht, The Netherlands, **1992**, 119).

through a clear identification of the relations between multi-functionality, deterministic dynamics, *and* stochastic dynamics (see also Chapter 15) *.

For a real understanding of phase transitions, we have to deal not only with the structure and function of elementary building blocks, but also with the properties which emerge in consequence of the complex organization which such simple entities may collectively yield when interacting cooperatively. And we have to realize that such emergent *high-level* properties are properties which – even though they can be exhibited by complex systems only and cannot be directly observed in their component parts when taken individually – are still amenable to scientific investigation.

These facts are generally accepted and easily recognized with respect to crystallographic symmetry; here, the mathematics describing and classifying the emerging structures (e.g., the 230 space groups) is readily available [9]. But the situation becomes more difficult when complex biological systems are to be investigated where no simple mathematical formalism yet exists to classify all global types of interaction patterns and where molecular complexity plays a key role: The behaviour of sufficiently large molecules like enzymes *in complex systems* can, as yet, not be predicted computationally nor can it simply be deduced from that of their (simple chemical) components.

Consequently, one of the most aspiring fields of research at present, offering challenging and promising perspectives for the future [2] is to learn experimentally and interpret theoretically how relevant global interaction patterns and the resulting high-level properties of complex systems emerge by using a stepwise procedure, to build ever more complex systems from simple constituents. This approach is used, in particular, in the field of *supramolecular chemistry* [10] (for further details see [5]) – a basic topic of this book – where some intrinsic propensities of material systems are investigated. By focusing on phenomena like non-covalent interactions or multiple weak attractive forces (especially in the case of molecular recognition, host/guest complexation as well as antigene-antibody

---

*   During cosmological, chemical, biological, as well as social and cultural evolution, *information* increased parallel to the generation of structures of higher complexity. The emergence of relevant information during the different stages of evolution is comparable with phase transitions during which structure forms from unordered systems (with concomitant entropy export). Although we can model certain collective features in natural and social sciences by the complex dynamics of phase transitions, we have to pay attention to important differences (see Section 1.4 and Chapter 15.4). In principle, any piece of information can be encoded by a sequence of zeros and ones, a so-called $\{0,1\}$-sequence. Its *(Kolmogorov)* complexity can thus be defined as the length of the minimal $\{0,1\}$-sequence in which all that is needed for its reconstruction is included (though, according to well-known undecidability theorems, there is in general no algorithm to check whether a given sequence with such a property is of minimal length). According to the broader definition by C. F. von Weizsäcker, *information* is a concept intended to provide a scale for measuring the amount of *form* encountered in a system, a structural unit, or any other information-carrying entity *("Information ist das Maß einer Menge von Form")*. There exists, of course, a great variety of other definitions of *information* which have been introduced within different theoretical contexts and which relate to different scientific disciplines. Philosophically speaking, a *qualitative* concept is needed which considers information to be a property neither of structure nor of function alone, but of that inseparable unit called *form* (εἶδος), which mediates between both (see also [2] and Chapter 15).

interactions), (template-directed) self-assembly, autocatalysis, artificial, and/or natural self-replication, nucleation, and control of crystal growth, supramolecular chemistry strives to elucidate strategies for making constructive use of specific large-scale molecular interactions, characteristic for *mesoscopic* molecular complexes and nanoscale architectures.

In order to understand more about related *potentialities* of material systems, we should systematically examine, in particular, self-assembly processes. *A system of genuine model character* [2, 5, 12], exhibiting a maximum of *potentiality* or *disposition* "within" the relevant solution, *contains very simple units with the shape of Platonic solids* – or chemically speaking, simple mononuclear oxoanions (see also Chapter 11 and [11]) – *as building blocks* from which an extremely wide spectrum of complex polynuclear clusters can be formed according to a type of unit construction. In this context, self-assembly or condensation processes can lead us to the fascinating area of mesoscopic molecular systems.

A significant step forward in this field could be achieved by controlling or directing the type of linkage of the above-mentioned fragments (units), for instance *by a template, in order to obtain larger systems and then proceeding accordingly to get even larger ones (with novel and perhaps unusual properties!) by linking the latter again, and so on. This is possible within the mentioned model system. Basically, we are dealing with a type of emergence* due to the generation of ever more complex systems. The concept of *emergence* should be based on a pragmatically restricted reductionism. The dialectic unit of *reduction and emergence* can be considered as a "guideline" when confronted with the task of examining processes which lead to more and more complex systems, starting with the most simple (chemical) ones [2].

Fundamental questions we have to ask are whether complex near-equilibrium systems were a necessary basis for the formation of dissipative structures during evolution and whether it is possible to create molecular complexity stepwise by a conservative growth process corresponding to the following schematic description [12]:

$$\mathbf{I} \quad \Rightarrow \quad \mathbf{III} \quad \Rightarrow \quad \mathbf{V} \quad \Rightarrow \quad \mathbf{VII} \; \textit{(2N-1)}$$
$$\Uparrow \qquad\quad \Uparrow \qquad\quad \Uparrow$$
$$\mathbf{2} \qquad\quad \mathbf{4} \qquad\quad \mathbf{6} \qquad\quad \textit{(2n)}$$

Here, the uneven Roman numerals, *2N-1,* represent a series of maturation steps of a molecular system in growth or development and the even Arabic numerals *2n* stand for ingredients of the solution which react only with the relevant "preliminary" or intermediate product, *2N-1*. The species *2n* can themselves be products of self-assembly processes. The target molecule at the "end" of the growth process would be formed by some kind of (near-equilibrium) symmetry breaking steps. The information it carries could, in principle, be transferred to other systems [12].

Very complex and large molecular systems are also produced in *Nature* by stepwise gene-directed processes. Of interest in this context is, for instance, the FeMo protein of nitrogenase, where the FeMo cofactor, a rather complicated nonequilibrium metal sulfur cluster, is also generated stepwise with the help of seven genes [13]. Basically, it is possible that rather complex – in our definition – multi-functional molecular systems with the size of small proteins can be generated from solutions of simple oxometallate building blocks according to the reaction scheme given above *or* by linking, stepwise as mentioned earlier, larger building blocks obtained by self-assembly processes. One example of a multi-functional molecular system is the giant polyoxometallate cluster [14] (see also M. T. Pope's article in Chapter 11) with 276 (non-hydrogen) atoms built up by three large transferable units.

An even larger polymolybdate cluster with more than 700 atoms(!) and a molecular weight of approx. 24 kDa, obtained by a self-assembly process, contains building blocks similar to those of the "smaller" 276 atom cluster and has the structure shown in Figure 1-1. *It is probably, as yet, the largest or at least one of the largest (inorganic) molecular systems formed by a self-assembly process and characterized by X-ray structure analysis* [15]. The previously-mentioned "smaller" cluster or a fragment of it *seems* to act here as a template, too (see Figure 1-1). Anyhow, the task of isolating these large anions within a crystal lattice presents a serious problem (see Supplement 1.5).

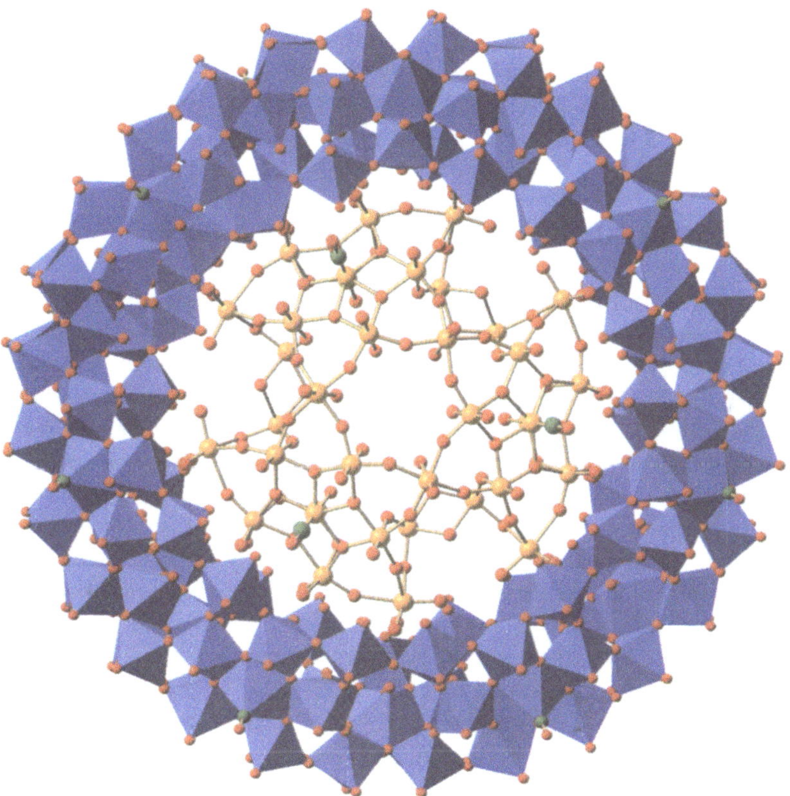

Figure 1-1   Representation of two giant polymolybdate clusters, the "smaller" one containing 276 atoms ($[Mo_{57}Fe_6(NO)_6O_{174}(OH)_3(H_2O)_{24}]^{15-}$) encircled (for demonstration) by the larger one ($[Mo_{154}(NO)_{14}O_{420}(OH)_{28}(H_2O)_{70}]^{(25\pm5)-}$; polyhedral representation) containing more than 700 atoms [14, 15] [*].

The large cluster anions can be called *complex* as they show complex behaviour or multi-functional properties:

---

[*]   First characterization of the mysterious "molybdenum blue".

(1) They contain cavities with nanometer-sized architectural features and, thus, can act as large hosts. (2) Their large surfaces resemble those of solids. (3) Paramagnetic centres can be deliberately placed stepwise on well-defined sites at the surface and it is possible to tune the magnetic properties (or the exchange coupling for molecular antiferromagnets and ferrimagnets). Under formal consideration, the species show (static) fluctuations of the stoichiometry. (4) The magnetic metal-centre-metal-centre interaction of proteins (according to (3)) as well as (5) the properties of active open-shell centres on solid surfaces can be mimicked (according to (2)). (6) The electronic structure resembles that of small-sized semi-conductors whereby the mobility of the electrons of the open-shell metal-centres is immense and, furthermore, it may be extremely sensitive to the type of cationic lattice in the corresponding compounds. (7) According to their reduction state, they act as electron storage systems so that (possible) guests can be reduced. (The term multi-property materials is now in common use in the field of materials science).

An important challenge would be to find systems which according to the reaction scheme given above would lead to quite complex, asymmetrical, nonequilibrium (metastable) structures with a high *information* content (for a rather rare example of a molecular system with low symmetry formed by a self-assembly process see [16]).

1.4   A *System* of High Complexity: Human Society

Obviously, the theory of complex systems and their phase transitions offers a successful formalism to model the emergence of order in *Nature*. The question arises how to select, interpret, and quantify the appropriate variables of complex models in the social sciences. In this case, the possibility to test the complex dynamics of the model is restricted: In general, we cannot experiment with human society. Yet, computer simulations with varying parameters may deliver useful scenarios to recognize global trends of a society under all sorts of conditions.

Evidently, human society is a complex multi-component system composed of diverse elements. It is an open system because there exist not only internal interactions through materials and information exchange ("ideas") between the individual members of a society, but also an interchange with the external environment, nature, and civilization. At the microscopic level (e.g., micro-sociology and micro-economy), the individual "local" states of human behaviour are characterized by different attitudes. Changes of society are related to changes in attitudes of its members. Global change of behaviour is modeled by introducing macrovariables in terms of attitudes of social groups.

In social sciences, one distinguishes strictly between biological evolution and the history of human society. The reason is that the development of nations, markets, and cultures is assumed to be guided by the intentional behaviour of humans, i.e., human decisions based on intentions, values, etc. From a microscopic viewpoint we may, of course, observe single individuals contributing with their activities to the collective macrostate of the society representing cultural, political, and economic order (and, hopefully, determined by the value of corresponding "order parameters").

Yet, macrostates of a society do, of course, not simply average over its parts. Its order parameters strongly influence the individuals of the society by orientating ("enslaving") their activities and by activating or deactivating their attitudes and capabilities. This kind of feedback is typical for complex dynamical systems. If the control parameters of the environmental conditions attain certain critical values due to internal or external interactions, the macrovariables may move into an unstable domain out of which highly divergent alternative paths are possible. Tiny unpredictable microfluctuations (e.g., actions of very few influential people, scientific discoveries, new technologies) may decide which of the diverging paths society will follow.

A particular *measurement problem* of sociology arises from the fact that sociologists observing and recording the collective behaviour of society are themselves members of the social system they observe. Sociologists strive to define and to record quantitatively measurable parameters of collective behaviour, using all sorts of "objective", that is, empirical and quantitative methods. But, while the world of macroscopic physical phenomena will certainly not be changed in a scientifically relevant way by the fact that it is being explored and investigated, this does not necessarily hold true for social systems – a further justification for the obvious fact that scientific procedures used in classical physics are not simply transferable to the study of human social behaviour. This well-known sociological phenomenon of "self-observation in a society" confirms the complex dynamics of a society, i.e., the nonlinear feedback between individual activities at the microscopic level and its global macroscopic order states.

While systems in physics and chemistry are often taken for granted and are considered to be arbitrarily delimitable units of consideration, social systems cannot even be defined (and much less analyzed and studied) without simultaneously considering their environment and delineating their boundaries from their interval dynamics as well as from their interactions with all those features that do not pertain to the system. The problems which obviously arise in this context are carefully analyzed by N. Luhmann in his well-known system theory approach. Problems of a similar nature arise when considering biological processes. In addition, it might be worthwhile to take into account also the epistemological aspects discussed by N. Luhmann even in connection with the study of chemical and physical systems. A case in point is, for instance, the well-known fact that the dynamics of a protein cannot be understood without studying it in solution.

Social migration, economic crashes, and ecological catastrophes are very dramatic topics today, demonstrating the danger of global world-wide effects. It is not sufficient to have good intentions without considering the nonlinear effects of single decisions. Linear thinking and acting may provoke global chaos although we act locally with the best intentions. In this sense, even if we are not able to quantify all relevant parameters of complex social dynamics, the cognitive value of an appropriate model will consist in useful insights into the role and effect of certain trends relative to the global dynamics of our society. In other words, the operational value of such an approach depends upon the possibility of using the model in order to examine hypothetical courses of our society.

Acknowledgements

We thank Prof. Dr. A. Dress and Prof. Dr. F. Vögtle for valuable discussions.

1.5 Supplement

Several *basic* methods available for modeling self-organization processes can be applied,
such as:
(1) Phenomenological kinematic models;
(2) Thermodynamical models (e.g., of irreversible thermodynamics);
(3) Models of deterministic dynamics (differential equations for the order parameters);
(4) Models of stochastic dynamics (*Chapman-Kolmogorov* equation for the probability
     distributions of order parameters);
(5) Models of statistical physics (probability distributions of the microstates of a system).
     While the thermodynamics of irreversible processes considers only time average values
     of physical quantities in nonequilibrium states, the modern theory of nonequilibrium
     fluctuations takes also deviations from these values into account. *Deterministic* and
     stochastic elements are included, for example, in Haken's concept of *synergetics* and
     order parameters. These two central planes (according to (3) and (4)) are framed by
     those of (2) and (5).

The modern stochastic theory is concerned with random processes that develop in time.
If $x(t)$ is taken as such, a complete description of all statistical properties of $x(t)$ demands
specification of an infinite number of *probability densities* p_n $(x_1,\ t_1;\ x_2,\ t_2;\ \ldots.;\ x_n,\ t_n)$
where p_n $(x,\ t)$ $dx_1\ dx_2\ \ldots\ dx_n$ is the joint probability that $x_1 < x(t_1) < x_1 + dx_1$, $x_2 < x(t_2)$
$< x_2 + dx_2$, and so on [17, 18]. It is not possible to deal with generalized problems of this
kind in practice and, thus, simplifying assumptions must be additionally introduced. It
should be noted that for independent processes, knowledge of $x(t)$ at one time t does not
imply knowledge about $x(t')$ at any other time t'. The simplest assumption that can be used
for the correlation is that provided by *Markov*, whereby single-step *transition probabilities*
form the important quantities in his model. A significant and widely used class of *Markov*
models is that of *random walks*, in which case a particle makes random displacements r_1,
r_2, ... at times t_1, t_2, ... [17, 18]. (The *excluded-volume random walk*, which is a non-
Markovian type, *plays a role in the theory of polymer configurations*). Two main procedural
possibilities are available to facilitate solving *Markovian* problems in continuous time, the
first procedure leads to the *master* and the second to the *Fokker-Planck* equation. But, both
techniques start out from an equation which is, basically, too general to tackle problems of a
specific physical nature. The *master* equation which starts out from an observation of the
transition probabilities in continuous one-dimensional state space can be reduced to a
Fokker-Planck equation which would be valid for a particular kind of *conditional
probability.* The related *Langevin* equation, which was successfully used for the
understanding of the Brownian motion, integrates a stochastic element with respect to the
dynamics of a system. *It provides a useful background for the understanding of complicated
unknown crystallization processes in which extremely large cluster anions – like those
mentioned above – formed in solution are involved* [12]. (A typical *Langevin* equation could
be given as $m\dot{u} + \gamma u = F(t)$ where m is mass, u the velocity, γu a *damping force* and $F(t)$ a
rapidly *fluctuating random force*). In general, crystal growth starts with a nucleation
process, where random fluctuations play a key role – and which is rather complicated in

cases where giant cluster species are involved [*]. For the understanding of the whole crystal growth, microscopic and macroscopic theories have to be taken into account.

The important *Fokker-Planck* equation allows us, for instance, to draw some very close and important analogies between phase transitions occurring *in thermal equilibrium*, and certain order-disorder transitions in *nonequilibrium systems of physics, chemistry, biology, and other disciplines.* (Some relevant philosophical aspects are also considered in Chapter 15). The equation for the *distribution function* of the laser amplitude A, for instance, ($f(A) = N\exp(-\lambda_1 A^2 - \lambda_2 A^4)$; $\lambda_{1/2}$ *Lagrange* parameter) is formally identical to that of the magnetization M as *order parameter* in the case of para/ferromagnets, whereby the corresponding *second order phase transition* can be treated by means of *Landau's* theory [19, 20].

In mathematical models of social dynamics, a socio-economic system is characterized on two levels, distinguishing the micro-aspect of individual decisions and the macro-aspect of collective dynamical processes in a society. The probabilistic macro-processes with stochastic fluctuations can be described by the *master equation of human socio-configurations.* Each component of a socio-configuration refers to a subpopulation with a characteristic vector of behaviour. Concerning the migration of populations, the behaviour and the decisions to rest in or to leave a region can be identified with the spatial distribution of populations and their change. Thus, the dynamics of the model allows us to describe the phase transitions between global macro-states of the populations. In *numerical simulations* and *phase portraits of the migration dynamics*, the macro-phenomena can be identified with corresponding *attractors* such as, for instance, a stable point of equilibrium ("stable mixture"), two separated, but stable ghettos, or a limit cycle with unstable origin [21].

In economics, the Great Depression of the 1930s inspired economic models of *business cycles*. However, the first models were *linear* and, hence, required *exogenous shocks* to explain their irregularity. The standard econometric methodology has argued in this tradition, although an intrinsic analysis of cycles has been possible since the mathematical discovery of *strange attractors*. The traditional linear models of the 1930s can easily be reformulated in the framework of *nonlinear systems* [22].

---

[*]   This leads to serious problems when trying to get compounds with these ingredients (particles) in a crystalline – or even microcrystalline – form. The average distance *the particle* moves decreases with increasing particle size (The corresponding process is now being systematically studied by us [12]).

2 Entropies and Lexicographic Analysis of Biosequences

Hanspeter Herzel, Werner Ebeling, Armin O. Schmitt, and Miguel Angel Jiménez-Montaño

Abstract

This paper is devoted to the statistical and linguistic analysis of biosequences. Information-theoretical tools (Rényi entropies, mutual information) are introduced and applied to selected DNA sequences (yeast chromosome III, Epstein-Barr virus genome). Moreover, several techniques for the detection of long-range correlations are reviewed, and possible sources of such correlations are discussed. Finally, we study grammar representations of sequences and exemplify this approach by studying a fragment of mouse DNA.

2.1 Introduction

Higher forms of life are unthinkable without the existence of carbon. It is its specific electronic structure that permits it to enter an uncountable number of bindings with almost all sorts of atoms. A whole branch of science is devoted to exploring this abundant wealth – organic chemistry.

So much the more one might be astonished that for the most important processes governing all forms of life – the storage, the replication, and the transmission of genetic material – carbon only plays a subordinate role. For some reasons, Nature renounces this reservoir and restricts herself to molecules of the simplest form: all genetic information is stored in linear molecules, the DNA (the DNA of some bacteria is closed to a loop). The advantage of linear molecules in comparison with more complicated shapes is that they suggest a natural order when reading information from them. Information stored on more complex structures, i.e. branched molecules, would require additional instructions on how to be retrieved so as to restore its original meaning unequivocally.

Linearity is an elegant fashion to circumvent this problem. All symbols are (normally) read step by step – much the same as in reading a book. We note in passing that Schrödinger gave cogent reasons in favour of a linear form of genetic molecules before its actual discovery in the 40s [1].

Also, in the choice of the alphabet, Nature clings to simplicity: four "letters" – A (adenine), C (cytosine), G (guanine) and T (thymine) – suffice to encode the genetic information. It is obvious by now that it is the *sequence* of these symbols solely that carries the meaning of the genetic text.

One will probably never be able to "read" genomes directly, i.e. recognize the function of proteins encoded in them as genes or even of the whole organism just by inspection.

Nevertheless, one can scrutinize DNA sequences by means of several statistical tools, each of them highlighting different aspects.

Maybe the most prominent signature of some structure is its order state. It can range from "over-structured" (periodic sequences) to "completely messy" (random sequences). Theoretical reasons exclude these two extreme cases. A periodic structure does not convey new information when one period is finished (except perhaps the total number of repeated periods).

In a thoroughly random sequence, on the other hand, all of the symbols and all of their possible combinations occur equally likely. Without redundancy, however, the slightest mistake in the replication procedure (for instance the deletion or insertion of one base, or the exchange of two symbols) could not be detected and would result in a completely wrong product. Therefore, we can expect DNA sequences to lie "somewhere in the middle" between these poles.

Our paper is devoted to the detection of statistically significant structures in DNA sequences. In view of the incredible complexity of the transcription and expression mechanisms, DNA sequences should exhibit deviations from random sequences on various scales.

We will quantify structures by information-theoretical tools and, in the final section, by "grammar complexity". Special attention is devoted to the detection and interpretation of "long-range correlations".

2.2   Entropies of Biosequences

One of the best established methods to quantify structure in symbolic sequences is to investigate their entropies [2–5]. The thermodynamic entropy function was established in statistical mechanics by Boltzmann, Gibbs, and Planck in the last century to measure the degeneracy of a macrostate [6], i.e. to count the number of microstates that are compatible with fixed variables of the macrostate. Only in the middle of the 20th century, Shannon recognized that this concept can be exploited to measure information, too.

The amount of information contained in a symbol on the average is best visualized by the question: "How many "intelligent" binary questions must one at most ask to determine the symbol unequivocally?" In the case of four equidistributed letters A, C, G, and T the following strategy would be "intelligent": First question: "Is the base a purine (R) or a pyrimidine (Y)?". The second question would be, depending on the answer to the first: "Is it A or G?" (if the answer was "R") respectively "Is it C or T?" (if the answer was "Y"). Two intelligent binary questions are enough to determine any symbol.

In general, the average number of questions is termed information. When the logarithm is taken to base 2 – as it is done throughout this work – information is measured in "bits". In the case of non-equidistributed symbols (i.e. some symbols have a greater chance to appear than others) one has to average over all probabilities p_i:

$$H_1 := \sum_{i=1}^{\lambda} -p_i \log p_i \; , \tag{2-1}$$

which is the well known *Shannon entropy*.

A straightforward generalization are the *block entropies*

$$H_n := \sum_i -p_i^{(n)} \log p_i^{(n)} \; . \tag{2-2}$$

Here, the summation is carried out over all n-words (n-tuples of letters) with non-vanishing probability $p_i^{(n)}$. H_n measures the *information content* of an n-word, i.e. the mean number of binary questions to guess an n-word generated by the underlying process. H_n becomes maximal if all symbols are equidistributed and statistically independent. For independent letters, any letter requires H_1 binary questions and, hence, $H_n = n \cdot H_1$. Consequently, the differences $H_1 - H_n / n$ can be regarded as a measure of correlations between symbols.

For stationary and ergodic sources, the limit

$$h := \lim_{n \to \infty} \frac{H_n}{n} \tag{2-3}$$

exists. It is called the *entropy of the source* and gives the information per letter with all statistical dependencies taken into account. The amount of new information of the $(n + 1)$th symbol given n letters is called *differential entropy*:

$$h_n := H_{n+1} - H_n \; . \tag{2-4}$$

This quantity measures the uncertainty for correctly guessing the $(n + 1)$th letter after the observation of the n previous letters.

For an n-th order Markov process,

$$h_n = h_{n+1} = \cdots = h \tag{2-5}$$

holds; therefore deviations of h_n from its limit $h = \lim_{n \to \infty} h_n$ indicate statistical dependency on the scale of $(n + 1)$-words.

In a certain sense, entropies H_n resp. h_n are the ideal candidates to detect structure in symbol sequences since they respond to any deviation from statistical independence. However, their estimation from finite samples appears to be problematic due to the combinatorial explosion: the number of possible n-words grows like λ^n which reaches astronomical numbers even for moderate word lengths n.

In several papers, finite sample corrections for entropies have been discussed extensively [7–10]. The results can be summarized as follows: If a word indexed by i occurs k_i times in a given sample of size N, one gets from this sample the *frequency vector* $\vec{k} = (k_1, \ldots, k_M)$. By taking the relative frequencies as estimates for probabilities, the so-called *observed entropy*

$$H^{obs} := \sum_i -\frac{k_i}{N} \log \frac{k_i}{N} \tag{2-6}$$

is obtained. Averaging over all possible realizations $\vec{k}$, the *mean observed entropy* results:

$$\left\langle H^{obs} \right\rangle := \sum_{\vec{k}} \sum_{i=1}^{M} -\frac{k_i}{N} \log \frac{k_i}{N} \cdot P(\vec{k}) \ . \tag{2-6a}$$

The symbol $(\vec{k})$ means that the summation has to be carried out subject to the restriction

$$\sum_{i=1}^{M} k_i = N \ , \tag{2-6b}$$

and $P(\vec{k})$ is the probability for one given realization $\vec{k}$.

Formula (2-6a) can be approximated by Taylor expansions in the cases of good and bad statistics [9]. Firstly, in the realm of relatively good statistics, i.e. for $M \ll N$, the expansion yields that the entropy estimates from realizations underestimate the real entropy by

$$\Delta H = \frac{M}{2N \ln 2} \ . \tag{2-7}$$

Sophisticated techniques based on the McMillan Theorem and the Maximum Entropy Principle allow for reliable estimations of entropies even in cases where $\lambda^n > N$ holds [10, 11].

Somewhat surprisingly, some interesting results can be obtained for large n, i.e., for very bad statistics [9]: For $\lambda^n \gg N$, most words of the sample appear only once, i.e., H^{obs} approaches $\log N$. The corresponding expansion of Equation (2-6a) reads:

$$\left\langle H^{obs} \right\rangle = \log N - N \cdot \sum_i p_i^2 + \frac{N^2}{2!}\left(2 - \frac{\log 3}{\log 2}\right) \sum_i p_i^3 + \ldots \tag{2-8}$$

We will discuss below how this expression can be exploited for the estimation of entropies.

The finite sample effect is demonstrated in Figure 2-1 by means of the yeast chromosome III [12]. Without correction, the observed block entropies H_n^{obs} bend down towards $\log N \approx 18.3$.

Checks with processes of known entropies suggest that the correction technique based on the McMillan Theorem and the Maximum Entropy Principle (see Reference [11] for details) allow reasonable estimations even for $n = 10$. The principle idea of these correction methods is to reconstruct the underlying probability distribution using variational principles rather than correcting the entropies themselves. This is remarkable since the number of possible 10-words ($4^{10} \approx 10^6$) already exceeds the sample size (315338).

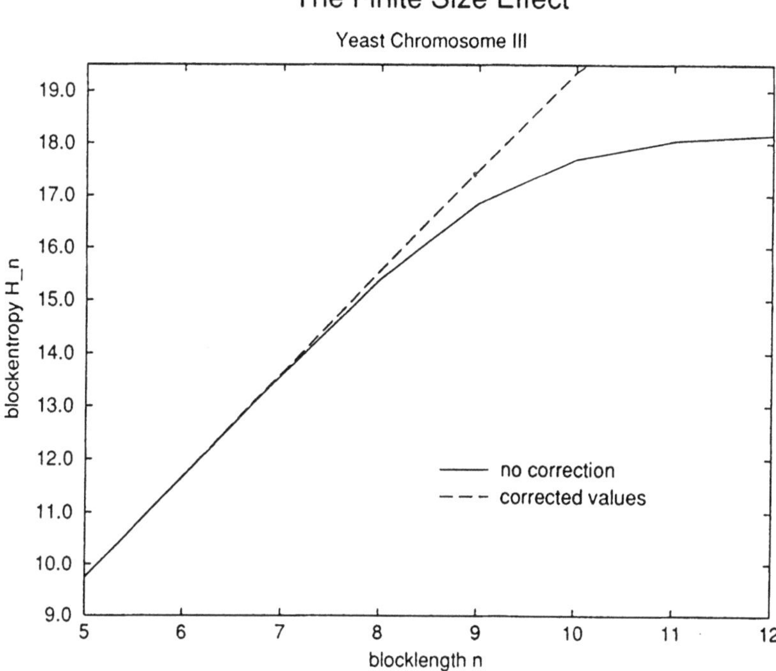

Figure 2-1   Word entropies for the DNA yeast chromosome III (315338 nucleotides). The full line corresponds to the naive estimation according to Equation (2-6). The dashed line refers to entropies obtained by the finite sample corrections discussed in Reference [11].

In Figure 2-2, differential entropies $h_n = H_{n+1} - H_n$ are shown for the Epstein-Barr virus (172281 base pairs). The decay of the uncorrected values is dominated by finite sample effects. In contrast, the corrected values display an interesting effect that repetitive sequences take on information content. This DNA sequence contains about 25% "repeat regions", i.e. subsequences occurring at least twice in the whole sequence. After removing these repeats, the information per letter described by the h_n is above 1.9 bit. However, the information of the complete sequence is reduced by about 0.5 bit due to repetitive subsequences. This seems plausible since repetitions induce redundancy and thus reduce the information content.

In Reference [13], a detailed theoretical model of DNA sequences with interspersed repeats is discussed. It is worth paying attention to the effect of repeats on the statistical properties of biosequences since they may constitute up to 50% of genomes of higher eukaryotes.

Besides the block entropies, the transinformation (or mutual information) $I(k)$ turned out to be a valuable tool in sequence analysis [14].

It is based on the probabilities $p_{ij}^{(k)}$ that, after k symbols, a letter i is followed by a letter j. If these events are statistically independent, we have

$$p_{ij}^{(k)} = p_i p_j \ .$$

(2-9)

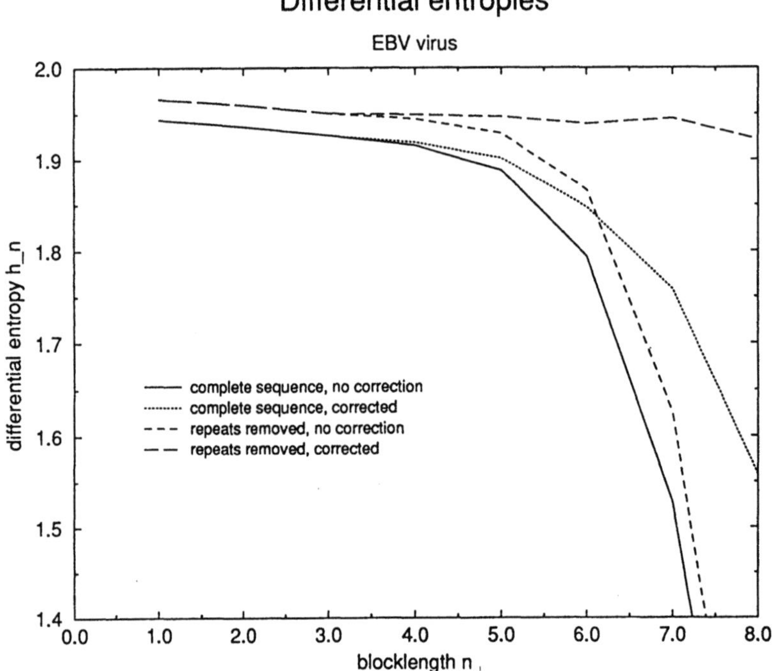

Figure 2-2   Entropies for the Epstein-Barr virus. Note, that the effect of repeats becomes visible only with the aid of finite sample corrections.

Hence, the *mutual information function*

$$I(k) := \sum_{ij} p_{ij}^{(k)} \log \frac{p_{ij}^{(k)}}{p_i p_j} \tag{2-10}$$

measures statistical dependencies of letters at distance k. As is easily seen, $I(k)$ can also be written as a difference of word entropies:

$$I(k) = 2H_1 - H_2^{(k)} , \tag{2-11}$$

with

$$H_2^{(k)} := \sum_{ij} -p_{ij}^{(k)} \log p_{ij}^{(k)} . \tag{2-12}$$

The estimation of the mutual information for long DNA sequences is only weakly affected by finite sample effects since 16 probabilities have to be estimated, only. More precisely, from Equations (2-7) and (2-11), the following finite sample correction can be found:

$$I(k) = I^{obs} - \frac{4}{N \ln 2} \; .$$

(2-13)

Here, I^{obs} denotes the mutual information estimated from observed relative frequencies of pairs.

Now we turn to a generalization of the entropy concept by using *Rényi entropies* [15].

Shannon entropies are a specific scalar functional of probability distributions. In order to characterize distributions more exactly, various moments or, alternatively, Rényi entropies are appropriate. These generalized entropies have been found to be very useful in dynamical systems theory [7, 16, 17].

Applications to texts and to pieces of music were given in a separate paper [18].

Rényi entropies of the order q are defined as follows:

$$H^q := \frac{1}{1-q} \log \sum_i p_i^q \quad (q \neq 1) \; .$$

(2-14)

In the limit $q \to 1$, we obtain the Shannon entropy. For large (small) q the generalized entropies H^q are dominated by the most (least) probable events. Variation of q allows to extract interesting features of the probability density. For example, for $q = 0$ simply the number of non-vanishing probabilities p_i is counted.

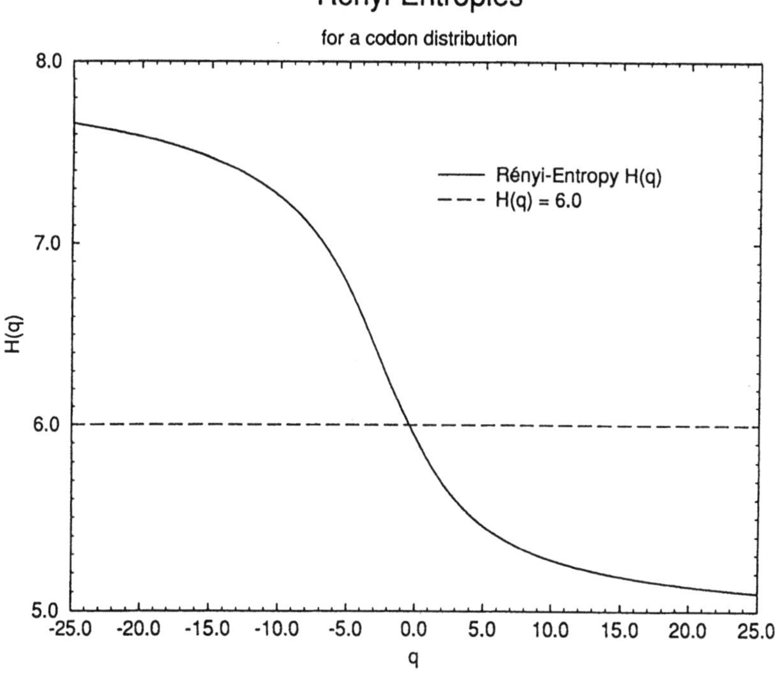

Figure 2-3   Rényi entropies from the codon usage table (see Reference [19]) of protein coding sequences.

Figure 2-3 displays the generalized entropies H^q for a specific distribution of 3-words. The corresponding p_i are from a codon usage table of protein coding regions [19]. It is well-known that the 20 amino acids are not used with the same frequency. Moreover, the different codons for a certain amino acid are often utilized in a non-random manner. Thus, the uneven codon usage is a central tool to identify protein coding regions [20].

The dashed line in Figure 2-3 corresponds to equidistributed codons ($p_i = 1/64$). The full line characterizes the deviations from an even usage. The Shannon entropy $H^{q=1} = 5.80$ bit is only weakly below the dashed line. For $q = 0$, we obtain

$$H^{q=0} = \log 61 \approx 5.93, \tag{2-15}$$

which reflects the fact that the three stop codons are absent. Since some codons are much more frequent than others, e.g. ($p(\text{CGT}) = 0.033$ and $p(\text{CGA}) = 0.004$), the deviation from 6 bit is more pronounced for large absolute values of q. This observation suggests that Rényi entropies with large absolute values q are better suited to identify protein coding regions than the Shannon entropy.

Now, we propose a rather general approach to the statistical analysis of biosequences which contains all the quantities discussed above as special cases. For this purpose, we discuss the joint probabilities $p_i^{(m,n,k)}$ that a certain word of length m is followed by an n-word after k letters. Then, the Rényi entropies

$$H^q_{(n,m,k)} := \frac{1}{1-q} \log \sum_i (p_i^{(m,n,k)})^q \tag{2-16}$$

allow to study virtually all statistical dependencies of subwords. Block entropies H_{m+n} are obtained for $k = 1$ and $q \to 1$, and the mutual information function results for $m = n = 1$ and $q \to 1$. Finite sample corrections have been calculated also for Rényi entropies [7, 9].

An interesting relation between finite sample effects and Rényi entropies is revealed by inspection of Equation (2-8). The convergence to the asymptotic value, $\log N$, is related to higher order Rényi entropies (Equation (2-14)):

$$\langle H^{obs} \rangle = \log N - N \cdot 2^{-H^{q=2}} + \frac{N^2}{2!}\left(2 - \frac{\log 3}{\log 2}\right) 2^{-2H^{q=3}} + \dots . \tag{2-17}$$

Consequently, by fitting the coefficients of N^k in this series, estimations of entropies can be obtained. For example, from the first terms in (2-17) one gets:

$$H^{q=2} = \log \frac{N}{\log N - H^{obs}} . \tag{2-18}$$

This formula essentially exploits repetitions of some words (which lead to a difference between $\log N$ and H^{obs}) as an indication of the underlying distribution of words.

2.3   Detecting Long-range Correlations in Symbol Sequences

Long-range correlations in a variety of fields such as semiconductors, highway traffic, music, and heart-variability attracted much attention during the past decades [21–24]. Usually, these phenomena are measured using power-spectra, correlation functions, or anomalous diffusion. Characteristic features of long-range correlations are power laws of the spectrum $S(f)$ for small frequencies f:

$$S(f) \propto f^{-\beta} \qquad (\beta \approx 1) \ , \tag{2-19}$$

or of the autocorrelation function for long time spans:

$$C(t) \propto t^{-\gamma} \ . \tag{2-20}$$

The latter function is intimately related to anomalous diffusion

$$\sigma^2(t) \propto t^\alpha \qquad (\alpha > 1) \ . \tag{2-21}$$

According to the Wiener-Khinchin Theorem and the Kubó Formula (see our relation (2-30)), the exponents β, γ and α are not independent.

In several papers, these techniques have been used to detect long-range correlations in DNA sequences [25–27]. Furthermore, applications to lengthy texts as, e.g., Moby Dick by H. Melville, are given in [28]. However, we will argue in the following that the mutual information function introduced in the preceding section is the most appropriate measure for symbol sequences.

In order to obtain correlation functions (or spectra), one assigns to each symbol A_i $(i = 1, 2, \dots , \lambda)$ a real number a_i. Then, the *autocovariance function* is defined as

$$C(k) := \langle a_i a_{i+k} \rangle - \langle a_i \rangle \langle a_{i+k} \rangle \ . \tag{2-22}$$

Using the probabilities p_i and $p_{ij}^{(k)}$ defined above we get

$$C(k) = \sum_{i,j} p_{ij}^{(k)} a_i a_j - \left(\sum_i p_i a_i \right)^2 \ . \tag{2-23}$$

Hence, the k-dependent part of the autocorrelation function $C(k)$ is just a quadratic form with the matrix of joint probabilities $p_{ij}^{(k)}$. This representation elucidates the main disadvantage of any identification of symbols A_i with numerical values a_i: There is a high degree of arbitrariness in assigning numbers to symbols A_i. For example, in [27, 28], the transitions between identical symbols are analyzed by means of an autocorrelation function which describes only correlations within the diagonal terms $p_{ij}^{(k)}$. Other correlations may remain hidden.

Consequently, we suggest to study the mutual information function which detects *any* deviation from statistical independence for pairs of letters at distance k. The transition probabilities are easily estimated, and systematic errors can be corrected using Equation (2-13).

Moreover, a power-law decay of the autocorrelation function can be related to a corresponding decay of mutual information: To illustrate this, let us assume that there are small deviations from statistical independence:

$$p_{ij}^{(k)} = p_i p_j + \varepsilon_{ij}^{(k)} \ .$$

(2-24)

Summation gives

$$\sum_{i,j} \varepsilon_{ij}^{(k)} = 0$$

(2-25)

in view of

$$\sum_{i,j} p_{ij}^{(k)} = \sum_{i,j} p_i p_j \qquad (=1) \ .$$

(2-25a)

Hence, expanding the expression for the mutual information yields

$$I(k) = \sum_{i,j} \left(p_i p_j + \varepsilon_{ij}^{(k)} \right) \log \frac{p_i p_j + \varepsilon_{ij}^{(k)}}{p_i p_j}$$

$$\approx \frac{1}{\ln 2} \left[\sum_{i,j} \varepsilon_{ij}^{(k)} + \frac{1}{2} \sum_{i,j} \left(\frac{\varepsilon_{ij}^{(k)2}}{p_i p_j} \right) \right]$$

$$= \frac{1}{2 \cdot \ln 2} \sum_{i,j} \left(\frac{\varepsilon_{ij}^{(k)2}}{p_i p_j} \right) \ .$$

(2-26)

Since $I(k)$ has an extremum at $p_{ij}^{(k)} = p_i p_j$, the linear terms vanish, and we get a quadratic dependency on $\varepsilon_{ij}^{(k)}$. Hence, if the autocorrelation function – which, in general, is proportional to the $\varepsilon_{ij}^{(k)}$ (see Equation (2-23)) – decays as $k^{-\gamma}$, we get for the mutual information

$$I(k) \propto k^{-2\gamma} \ .$$

(2-27)

This relation has already been established by Wentian Li for specific binary sequences [29]. It was mentioned above that autocorrelation functions are intimately related to the growth of the variance of the corresponding process. In order to define an analogue to the mutual information, we discuss this relationship in some detail: Let us define a random walk,

$$s(l) := \sum_{i=1}^{l} a_i \tag{2-28}$$

derived from the string of numbers a_i from the symbol sequence under consideration. For simplicity, we assume vanishing mean values

$$\langle a_i \rangle = \langle s \rangle = 0 \ . \tag{2-29}$$

Then, according to Reference [26], the variance of the process grows like

$$\langle \sigma_s^2(l) \rangle = \left\langle \left(\sum_{i=1}^{l} a_i \right)^2 \right\rangle = \sum_{i,j=1}^{l} C(i-j) \ . \tag{2-30}$$

Hence, the variance growth can be expressed by a finite sum over the autocovariance function defined in Equation (2-22).

Consequently, a slow decay of the autocorrelation function is reflected in anomalous "diffusion". In contrast, it is obvious that a sufficiently fast decay of the autocorrelations leads to normal diffusion

$$\langle \sigma_s^2(l) \rangle \propto l \ . \tag{2-31}$$

It was found by empirical studies of Peng et al. [26] that such a random walk analysis indicates long memories in DNA sequences. Compared with the autocorrelation function, fluctuations are substantially reduced due to summation (see Equation (2-30)).

We have already pointed out that for symbol sequences the mutual information has to be preferred to autocorrelations. The advantage of an integral quantity can be obtained for the mutual information function by defining

$$I_s(l) := \sum_{i,j=1}^{l} I(|i-j|) \ . \tag{2-32}$$

Analogously, the variance growth of the smoothed function $I_s(l)$ contains still information on the decay properties of $I(k)$.

In the next section, we apply the preferable measures $I(k)$ and $I_s(k)$ to a long DNA sequence and discuss possible sources of long correlations.

2.4   Long-range Correlations in DNA Sequences

In this section, we exemplify the search for long correlations by analyzing the yeast chromosome III, the first completely sequenced eukaryotic chromosome. First indications of long correlations over thousands of bases have already been reported in References [9, 30].

Figure 2-4 shows the mutual information for $k \in (1000, 1100)$. The signal is far above the estimated level for uncorrelated symbols (see Equation (2-13)). The most prominent feature is the pronounced period three which results from the triplet code of protein coding sequences.

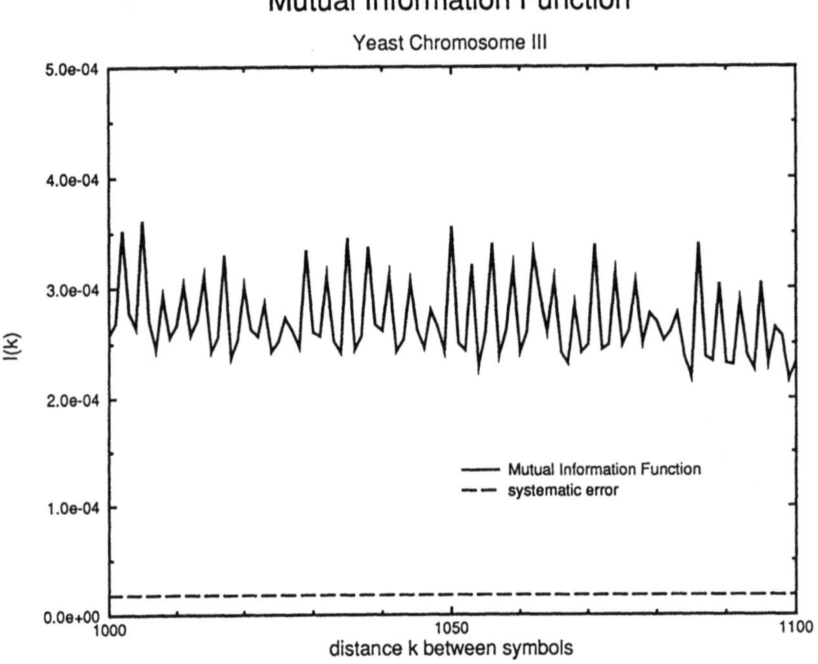

Figure 2-4   Mutual information as a function of k for the yeast chromosome III.

It was already presented in Section 2-2 that the uneven codon usage is a characteristic of protein coding fragments. 182 open reading frames longer than 100 bases have been found on the chromosome. Some of them are several kilobases long and, hence, the period three in Figure 2-4 is not unexpected even for distances $k > 1000$.

The interpretation of Figure 2-5 is less straightforward. In order to reduce the period three, we have plotted $I(k)$ only for multiples of three, i.e., we analyze the decay of the peaks in Figure 2-4. The log-log representation visualizes an interesting feature: there is a power-law decay over two orders of magnitude. Due to the strong period three in this k-range, we attribute this decay to the distribution of protein coding regions. The characteristic length scale of the open reading frames is indeed around 1000 bases.

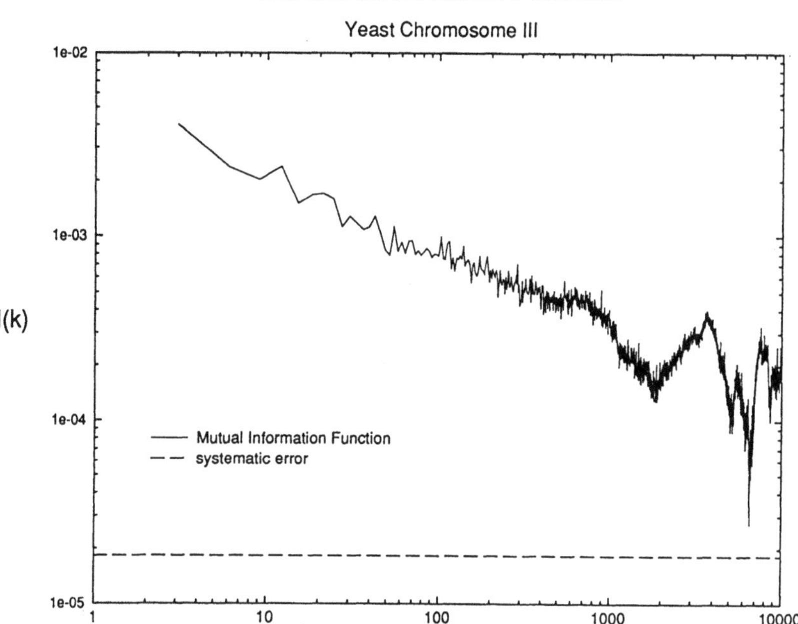

Figure 2-5   Log-log plot of the mutual information from yeast chromosome III.

As shown in Reference [9], there are also peaks in the mutual information function around $k = 3700$ and $k = 7400$ which are not yet understood. They may reflect a certain organization of genes on the chromosome or they might stem from similar nucleotide composition in regions separated by this distance.

Figure 2-6 shows the summed up mutual information $I_s(k)$ (see Equation (2-32)) for the same DNA sequence. In this representation, the correlations over more than five bases are clearly visible when $I_s(k)$ computed for the actual DNA is compared with the same function for a Markov process of 5th order. The 6-word probabilities of this Markov process are taken from the yeast DNA and, hence, the short-range correlations are the same.

Figures 2-4–2-6 illustrate that from mutual information various structures can be extracted efficiently. In the following, we discuss some possible sources of long correlations in DNA sequences.

As far as we know, there are no consistent biological interpretations of the long-range correlations which have been found in several empirical studies [9, 25–27]. Possible sources of correlations which are well-documented in the literature [31–34] are listed below.

In connection with Figure 2-4, we have already discussed the period three amplitude induced by the uneven codon usage of protein coding sequences. Trifonov [33] emphasizes that there are additional superimposed codes such as the translation framing code, responsible for correct triplet counting by the ribosome during protein synthesis, and the chromatin code which is related to the nucleosome structure and which induces periodical repetitions of, primarily, AA and TT dinucleotides at a distance of about 10.5 bases [35].

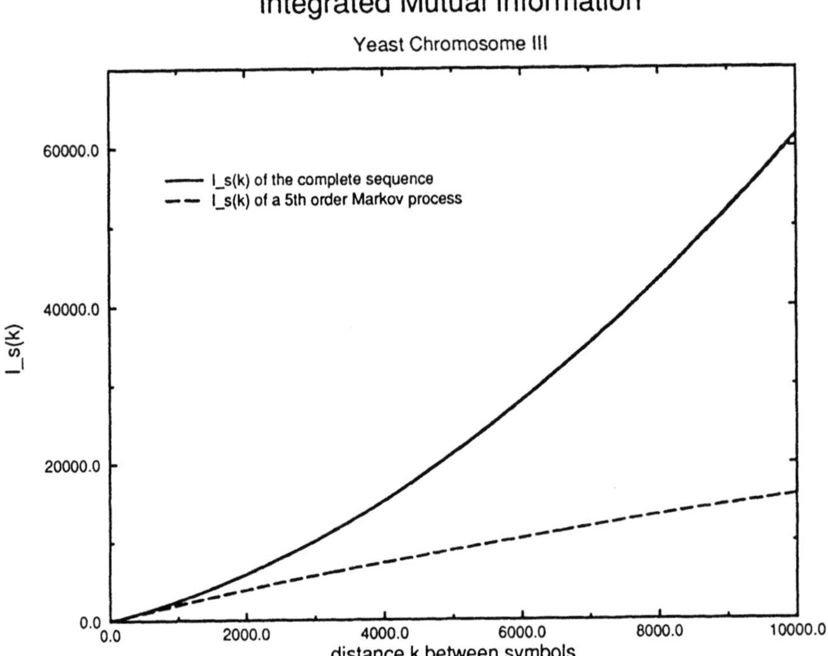

Figure 2-6   Summed up mutual information for yeast chromosome III and a Markov model of this sequence.

Periods of 10 to 11 bases may also be related to the secondary structure of the DNA molecule (B-DNA exhibits a periodic structure of 10.4 bases) and of proteins (α-helices). The mutual information shown in Figure 2-4 of Reference [14] exhibits indeed a peak at $k = 10$.

The effect of the tertiary structure of proteins (alternation of α-helices, β-sheets, and random coils) on the primary DNA structure is not yet fully explored. The characteristic length of protein domains is in the order of 100 bases, and some "footprints" of these structures in the DNA sequence should exist. However, they may be hidden due to the non-uniqueness of the triplet code (most amino acids are encoded by several codons).

Genes of higher eukaryotes are characterized by the alternation of protein coding segments (exons) and intervening noncoding sequences (introns). The typical length scale of genes is in the order of 1000 bases. However, there are genes as large as 1000 kilobases [34]. The organization of exons and introns may cause, therefore, correlations ranging from one hundred to several thousand bases. An example of such correlations has been detected by Beckmann and Trifonov [36]: The exon-intron boundaries, termed splice sites, prefer a "ladder" with a period of about 205 bases. The authors suggest that this periodicity might be related to the folding of the DNA molecule around nucleosomes which has a similar periodicity [36]. Another hint for the important role of the exon-intron organization is given by Borštnik et al. [37]: They study introns and flanking sequences separately and find that there are less long-range correlations within these groups than in the entire sequences.

Besides the internal structure of genes, also the distribution of genes within genomes may exhibit some organization which can generate correlations over thousands of base pairs. For example, genes for histones, rRNA and tRNA are often clustered [34, 38]. Sea-urchin DNA contains, e.g., hundreds of tandemly repeated histones within repeat units of about 6300 bases [34].

It has been found by reassociation experiments of single-strand DNA to double-strand DNA that genomes of higher eukaryotes contain up to 50% repetitive sequences. Such a large amount of repeats can, obviously, affect the statistical properties of DNA sequences drastically. In Reference [13], the effect of repeats on entropies is studied in detail. It turns out that the information content is significantly reduced by repetitive sequences (see also Figure 2-2).

In order to discuss the contribution of repeats to long-range correlations, we distinguish two classes:

- First, there are "simple repeats" consisting of units of a few bases which are tandemly repeated up to 10^6 times. Such highly repetitive sequences are often found near the centromer of chromosomes. It is obvious that such long runs of tandem repeats lead to strong periodicities. For example, repetitions of "ACAAACT", "ATAAACT", "ACAAATT" and "AATATAG" constitute about 40% of the genome of *Drosophila virilis* [38].

- Another class of repeats are relatively long repeats, termed LINES (about 6000 bases long) and SINES (a few hundred bases) which are dispersed over the whole genome. A prominent example is the family of Alu repeats of an approximate length of 300. It is estimated that the human genome contains about 900 000 copies of this repeat [34]. These dispersed repeats have been distributed, presumably, by replicative transpositions [38]. Such interspersed repeats may contribute to long-range correlations in two ways: Firstly, there are internal repetitions within LINES such as "long-terminal repeats" (LTRs). Keeping in mind that genomes may contain more than 20 000 LINES, these LTRs separated by several thousand base pairs may have considerable influence on measures of long correlations. Secondly, non-random distribution of dispersed repeats may induce correlations. Lewin mentions, for example, "short-period interspersions" as an alternation of repeats of length 300 and nonrepetitive segments of length 1000 [38].

In summary, the internal structure and the distribution of exon, introns and repeats constitute sources of long-range correlations over hundreds and thousands of bases.

There are, moreover, indications that the hierarchy can be extended up to the scale of whole chromosomes ($10^5 - 10^8$ bases). Besides functional units, such as origins, centromers, and telomers, thousands of "chromosome-bands" are well-known [32, 38]. These bands are due to AT- or CG-rich regions, which again are related to the distribution of repetitive segments. Since the resolution of the band-structure is above 10^6 bases, these bands reflect structures even on such huge scales.

Thus, it is likely that DNA sequences are organized in a hierarchical manner which is associated with long-range correlations over many orders of magnitude. Obviously, the detection of statistical dependencies is only the first step towards an understanding of such structures.

2.5   The Linguistic Concept

In this section, we discuss a linguistic-oriented approach to biosequences [2, 39–41] and continue earlier studies on context free grammars and grammar complexity [42, 43]. Our approach is based on the investigation of the frequency of subwords. Since physico-chemical discernability (molecular recognition) governs biochemistry, we introduce also descriptions using distinctive features that embody the relevant physico-chemical constraints, i.e., we go to attribute-valued grammars.

We start with several definitions along the line of Chomsky's approach to languages [44]:

Definition 1: An alphabet X is a finite set of λ symbols: $X = \{A_1, A_2, \cdots A_\lambda\}$.

For DNA molecules, the four nucleotides A, C, G, T – and for RNA molecules A, C, G, U – play the role of the alphabet. In the case of proteins, the alphabet consists of 20 letters, which are A, C, G, F, ... V, W, Y in Dayhoff's notation. However, other choices of alphabets are possible as well. In the case of polynucleotides, we may restrict ourselves to just two letters, say R standing for purines A, G, and Y for pyrimidines C, T, or U. For polypeptides, a categorization into 4 groups is a possible choice:

A: neutral amino acids (A, P, G, S, T)

D: hydrophilic amino acids (D, E, Q, N, D, H, K, R)

C: small hydrophobic amino acids (C, L, I, V, M)

F: big hydrophobic amino acids (F, Y, W)

The principle of the abbreviated notation is that each of the amino acids in the four groups is denoted by the letter corresponding to its first representative. An even more drastic reduction is given by splitting them into only two groups, the hydrophobic and the hydrophilic amino acids. Reduced alphabets may be very useful in linguistic studies.

Definition 2: A finite string of length n consisting of symbols of X is called a subword of length n. The set of all finite strings is called free semigroup X^*. The set of all subwords of length n (which is called X_n) has $N_n = \lambda^n$ elements, a number growing exponentially with n.

Definition 3: A language S is any subset of X^*. This subset can be specified in many ways, by a set of generating rules, an automaton, a nonlinear map etc. A language may be given by an infinite string over the alphabet X which has the properties of stationarity and ergodicity.

Definition 4: The S-vocabulary of a language is the set of all subwords ordered by increasing length n and with respect to their relative frequency. The S-vocabulary consists of a rank-ordered list of letters S_1, a rank ordered list of pairs S_2, triplets S_3, etc..

Definition 5: The K-vocabulary of a language is the set of all subwords (independent of their length) which are constitutive for the language. "Constitutive" means here that any string of the language may be constructed by concatenation of keywords. The keywords are those elements of a language which may be associated with meaning. The fuzzy character of "meaning" does not allow so far to give a sharper definition of the term "keyword". It is one of the main purposes of this approach to find a way to get at least candidates for keywords being the constitutive elements of the language of biomolecules.

Definition 6: A context-free grammar $G = (N, T, P, S_0)$ is a quadruple consisting of a start symbol S_0, a set of terminal symbols $T \in X$, a set of nonterminals $N = \{S_1, S_2, S_3, \cdots\}$ which is also called variables or categories, and the set P of production rules $S_i \to p$, where $p \in (N \cup T)$.

Now we turn to the complexity measure of a grammar representation as proposed in Reference [42] in 1980; it has been applied recently in Reference [45].

Unfortunately, grammar representations are not unique, nor does there exist a unique way to evaluate their complexity. In earlier papers [2, 42], terminals (letters from the original alphabet) and nonterminals (letters of the auxiliary alphabet) have been treated equally. Thus, in the program GRAMMAR 80, the complexity measure was

$$K_{G80} := n_\lambda + n_\sigma \, , \tag{2-33}$$

where n_λ and n_σ denote the number of terminals and nonterminals, respectively. Implementation of this algorithm (GRAMMATICA, written in 1987 by L. Z. Cortina) has shown that the minimization of K_{G80} introduces too many nonterminals and, consequently, the representations are rather long when measured by the goals of data compression [46].

As a first generalization, we propose to ascribe different scores S_λ and S_σ to terminals and nonterminals leading to

$$K_{G93} := S_\lambda n_\lambda + S_\sigma n_\sigma \, . \tag{2-34}$$

A possible choice may be

$$S_\lambda := \log \lambda$$

$$S_\sigma := \log \sigma \quad , \tag{2-35}$$

which can be motivated as follows: In order to generate a string from terminals and nonterminals, one has to find the corresponding letter in a list of λ or σ elements and search times for finding this letter scale typically with the logarithm of the length of the corresponding list.

A second, rather general proposal is

$$K_{G94} := \sum_{i=1}^{\lambda} S_\lambda^{(i)} + \sum_{j=1}^{\sigma} S_\sigma^{(j)} \, . \tag{2-36}$$

Such arbitrary choices of weights, $S_\lambda^{(i)}\,(i = 1, 2, \ldots, \lambda)$ and $S_\sigma^{(i)}\,(j = 1, 2, \ldots, \sigma)$, introduce the flexibility to adapt the complexity measure to standard methods of data compression [46] as well as to specific biologically motivated valuations.

At the end, we apply our concepts to a fragment of length 234 bp of the satellite DNA from the mouse *M. musculus* [38]:

GGACCTGGAAAAAATATGGCGAGAAAAAAAATGAAAATCACGGAAAATGAGA
AATACACACTTTAGGACGTGAAATATGGCGAGGAAAACTGAAAAAGGTGGAAA
ATTTAGAAATGTCCACTGTAGGACGTGGAATATGGCAAGAAAACTGAAAATCAT
GGAAAATGAGAAACATCCCTTGACGACTTGAAAAATGACGAAATCACTAAAAA
ACGTGAAAAATGAGAAATGCACACTGAA.

The estimated block entropies of this sequence are

$$H_1 = 1.85 \text{ bit}$$
$$H_2 = 3.58 \text{ bit}$$
$$H_3 = 5.07 \text{ bit} .$$

(2-37)

Here, the finite sample correction according to Equation (2-7) has been applied. The above entropies indicate that there are significant deviations from randomness. Further indications of correlations are provided by the mutual information function in Figure 2-7. Beside correlations over about 20 bases, pronounced peaks appear around $k = 58$ and $k = 117$.

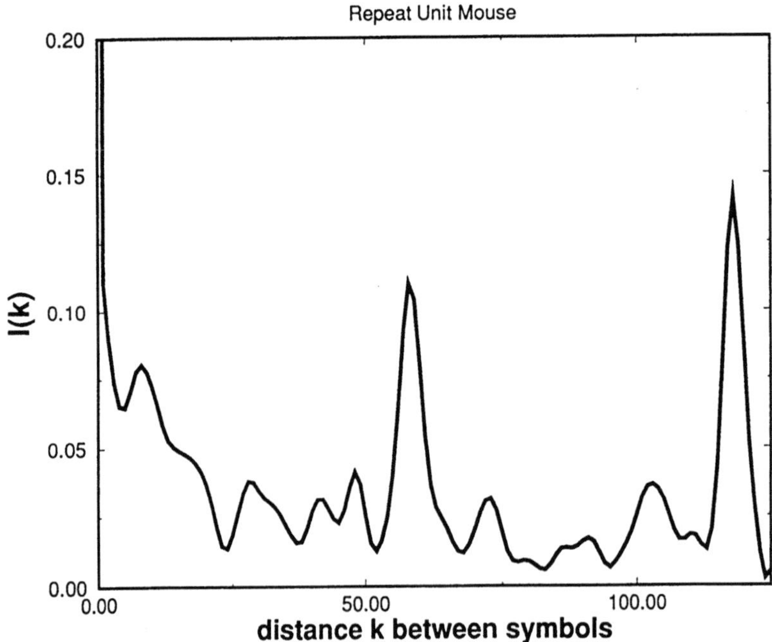

Figure 2-7 Mutual information function of the fragment of mouse DNA. Finite sample corrections using Equation (2-13) have been used. Moreover, the function has been smoothed by a moving-average filter.

These peaks are manifestations of a hierarchical organization of the sequence revealed by homology analysis [38]: There is an 81% homology of the first and second half and about 60% homology of the quarters which have a length of about 58 bases. As Figure 2-7 demonstrates, a strong correlation in each repeat unit is easily recognized by mutual information.

Such correlations imply redundancy and, hence, it is no surprise that the DNA fragment can be effectively compressed by introducing nonterminals a, b, ... m. From the relatively short representation

$$S \rightarrow \text{GmCcGaCbTACAkTeTiATGGCGAjflGGTjATTTAghGecAaTbCAhTmmt dCikllCGdgCACf}$$

the original sequence can be easily reconstructed using the auxiliary symbols

$$
\begin{aligned}
a &\rightarrow \text{AGAAAACTGAAAATCA} \\
b &\rightarrow \text{GGAAAATGAGAAA} \\
c &\rightarrow \text{TGGAATATGGC} \\
d &\rightarrow \text{TGAAAAATGA} \\
e &\rightarrow \text{TAGGACG} \\
f &\rightarrow \text{ACTGAA} \\
g &\rightarrow \text{GAAATG} \\
h &\rightarrow \text{TCCACT} \\
i &\rightarrow \text{GAAAT} \\
j &\rightarrow \text{GGAAA} \\
k &\rightarrow \text{CACT} \\
l &\rightarrow \text{AAA} \\
m &\rightarrow \text{GAC} \, .
\end{aligned}
$$

This representation corresponds to a relatively low grammar complexity

$$K_{G80} = 161$$

which reflects some degree of order within the sequence.

2.6   Summary and Discussion

We discussed in this paper various aspects of the statistical analysis of DNA sequences. The main results can be summarized as follows:

- Block entropies H_n or differential entropies h_n measure the information content of symbolic sequences. Since these entropies are heavily affected by finite sample effects, appropriate corrections are required.
- Repetitive fragments reduce the information content drastically (see Figure 2-2).
- Rényi entropies may be used to characterize the distributions of words comprehensively.

- Long-range correlations are efficiently analyzed by the mutual information function. In contrast to methods based on autocorrelations, this function detects any deviation from statistical independence for pairs of letters at any given distance k.
- Observed long-range correlations can be related to a hierarchical organization of genomes. Particularly, the alternation of exons and introns and repetitive sequences are, presumably, sources of long correlations.
- By studying the frequencies of "words" in DNA sequences in detail, grammar representations can be derived. The grammar complexity is a measure of compressibility of biosequences.

Regarding the ever accelerating growth of published DNA sequences, theoretical approaches as discussed in this paper are helpful to understand the basic mechanisms of translation, replication and evolution. We emphasize that information-theoretical tools, finite sample corrections and grammar representations can contribute to the classification of sequences and reveal their functional and evolutionary relatedness.

Acknowledgements

Our studies have been supported by CONACYT (project 1932-E9211, Mexico) and the Stiftung Volkswagenwerk. We acknowledge enlightening discussions with Oscar Chavoya, Edward Trifonov, and Jacob Engelbrecht.

3 Coding Coenzyme Handles and the Origin of the Genetic Code

Eörs Szathmáry

Abstract

The coding coenzyme handle (CCH) hypothesis for the origin of the genetic code is presented in a series of postulates and theorems. The hypothesis argues that (i) useful coding preceded translation; (ii) amino acids were first utilized as coenzymes of ribozymes in an RNA world; (iii) coenzymes were equipped with trinucleotide handles in a coded fashion; (iv) aminoacylation of handles was carried out by specific synthetase ribozymes (primordial assignment catalysts); (v) handles were essentially anticodonic in nucleotide composition; (vi) handles and complementary trinucleotides were part of the amino acid recognition site of ribozymes; (vii) the facts that the first letter of the codon is correlated with amino acid kinship in biosynthesis and that the middle base is correlated with amino acid polarity (the "code within the codons") follows from the theory; and (viii) Group I self-splicing introns are descendants of the primordial synthetases. The hypothesis is experimentally testable.

3.1 Introduction

The origin of the genetic code has puzzled biologists ever since the discovery of its nature. Many interesting ideas have been presented, which do not yet yield a nearly complete picture. Essentially, there are four ways to approach the problem:
1. Building of biochemical, "bottom-up " scenarios which are meant to explain the evolutionary pathway from a non-coding system to one of coding.
2. Collection of data about the recent evolution of the genetic code in the hope that this may reflect on the primeval evolution of the code.
3. The analysis of patterns in the genetic code by looking for correlations between amino acid and codon/anticodon properties.
4. Formulating hypotheses about the logic of extensions of the genetic code through the incorporation of novel amino acids.

Of course, concrete works quite often take more than one approach to the problem. Approaches 2 and 3 belong to the "top-down" category. Success can be expected when the bottom-up and the top-down approaches meet somewhere in between, like stalagmites and stalagtites in a cave when forming a column. Approach 4 need not be tightly linked to approach 1: theories belonging to this category often assume that some coding/translation system is already present.

It is not my aim to give a comprehensive review of the literature here, such can be found elsewhere [1–4]. Rather, I shall present the so-called coding coenzyme handle (CCH) hypothesis in a form which is different from previous expositions [3, 5, 6] in that it has a deductive structure: theorems or theses are deduced from other theses and postulates. Among the latter one finds pieces of experimental evidence as well as theses which work as axioms. In order to aid the reader, a postulate appears after the first thesis in which it is referred to. Theses and postulates are numbered, e.g. T1 and P1 refer to the first thesis and first postulate, respectively. This is necessary since the structure of the theory is not strictly sequential.

A note on terminology is in order. Whenever a presumptive component of genetic coding is addressed, it is referred to by the name of the current component, put in quotation marks. "Anticodon" thus refers to a triplet, the descendent of which is today's anticodon, despite of the fact that the direct role of the former is different from that of the latter.

In Section 2 the theory as such is presented, and Section 3 discusses some further issues, including suggestions for experimental testing.

3.2   Postulates and Theses

3.2.1   Origin of Coding Amino Acid Handles

T1. It follows from P1 and P2 that *protocells, utilising amino acids which reliably complemented the regular functional groups of ribozymes, were at a selective advantage.*
　　P1. The genetic code originated in an RNA world [7, 8] which was metabolically complex [9, 10]: ribozymes catalyzed the reactions of a fairly rich metabolic network.
　　P2. Ribozymes with an increased catalytic efficiency were selected through the competition of protocells containing these ribozymes (cf. Reference [11]).
T2. Itb follows from P3 and T1 that amino acids must have been utilized in a form which was different from a covalent modification of ribozymes. *Re-usable, specifically recognizable amino acid modules* would have been positively selected. This amounts to saying that *amino acids were introduced into the catalytic machinery as coenzymes of ribozymes.* Availability was guaranteed by P4.
　　P3. The size of the genetic alphabet was, as it is now, four (i.e. two base pairs: [12, 13]). Although some bases can be modified with aliphatic amines, carboxylates, sulfur and hydroxyl groups [9], this could not have been a widespread phenomenon because of the need of specific modifying enzymes. The need for enzymes which modify other modifying enzymes would have resulted in an infinite regress [3. 9].
　　P4. At least some amino acids were part of metabolism before they were used for anything else (cf. Reference [14]).This must be so because glutamine, glycine, aspartate and serine are present in the biosynthesis of purines and aspartate is needed for the biosynthesis of pyrimidines [15]. Ribozymes were able to recognize amino acids in reactions where the latter were among the reactants. It was shown experimentally that RNAs can recognize and bind Arg and citrulline [16, 17].
T3. It follows from P5 and T1 that *an organism using a larger number of amino acids had an advantage.*
　　P5. The number of biochemically occurring amino acids is large.
T4. From this, P6 and T2 follows that *many specific handles were needed.*

P6. Ribozymes utilized coenzymes with nucleotide-like handles, by which they could grab them [18, 19]. Many contemporary coenzymes are survivors of this ancient metabolic state.

T5. From this and P7 follows that the easiest solution was *to equip the amino acids with specific oligonucleotide handles.*

P7. A common theme of biology is the use of combinations of a finite number of building blocks to encode enormous amounts of sequential information. Nucleic acids are built on this principle. It turns out to be useful in their genetic as well as enzymatic role.

T6. Such *oligonucleotides were able to pair with ribozymes by conventional Watson-Crick type base pairing.* Thus an evolutionary bifurcation took place: amino acid handles could not have non-standard links between nucleotides such as is found in NAD and FAD.

T7. From this and T2 it follows that the lengths of these oligonucleotides must have been set by the conflicting needs of specificity (which increases with length) and reversibility of binding (which decreases with length). P8 suggests that the optimal number was 3: *handles were made of triplets.*

P8. It has been found that a trinucleotide can act as a cofactor of an existing RNA enzyme [20]. Data on association-dissociation between oligo- and polynucleotides are consonant with this view [21–23]. Oligomers of 9, 8, 6 and 4 nucleotides worked as efficient cofactors of enzymatic aminoacylation of single-stranded RNA derived from tRNA^{Ala} [24]. Note that in this latter case the synthetase is an enzyme requiring a stable duplex in its usual environment, and thus it is not surprising that the dissociable cofactor cannot be shorter than 4 nucleotides. Even so, the yield with a ribohexamer is greater than that with the more tightly bound nanohexamer.

T8. From this and P9 follows that *the number of required amino acids could not have been larger than the number of triplets.*

P9. The number of currently encoded amino acids is 20 (selenocysteine would be the 21st [2]). This shows that the number of triplets (64) is large enough to encode a highly evolved catalytic alphabet.

T9. From this follows that only one amino acid was assigned to a given triplet, but several triplets could have been assigned to the same amino acid. This was *the primordial "genetic code":* like its successor, it *was unambiguous and degenerate.* Thus *useful coding preceded translation.*

3.2.2 Specific Aminoacylation of Handles

T10. It follows from P4, P10 and P11 that *the attachment of amino acids to their respective handles must have been catalyzed by* specific assignment catalysts: *aminoacyl-handle synthetase ribozymes.*

P10. In biological systems specific reactions between small molecules are catalyzed by enzymes. Enzymatic catalysis renders slow and low-specificity reactions fast and discriminating.

P11. It has been shown that Arg is specifically recognized by an existing ribozyme, and all binding sites have a codonic sequence [16, 25]. tRNAs binding to amino acids (Arg and citrulline) were generated by *in vitro* selection [17]. The *Tetrahymena ribozyme* was shown to have aminoacyl esterase activity [26].

T11. From this and P12 it follows that *a C4N-like (anticodonic) part was* very likely *an essential part of synthetase ribozymes.* An exception (maybe not the only one) to this rule was *the synthetase for Arg, where the recognition site was codonic* (P11). Thus *coding was* *essentially stereochemical,* but *it was impossible* – due to the weakness of the association (P12) – *without the appropriate catalysis by ribozymes.*

> **P12.** It was shown that there is a weak, but measurable and specific association between certain amino acids and the first one or two nucleotides from their cognate anticodons [27]. Moreover, stereochemical models suggest that there is a lock-and-key relationship between amino acids and a complex of *four* *n*ucleotides (C4N) comprising the cognate anticodon and the so-called discriminator base (immediately next to the CCA 3'-end) of the respective tRNA [28].

T12. Since the *synthetases* must have been able to bind amino acids and triplet handles reversibly (T10), they *must have contained a triplet at the recognition site which was complementary to the respective handle.* Thus, during the charging reaction of the handle with the cognate amino acid, *a transient, formal "codon-anticodon" complex was formed.* From T9 follows that one part of the complex (presumably, mostly the "codon") was a triplet in the sequence of the ribozyme.

T13. It follows from P13 and T10 that the preferred attachment site of the amino acids to the handles was the 3'-end of the latter. Thus the structure of *the amino acid coenzymes was 5'-NNN-aa,* where N is any base and aa is an amino acid.

> **P13.** There is experimental evidence that spontaneous aminoacylation of oligonucleotides prefers the 3'-end (reviewed in Reference [29]).

T14. From this and T9 follows that wobble pairing between one base of the handle and the opposite base in the recognition site of the enzyme (T11) was *free to evolve.* This could not have been the middle one, since that would have rendered the binding energy of the "codon-anticodon" complex too low. Since the ribozymes attached the amino acid to the 3'-end (T13), wobbling at this end was presumably tolerated much less than on the other end. This suggests that *wobble pairing evolved at the 5'-end of the "anticodonic" handle, i.e. at the 3'-end of the "codonic" synthetase binding site.*

3.2.3   Extension of the Amino Acid Vocabulary

T15. From P14, T8 and T9 follows that *the number of nonsense triplets has decreased through evolution* as novel amino acids were recruited and the vocabulary was extended. Extension amounted to the takeover of nonsense "codons" – similar to the logic of stop-codon takeover advocated by Lehman and Jukes [30].

> **P14.** It is plausible to assume that the amino acid vocabulary was extended partly through the successive recruitment of amino acids from metabolism to a coded function [14, 30, 32].

T16. From this and T9 follows that *nonsense triplets did not mean termination signals:* in the absence of translation, there was nothing to terminate. Consequently, the mutational load was not increased to unacceptable levels due to frequent mutations to numerous stop codons.

T17. *Ambiguity reduction* [33, 34] *must have accompanied vocabulary extension* for the following reason. Stereochemical complexes have finite interaction energies. Therefore, the specificity of synthetases (T10, T11) was always finite. Synthetases were selected for increased specificity through selection for unambiguous coenzyme handles (T4).

> Indeed analysis of tRNA sequences revealed some traces of ambiguity reduction [35].

T18. *Amino acids with similar "codons" should tend to have similar polarity values.* This follows from the stereochemical nature of amino acid recognition (T11) which requires that similar structures should have similar functional groups in similar positions.

A significant correlation was found between the polarity distances of amino acids and their codonic Hamming distance in the genetic code [36], in line with earlier observations [34, 37].

T19. *Novel amino acids were successfully introduced into the coenzyme repertoire by the action of mutant synthetase ribozymes,* following from T9 and T10.

T20. *Vocabulary extension* must have *often followed a* so-called *Woese pathway:* the novel amino acid turned out to have a polarity value similar to the one whose mutant ribozyme was responsible for the extension. This follows from T18 and T19.

T21. *Amino acids sharing the same middle base must have, on the average, similar polarity values.* By "wobbling", the first "anticodon" base is excluded from determining amino acid identity (T14). The third nucleotide becomes aminoacylated, thus it is likely to be more specific for this synthetic step (T13) rather than recognition. More importantly, P15 says that this must be so.

> **P15.** The middle anticodon base of the C4N complex was shown in models to have paramount importance, whereas the third anticodon base preferentially binds to non-specific parts (NH_3^+ and COO^-) of the amino acids [28].

T22. *Vocabulary extensions by the Woese pathway were performed by mutant ribozymes coding for an altered third "anticodon" base.* This follows from T19–T21.

T23. *Amino acids must have occasionally been transformed into novel ones by metabolism while linked to their handles* (P1, P4). This is called a homeotopic transformation.

Two current examples are the amidation of glutamic acid on $tRNA^{Gln}$ and the addition of hydrogen selenide to an activated $tRNA^{Secys}$. Charged $tRNA^{Lys}$ is used in the synthesis of lipids, and $Glu\text{-}tRNA^{Glu}$ is important in the first step of chlorophyll synthesis in chloroplasts (see review, Reference [15]).

T24. *In several cases homeotopic novelties must have had polarity values rather different from those of the source amino acids,* following from P16.

> **P16.** It was shown that two amino acid characters, polarity and biosynthetic kinship, are uncorrelated [38]. This is understandable considering that the attachment of a single charged group to a molecule can change its polarity considerably: this transformation may require only one enzymatic step.

T25. Condition for successful recruitment of a homeotopically arisen novel amino acid (by a so-called *Wong pathway*) was that the transformation must be carried out on a *mutant handle, accepted by a mutant synthetase of the original amino acid.* If *the polarity of the new amino acid strongly differed from that of the source one,* this mutation must have occurred predominantly *at the middle position of the "codon/anticodon".* This follows from T21 and T24.

T26. It follows from this that *amino acids related in biosynthesis but unrelated in polarity should tend to share a common first "codon" base.*

T27. By combining T22 and T26 one can deduce *the "code within the codons",* first observed and named by Taylor and Coates [39]: *the first codon base is correlated with amino acid biosynthetic groups, the second with polarity.*

Thus anticodon distances should *independently* correlate with amino acid biosynthetic and polarity distances, respectively: this was indeed found [38].

3.2.4   Evolution of the Coding Machinery

T28. The *first synthetases must have been mutant forms of ribozymes manipulating other nucleic acids.* The first *amino acid to be charged to an oligonucleotide was replacing a nucleotide in the original reaction,* as suggested by Weiner and Maizels [40]. This is consistent with P1, P4.

T29. A further step in evolution was the *fortuitous application of two (or more) amino acid coenzymes at neighbouring positions* of the ribozymes. It follows from P2 that such transitions were often selectively advantageous.

T30. A *ribozyme* that could catalyze the *peptidyl transfer* between adjacent handles could be selected, provided *the peptide thus arisen could specifically bind to (an altered form of) the ribozyme on its own.* This follows from P17. Such an enzyme was a *presumptive "ribosome".* Peptidyl transfer is catalyzed by rRNA rather than by a protein, even today (P11).

> **P17.** RNA parts of enzymes were gradually replaced by coded oligo- and polypeptides, because of the enhanced catalytic efficiency of the latter [41]. Such RNA-peptide complexes were stable and specific. It was found that a protein derived from the p7 nucleocapsid of HIV enhances the catalytic activity and selectivity of a hammerhead ribozyme 10- to 20-fold by its strand exchange activity [42].

T31. By this evolution the *handles turned into "adaptors"* and became reusable (P18) in the sense that peptide cofactors did not need handles permanently, in contrast to amino acids.

> **P18.** It is advantageous for an organism to recycle its component molecules: a catalytic role is often preferable to a stoichiometric one.

T32. An evolutionary bifurcation took place: *some RNAs remained active as* enzymes, or rather *cores of RN peptide enzymes,* while *others became more and more efficient templates for oligopeptide synthesis.* This follows from P19 and P20. The latter became *presumptive "messenger RNAs".* Their DNA copies became *the first "protein" (micro)genes.*

> **P19.** The nature of the triplet code and the geometry of amino acids and nucleotides in general prevent the same RNA from being a ribozyme and a messenger for an isofunctional protein enzyme at the same time [3].

> **P20.** When adaptations of the same object serving different roles are in conflict, division of labour by the appearance of specialist objects is advantageous. There are several examples of this in evolution: e.g. the origin of many specific, efficient enzymes from a few, non-specific, weakly catalytic ones [11], or the origins of tissues in organisms, or the appearance of castes in insect societies [54].

T33. Evolution of presumptive "ribosomes" must have occurred through the *recruitment of further RNAs, peptides and (later) proteins.*

T34. Evolution of adaptors must have gone in a way that made them more efficient in binding to (i) the message, (ii) the ribosome and (iii) maybe to each other (see [44] for the latter). *Anticodons are the direct descendants of the handles.*

T35. It is likely that binding of handles to "mRNA" was the first adaptor function to evolve further. This could have been achieved by *replacing the trinucleotide handle with a structure essentially corresponding to the current anticodon stem-and-loop structure of tRNA.*

Similar structures must have been available in an RNA-based metabolism (P1). Moreover, P21 says that having a definite conformation for not only the messenger (first by being part of a ribozyme, then fixed by the ribosome) but also for the adaptor increases the strength of binding.

P21. It was demonstrated that binding between tRNAs with complementary anticodons is considerably stronger than between two triplets, or one triplet and one tRNA; e.g. between tRNA^{Phe} and tRNA^{Glu} the association constant is $3{,}6 \times 10^5$ M^{-1} at 25 °C. This is due to the loop constraint in both RNAs [45]. The association constant between tRNA^{Val} and tRNA^{Tyr} is 4×10^5 M^{-1} at 0 °C, which is three orders of magnitude larger than that between tRNA^{Tyr} with its codon triplet [46]. It was suggested on this basis that interaction between two RNA loops may have been important in the origin of the code [47].

T36. A necessary condition for the replacement of the primordial handles with stem-and-loops was the concomitant evolution of the synthetases. It was essential to match the amino acid against the anticodon, either directly or indirectly. It was perhaps at this stage that *the original C4N complex was broken up and the one-armed tRNA and the amino acid were recognized separately by the same synthetase.*

T37. *The relative frequency of the amino acids in proteins should deviate from what would follow from their codon multiplicities alone.* This follows from the stereochemical nature of primordial coding (T11) and the fact this assignment had nothing to do with amino acid functionality in proteins *later* in evolution.

This expectation is borne out: selection against the code, as observed by Jukes et al. [48], is apparent: lysine, aspartic acid, glutamic acid, and alanine are overrepresented whereas arginine, serine, leucine, cysteine, proline, and histidine are underrepresented in proteins relative to their respective codon multiplicities.

3.2.5   The Role of Complementary Strands in Coding and Translation

T38. When the original handles were replaced with one-armed tRNAs (T35), the easiest solution was to *adopt the positive as well as the negative strands* (P22), since they were replicationally available anyway.

It is very interesting that consensus sequences of the anticodon arms of tRNAs with complementary anticodons are still very similar to each other [49].

P22. Sequence evolution is opportunistic: it tries out all available sequences, and retains the successful combinations.

T39. The structure of *the genetic code allowed the recruitment of the "sense" as well as the "antisense" strands of ribozymes for messenger function.* This follows from P22, T21 and P23:

P23. As Konecny et al. [50] demonstrated, the fact that silent substitutions on one strand correspond to chemically conservative mutations on the other, and *vice versa,* is a consequence of the middle base determining amino acid polarity. As was shown by Zull and Smith [51], secondary structures are also conserved.

T40. Following from P22 and T38, protein enzymes of metabolism are expected to show not only sense-sense, but also sense-antisense sequence similarities. *Sense-antisense similarities should predominantly be found between genes of proteins of unrelated functions,* since the genetic code does not allow isofunctionality of proteins translated from complementary strands.

This follows also from the role of the codonic mid-base (T21): hydropathy of amino acids with complementary codons must be different. Remarkably, it has been shown that genes of many enzymes and even RNAs from various parts of metabolism show sense-antisense similarity. Enzymes are included from glycolysis, the TCA cycle, and the pentose phosphate cycle [52].

3.2.6   Avoiding the Error Catastrophe of Translation

T41. The serious *problem of an error catastrophe of translation is avoided in the CCH model.* The number of sites for amino acids with a coding handle, necessary for this coding itself, must have been very low. This ensured that a self-coding system could arise without erroneous assignments leading to more erroneous assignments.

This follows from P24 and the fact that the bulk of enzymatic activity was carried out by RNA (P1), and that a definite stereochemical bias of assignment was present (T11):

> **P24.** Models of error propagation of translation (coding) show that the probability of nucleation of a coded system increases with the degree of stereochemical bias and decreases with the number of sites that should be occupied by coded amino acids for sufficient specificity [53, 54].

3.3   Discussion

It has been assumed that coding was after all a stereochemical process, aided by ribozymes, in which the anticodon played a crucial role. Although conclusive evidence for this is by no means present, it seems to be the best-supported model to date. That assignments of amino acids to codons were made by ribozymes was assumed by Weiner and Maizels [40], though without presenting any stereochemical detail on how this could be done. For the time being, it seems not unwise to assume the validity of the C4N model and see how far one is able to get with it.

It has not escaped my attention that a detailed stereochemistry for codon-amino acid interactions was also worked out recently [55], but it holds that there are many other pieces of evidence in favour of the anticodon hypothesis (reviewed in Reference [56]). For example, there is a significant correlation between the retention factor (in a certain system) of the first two anticodonic dinucleotides and a combination of a polarity measure and bulkiness of the cognate amino acids [57, 58].

The treatment of binding Arg by its codons (T11) may seem muddled, but it is extremely relevant that, as observed by Lacey et al. [56], these codons, thought to be essential for the interaction, are actually base paired with Arg anticodons in the structure, so it is right to wonder what the anticodon is doing there. This becomes even more important if one considers the deduced significance of such a pair for the present model (T12). In fact I remarked earlier that formation of such a pair would have allowed Nature to decide whether the codon or the anticodon should play a more important role in amino acid recognition [3]. Note, however, that the Arg-binding site of RNAs generated by *in vitro* selection seems codonic, and *not* base paired [17].

If the anticodon did play an important role in amino acid assignments from very early on, it would be reassuring if it were still important in the recognition of tRNA identity by current synthetases. The past ten years changed our knowledge dramatically: the anticodon is important for 17 out of the 20 *E. coli* isoaccepting groups [59]. Lacey et al. [56] call attention to the fact that the anticodon-binding site of the Gln-tRNA synthetase consists of hydrophilic amino acids, which is remarkable since Gln is also hydrophilic. Moreover, among them one finds Gln itself! This is certainly consistent with the anticodon hypothesis.

In this paper I have not dealt with a realistic model of the synthetases, but I suggested previously that they may have been the ancestors of present-day self-splicing Group I introns. This is supported, among other things, by the facts that (i) these introns are found in tRNA genes of eubacteria; (ii) they are inserted immediately downstream of the anticodon;

(iii) they form a codon-anticodon complex with the end of the upstream exon; (iv) Group I ribozymes were shown to have aminoacyl esterase activity (P11) (Reference [3] and further refs. therein).

Table 3-1   Summary of critical transitions in the CCH hypothesis

Initial state (preadaptation)	Final state
coenzymes of ribozymes	coenzymes of proteins
	amino acid residues
amino acid handle	tRNA
triplet of handle	anticodon
metabolic ribozymes	mRNAs of protein enzymes
	disappeared RNA cores
handle binding sites of ribozymes	codons
peptidyl transferase ribozyme	ribosome
peptides bound to ribozymes	proteins
handle charging ribozyme	group I self-splicing intron

Where two final states are indicated, an evolutionary bifurcation is understood.

Another hypothesis, somewhat similar to mine [3, 5], was published at about the same time in which Gibson and Lamond argued that almost all *substrates* of ribo-organisms were adaptor-mounted, including amino acids [60]. The ancestors of ribosomes were "metabolosomes", binding metabolites by their adaptors to orchestrate biosynthetic processes. The crucial difference in my hypothesis is that amino acids have a predominantly catalytic role as cofactors. In addition, they did not discuss the logic of vocabulary extension, or the evolution of adaptors themselves, or the mechanism of specific aminoacylation (how do the adaptors become mounted?). Moreover, I assume that coding evolved after, and not before, cellularisation.

Experimental tests are crucial for the ultimate fate of the CCH theory. The first thing to do is to select for amino acid binding RNAs, and to look for the composition of the specific binding sites [61]. This has been started by Famulok [17]. At a later stage it would be useful to select for RNAs that have a built-in C4N part [3]. Next, the role of amino acid coenzymes in ribozymatic reactions should be investigated.

The major transitions from preadaptations to adaptations as conceived in the CCH model are listed in Table 3-1. A graphical representation of the same ideas is given in [6]. The major theoretical work to be done in the foreseeable future is:
1. Elaboration of a detailed model for the coevolution of synthetases with adaptors.
2. Presentation of a scenario for amino acid vocabulary extension.
3. Drawing of a version of the CCH model which does not rest on the assumption of a direct amino acid-(anti)codon interaction.

Acknowledgements

I thank Andreas Dress and John Maynard Smith for encouragement and support, Michael Famulok for discussion and Rüdiger Wehner for his hospitality in his department. This work was partly supported by the Hungarian National Scientific Research Fund (OTKA), grants no. I/3 2046 and T44-62.

4 RNA Replication and Evolution

Christof K. Biebricher

Abstract

RNA molecules replicating *in vitro* with the help of an RNA replicase are hitherto the most simple system capable of Darwinian evolution. RNA molecules reproduce, mutate and compete as organisms do; moreover, in this system a quantitative description of the evolution is possible: mutation rate and selection rate values can be calculated from physico-chemical parameters which predict quantitatively the evolution behaviour. A precise derivation of the phenotypic properties, e.g., the replication efficiency, from the genotype, the nucleotide sequence of the RNA, is not yet possible; however, structural elements required for the replication have been identified. Artificially designed short RNA sequences having this structural property were synthesized and have been found to replicate.

4.1 Introduction

Most scientists agree that the enormous complexity of life developed from simpler systems by a most powerful and efficient self-organization process, the biological evolution: more or less random mutations occur during reproduction of organisms; selection culls out the most efficient mutants while discarding inefficient ones. The selection success is evaluated globally, i.e., by the functioning of the whole system under the prevailing conditions. In view of the complexity of even the most primitive organisms, it seems obvious that quantitative evolutionary studies require much simpler systems.

Evolution is not bound to life itself. Three conditions suffice for observing Darwinian evolution [1]: i) Metabolism, i.e., the system must be far from equilibrium; ii) Autocatalysis, an information carrier must reproduce its information with high fidelity; iii) Mutation, the reproduction fidelity must have a certain limit.

While the first two conditions may be realized by autocatalytic reproduction of any chemical system, the third condition requires the presence of information defining certain "types" of the species, in biology realized by the sequence of nucleotides in a nucleic acid sequence. Also needed is an evaluation system for the reproductive success, conveniently quantified by the relative excess production rate of the "type", i.e., its relative synthesis rate minus its relative decomposition rate [1]. Molecular evolution can be studied in the test tube if a suitable system fulfilling these conditions can be found. Despite considerable progress in the last decade, a purely chemical system for autocatalytic amplification of molecules carrying information has not yet been developed [2]. The simplest system that meets the criteria is still one developed by Spiegelman and collaborators more than 25 years ago [3, 4]: enzymatic RNA replication by RNA replicase. The RNA replicase is a sophisticated enzyme derived from a virus, the coliphage Qß; as templates serve certain short-chained

RNA species [5]. The enzyme is purified from biological material to homogeneity and then used as an environmental factor catalysing the RNA replication in a defined chemical system containing replicating RNA and the precursor nucleoside triphosphates. During an experiment, the enzyme is neither produced nor consumed, its concentration remains constant. In this system, an RNA molecule capable of replication multiplies virtually indefinitely, i.e., until the resources are used up.

The molecular basis of both RNA and DNA replication is Watson-Crick type base pairing. However, while DNA replication involves separation of the two strands of the double helix and completion of the single strands to double strands, the features of RNA replication are different and shall be discussed in this paper in more detail: (i) A single-stranded template is used to synthesize and liberate a single-stranded complementary replica [6–8]. No proteins other than the replicase are involved in the reaction and no nucleoside triphosphates are consumed except the ones that are incorporated into RNA. (ii) The replicase is highly discriminating in accepting RNA as template [5, 9–11]. Therefore, the RNA template shares the catalytic properties with the enzyme. Probably, the defined secondary and tertiary structure of the RNA is recognized by the replicase and takes part in catalysing the many steps of the replication mechanism.

4.2   The Mechanism of RNA Replication

The RNA replication mechanism of Qβ replicase is now well understood [3, 6–8, 12–16]. The replicase binds in a first step to the RNA template. The binding strength, however, is not strongly correlated to the efficiency of replication: Several RNA templates unable to replicate, e.g., rRNA, bind replicase quite strongly [17, 18], while some short-chained replicating RNA species bind rather weakly [19, 20]. Replication is initiated by geminal association of two GTP molecules at the 3' terminus of the template followed by phosphodiester formation. From the appropriate nucleoside triphosphate substrates an antiparallel and complementary replica strand is synthesized. Replication termination involves replica liberation and eventually the slow dissociation of the resulting inactive template-replicase complex into its components.

When the synthesized replica is also able to serve as a template, the two complementary strands act cross-catalytically and an exponential (autocatalytic) amplification of the RNA takes place. Exponential amplification is only possible if excess replicase is available; when replicase is saturated with template, a linear synthesis rate of strands is observed [12, 13].

The two complementary strands of an RNA species can react with each other to form a double strand, which is unable to serve as a template [8]. Therefore, the rate of double strand formation has to be taken into account as a destruction rate of template. The destruction term is dependent upon the square of the RNA concentration; thus, a steady state in the production of template is obtained where the synthesis of new strands is balanced by the loss of double strand formation [14].

The concentration profiles can be calculated by numerical integration of the pertinent differential equations specific for the growth phase; the agreement between the predicted and the measured growth profiles is excellent [10, 13–15].

4.3   Quantitative Determination of Selection Rate Values

Different RNA species in the same incubation medium share the same environment and compete for resources; hence, their relative population sizes change by selection forces. In this system, quantitative selection rate values (defined as relative change of the relative population size of a type) can be calculated when the pertaining rates in the replication mechanism are known. However, the selection values are constant and independent of the RNA concentration only at an infinite dilution of the RNA; under these conditions, competition is excluded and the selection rate values are simply equal to the relative growth rates. At higher RNA concentrations, however, in particular in the linear growth phase, the selection rate values are not correlated to the growth rates at all: often a species with a lower growth rate takes over rapidly. This is not surprising, because under these conditions the RNA species must compete for enzyme molecules, and RNA species that bind these molecules most rapidly have the highest selection rate values. At still higher RNA concentration losses by double strand formation become significant for the selection rate values and even negative selection rate values can occur. Quite complicated RNA profiles are obtained, because the growth of the different RNA species changes the environment and influences the rate values. The details can not be discussed here, but it has been shown that the selective rate values of every RNA species can be *quantitatively predicted* from the synthesis rates, the loss rates, the enzyme binding rates etc. of the different species. Again, the growth profiles can be calculated by numerical integration of the rate equations and an excellent agreement with the complicated experimental profiles is obtained [10, 11, 15, 16].

4.4   Mutation Rates and Mutant Frequencies

The nucleotide misincorporation rates in RNA replication *in vitro* and *in vivo* are several orders of magnitude higher than in DNA replication, because RNA replication lacks a proof-reading mechanism [21]. The error rate of Qß replicase is about 3×10^{-4} [22, 23]. The probability of producing a mutant copy is calculated to be less than 3% per strand and replication round, and one would expect that the majority of those mutants would have only one sequence position altered.

However, the mutant frequency, i.e., the relative proportion of a mutant to the total population, does not depend on the error rates alone: mutants produced during replication compete with each other and the mutant frequencies are also functions of the selection values. After many replication rounds, a steady state mutant distribution – the quasispecies – is obtained where each mutant type achieves a constant mutant frequency which is a function of its formation-by-mutation rates as well as of its selection rate values [1, 16, 24].

Mutant distributions of replicating RNA species were determined by sampling the sequences of single strands. Direct sequencing of single strands, however, is not yet possible and single strands have to be amplified ("cloned") before sequencing. Amplification was done at the DNA level, because DNA replication is much more accurate than RNA replication and lacks a selective bias. Only 40% wild type was found, accompanied by mutants with several base substitutions. We conclude that the mutant distribution is largely determined by selection forces; indeed, the multi-error mutants in the population were found to be neutral or nearly neutral [25], probably by compensating mutations, because the secondary structure of the RNA was conserved. Regions where the primary sequence was conserved suggest that they play an important role in recognition.

4.5   Structural Requirements of RNA Replication

The phenotypic expression in the RNA replication system is reduced to the efficiency of an RNA type to direct its own synthesis by Qß replicase. How can we correlate genotypes and phenotypes? Suggestions are derived from sequence comparison of a large number of different RNA species replicated by Qß replicase, particularly of the very short (35–45 nucleotides) "minihelices" that are formed independently in template-free incorporations [11, 20, 26, 27]. However, no consensus sequences were found except for the invariant termini: the sequences of both complementary strands begin with pppGG(G) and end with (C)CCA, whereby the 3'-terminal A is attached by the replicase to the synthesized replica without template instruction. Quite large pieces of RNA could be inserted into replicating RNA species with little loss of template activity, as long as the ends were preserved and the RNA inserts met some structural criteria [28].

The lack of a "consensus sequence" for recognition is not unusual in RNA. In tRNA-synthetase binding, certain chemical side groups in a defined sterical arrangement are recognized [29]. Since it is still not possible to calculate tertiary structures from the nucleotide sequence of an RNA, we used algorithms for calculating secondary structures [30, 31], to compare the secondary structures of replicating RNA species.

The calculated secondary structures of the replicating RNA species show striking similarities. While their 5' termini are usually involved in double-helical structures, their 3' termini are unstructured. This is an unusual feature: one would rather expect that a double-helical 5' end in one strand would correspond to a 3'-terminal double helix in the complementary strand, since the complementary strands are antiparallel. The fact that both complementary strands show 5' terminal double helices is caused by the contributions of G:U base pairs at strategic positions [20].

Some replicating sequences are short enough for a synthetic approach to test whether the 5' terminal structure is required. A synthesized nucleotide sequence of a replicating RNA species was replicated normally by Qß replicase. After 20 replication rounds, the progeny RNA was found still to be a quasispecies distribution around the starting sequence. This was expected, because the sequence had already reached the local optimum in the fitness landscape. However, when a mutant of this sequence was synthesized where a 3'-

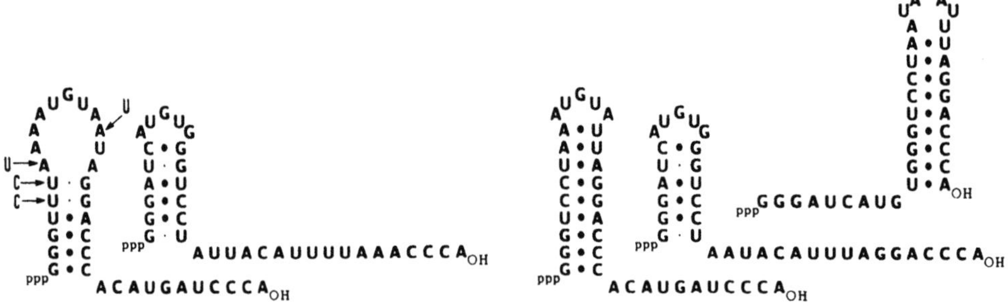

Figure 4-1   Tentative structures of plus and minus strands of a short replicating RNA species (left) and of a mutant unable to replicate (right). The altered positions are indicated in the left part. Two structures are shown for the mutant minus strand, the rightmost being the most stable.

terminal double helix was favoured in one of the strands (Figure 4-1) the RNA was indeed unable to replicate.

To test whether the proposed condition might also be sufficient for replication, a few RNAs that have the required structural features but were not found in previous experiments were devised, synthesized and investigated for their ability to replicate with Qß replicase. Replication was indeed observed, but proceeded initially at a very low rate. The progeny RNA produced by replication was cloned and sequenced. A sequence drift with a few base exchanges was observed (Figure 4-2). It was shown by another few rounds of replication followed by cloning and sequencing that this mutant RNA was evolutionary stable and had thus reached its local fitness maximum. Figure 4-2 shows that the structural feature we consider as being required for replication was improved. In other examples, optimization by mutation included also duplication of sequence parts by recombination processes [32].

Minihelices replicate much more slowly than optimized RNA species with longer chain lengths. Obviously additional sequence features also contribute to the replication efficiency of RNA. In particular, the short sequences bind replicase only weakly and are thus inefficient templates; optimized RNA species with longer sequences contain some additional sequence parts specific for replicase binding. A highly conserved interior stem in the RNA species MNV-11 is probably responsible for replicase binding.

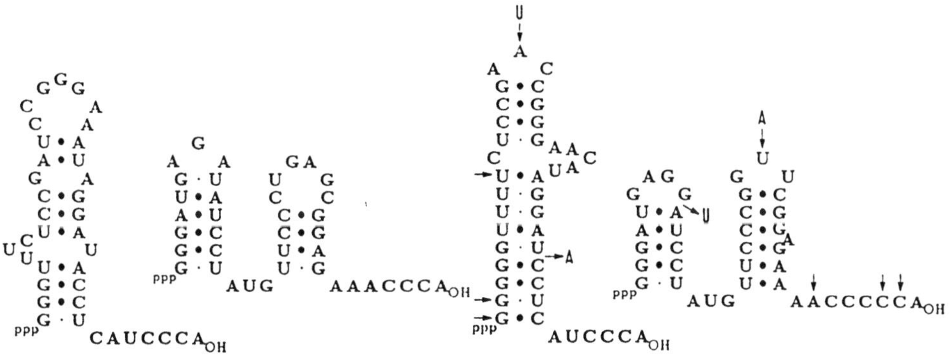

Figure 4-2   Tentative structures of plus and minus strands of an artificially devised short RNA species (left) and the optimized mutant formed during amplification (right). Arrows between two symbols designate base exchanges, arrows from internucleotide spaces mean base deletions and arrows pointing to symbols base insertions.

Optimized replicating RNA species are also selected for minimizing their rate of double strand formation. In large RNA molecules, regions sensitive to double strand formation can be buried in the interior of the RNA bulk. Medium-sized replicating RNA species were found to contain particularly stable RNA stems [20, 33]. Short-chained RNA species can minimize the rate of double strand formation by the described folding of the 5' termini, because in this way the maximum number of bases is made unavailable for contact in either of the two strands; loops in one strand correspond to stems in the complementary strand.

4.6   RNA Replication Catalyzed by RNA Polymerase

Surprisingly, DNA-dependent RNA polymerase, e.g., from E. coli or coliphage T7, may behave also as RNA replicases when supplied with appropriate RNA templates [34, 35] even though their role *in vivo* is transcription, not replication. The mechanism of RNA amplification by RNA polymerase has not yet been studied in detail, but what is known suggests that the reaction proceeds essentially as with Qß replicase. Therefore, the structures of template and replica RNA probably aid strand separation to a large extent. The extraordinary specificity of all RNA replicases suggests that much of the replication potential is provided by the RNA itself. However, we observed that RNA species replicated by Qß replicase are not accepted by T7 RNA polymerase, and vice versa. Replication is thus a concerted action of the replicase and the template and replica RNA strands.

4.7   Template-free RNA Synthesis

In the absence of extraneously added template, Qß replicase is able to synthesize a template *de novo*. To observe this remarkable reaction, extreme care has to be taken to remove all RNA impurities from the solution [36]. The reaction requires high concentrations of enzyme and substrates and occurs only after long lag times. In a first step, Qß replicase condenses – at a rate 5 orders smaller than in template-instructed synthesis – nucleoside triphosphates to oligonucleotides with chain lengths of 5–50 and composition $pppGpR(pN)_n$. The sequence of the oligonucleotides is random except for the 5' terminus [27]. If one of the sequences produced can replicate, no matter how slowly, it gets amplified and eventually grows out of the solution. The template-free RNA synthesis is kinetically clearly distinguishable from a template-instructed process [12]. The first templates that can be isolated are different in each experiment [11, 26], have short chain lengths (30–45) and replicate at relatively low rates [20]. Preliminary experiments indicate that T7 RNA polymerase shows analogous reactions.

The reaction is a remarkable example of how biological information is gained by trial and error.

4.8   Summary and Conclusions

The results may be summarized as follows:
- RNA replication is a system exceptionally suitable for the quantitative study of evolution processes *in vitro*.
- RNA molecules reproduce, mutate and compete like organisms. Depending on environmental conditions, the selection pressure and the selection rate values change, but can be calculated from physico-chemical parameters.
- The sequence complexity and thus the information content of an RNA molecule often increases during its evolution.
- The RNA is not merely the template of the replicase, but shares the catalytic role with the replicase.
- The synthesized RNA is liberated in single-stranded form; the RNA structure participates in strand separation.

- RNA species suitable for replication have common structural features: Their 5' termini are involved in a double-stranded form, while their 3' termini are not.

One may speculate why this structure might be important. During replication, a single-stranded complementary replica strand is synthesized and then template and enzyme are recycled. The synthesis of the complement suggests the transient formation of a double helix between replica and template at the replication fork. During chain elongation the replica must be peeled off the template by stepwise strand separation and the separated strands must be protected from re-forming a double strand. Stem structures within the replica (and probably also the template) could serve for this purpose and are particularly important at the 5' terminus of the replica where strand separation starts. Indeed, when strands can not separate, e.g., in the terminal elongation reaction producing hairpins [32], replication stops after incorporation of a few nucleotides. Furthermore, medium-length replicating RNA species are usually highly structured [8], but formation of branched helices, e.g., cloverleaf structures, is not observed, since they could form only after liberation of a large single-stranded stretch. Inserting unstructured RNA stretches into the sequences of replicating RNA species diminishes the replication efficiency [28].

Primitive RNA replication could have been mediated by ribozymes [37, 38]. All known ribozymes are highly specific for their RNA substrates [39, 40]. It seems plausible that primitive RNA replication (i) was highly selective in accepting RNA templates and (ii) proceeded by strand separation and not by completion of a template-primer complex to a perfect double strand.

Acknowledgements

I thank R. Luce for many experimental contributions and Prof. M. Eigen for his interest and support. This work was supported in part by a grant of the European Community (PSS*0396).

5 DNA Binding and Bending by two different Proteins: Factor for Inversion Stimulation (FIS) and Tetracycline Repressor (TetR)

Wolfram Saenger, Dirk Kostrewa, Joachim Granzin, Frank Cordes,
Claus Sandmann, Caroline Kisker, and Winfried Hinrichs

Abstract

The two proteins, Factor for Inversion Stimulation (FIS) and Tetracycline repressor (TetR),
occur as homodimers. They contain similar DNA-recognition modules composed of helix-
turn-helix motifs which bind to DNA by insertion into adjacent major grooves. FIS recognizes a large number of DNA sequences with "weak and bendable" pyrimidine-purine steps
which are seven base pairs apart. It bends the DNA upon binding because the separation of
the helix-turn-helix motifs is 5 Å shorter than required for binding to adjacent major
grooves in undistorted B-DNA which are 34 Å apart. TetR, in contrast, binds to a very
specific sequence of DNA. Upon binding of the antibiotic tetracycline, TetR is conformatio-
nally changed so that the separation between the helix-turn-helix motifs increases to 39 Å
and binding to DNA is abolished.

5.1 Introduction; Biological Significance of FIS and TetR

In recent years, a number of different DNA-binding proteins have been characterized with
respect to their functional and structural properties. In this article, we compare two of the
DNA-binding proteins, Factor for Inversion Stimulation (FIS), and Tetracycline repressor
(TetR). These two proteins have totally different functions and three-dimensional structures,
but share a comparable helix-turn-helix motif that recognizes and inserts itself into adjacent
major grooves in B-form DNA.

FIS binds to a number of seemingly unrelated DNA-sequences and bends DNA by 90°
so that it adopts a configuration which is required for certain processes in the regulation of
gene expression. These involve gene inversion; gene excision and gene insertion in several
phages; activation of transcription of tRNA and rRNA genes; and the regulation of the
biosynthesis of FIS itself. The number of different processes in which FIS is involved
requires that FIS recognizes a number of different DNA sequences. Of these, 17 have been
identified so far by footprint experiments. They have a very loose consensus sequence T/G
N N Py Pu N N A N N Py Pu N N C/A., where N denotes the presence of A, G, T or C; Py a
pyrimidine (T or C); Pu a purine (A or G); the slash / means "or".

In contrast, TetR recognizes a certain DNA sequence, 5'-CTATCATTGATAG-3'. It is
the protein that regulates the resistance against the antibiotic tetracycline. In the absence of
tetracycline, TetR binds to two operators (DNA of above-given sequence), one regulating
the expression of a membrane-intrinsic resistance protein that exports tetracycline out of the

cell, thereby causing resistance, and the other regulating the expression of TetR itself. If tetracycline enters the cell, it binds to TetR and induces a conformational change so that TetR releases the operators and transcription of the two genes can proceed.

FIS and TetR are compared in this article. Since the helix-turn-helix motifs in dimeric FIS are separated 5 Å shorter than required for the insertion into adjacent major grooves in straight DNA which are 34 Å apart, the DNA is bent to reduce this distance. In contrast, the TetR dimer in the absence of tetracycline binds to straight DNA; if tetracyline forms a complex with TetR, it induces a conformational change in the repressor. This increases the separation of the helix-turn-helix motifs up to 39 Å, so that binding to B-DNA is no longer possible.

5.2   General Principles of Protein and Nucleic Acid Structure

5.2.1   Protein Structure

A protein is a linear polymer consisting of mostly 20 different amino acids with different side chains which are linked by peptide bonds in a protein-specific sequence. The polypeptide is folded into a certain three-dimensional structure to provide the functionality of the respective protein [1]. Since the peptide bond is planar due to partial π-character (Figure 5-1), the bonds about which rotation of atomic groups and consequently folding of the polypeptide chain into a three-dimensional structure can occur are the N-Cα and Cα-C

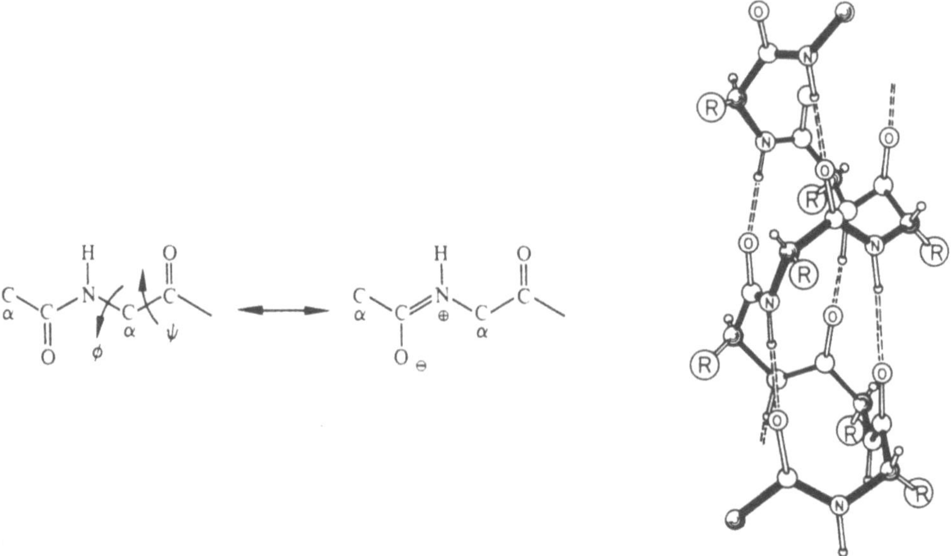

Figure 5-1   (left) The peptide bond is planar due to partial π-character, and folding of the polypeptide chain can only occur around Cα-C and Cα-N bonds. The 20 naturally occurring amino acids are characterized by different side chains attached at Cα. (right) Folding of a polypeptide segment into an α-helix; R denotes the side chains, the N-terminus is at the bottom, the C-terminus at the top.

bonds, with torsion angles ϕ and ψ respectively. The rotation around these two bonds is not totally free but restricted to certain ranges because there is steric hindrance between the amino-acid side chains attached at the Cα atoms. As a consequence, there are two main domains, which are characteristic for the formation of α-helices and of β-pleated sheets. These "secondary structure" elements are internally stabilized by hydrogen bonds $N-H\cdots O=C$ between the peptide groups. The arrangement of secondary structure and unstructured segments in space is called "tertiary structure". It is characteristic for each protein and determines its functional properties. The polypeptide chains of both, FIS and TetR, are folded such that only α-helical structures are found; they are connected by short sequences that are not α-helical.

5.2.2   Structural Principles of Nucleic Acids

DNA is the material that encodes the biological information and is inherited from one generation to the next. Chemically, it is a polynucleotide consisting of four different nucleotides which are arranged in a certain sequence and define the content of biological information. Two of these nucleotides are purine nucleotides, adenosine (A) and guanosine (G), and two are pyrimidine nucleotides, thymidine (T) and cytosine (C) (see Figure 5-2). According to the concept developed by Watson and Crick, there is complex formation between A and T to form a base pair A-T, and G and C to form a second base pair G-C, which is structurally isomorphous to A-T [2]. In DNA, two strands of the polynucleotides are bound together by complementary base-pairing of the Watson-Crick type, i.e. complexation by base-pairing A-T, T-A, G-C, C-G. The two strands are in antiparallel orientation and twisted in a righthanded double helix with a ten base-pair repeat at 34 Å. Since the sugar-phosphate backbones are not attached at diametrically opposite sides of the base pairs but on one side, the DNA double helix has two grooves, the minor groove formed between the sugars and the major groove on the opposite side.

5.2.3   The Interaction between DNA and Helix-Turn-Helix Motifs

A family of DNA binding proteins contains helix-turn-helix motifs to recognize and bind the DNA double helix. The helix-turn-helix motif consists of two parts [3]. The first helix is called the "supporting helix", and the second helix is the "recognition helix". As these names suggest, the "supporting helix" is necessary for the geometrical stabilization of the helix-turn-helix motif, which is due mainly to hydrophobic interactions between several amino-acid side chains of the two helices. The "recognition helix" inserts itself into the major groove of the DNA double helix. It forms specific contacts between amino acids on this "recognition helix" and functional groups which are located in the major groove of DNA. These interactions are mainly of the hydrogen bonding type, i.e. $N-H\cdots N$, $N-H\cdots O$, $O-H\cdots O$, $O-H\cdots N$. In addition, there can be salt bridges between positively charged amino acids such as lysine and arginine, and the negatively charged phosphate groups, i.e. $N^+-H\cdots {}^-O-P$.

There is no definite scheme how the "recognition helix" of the helix-turn-helix motif inserts itself into the major groove of DNA. This can occur with the N-terminal end, or with the C-terminal end, or the "recognition helix" can insert itself as a whole into the major groove. Since for FIS as well as for TetR, the structures of only the proteins are known and not of protein-DNA complexes, it is uncertain how the interaction between "recognition helix" and major groove works in detail. For the FIS-DNA complex, a model has been built that gives an idea of the bending of DNA around FIS.

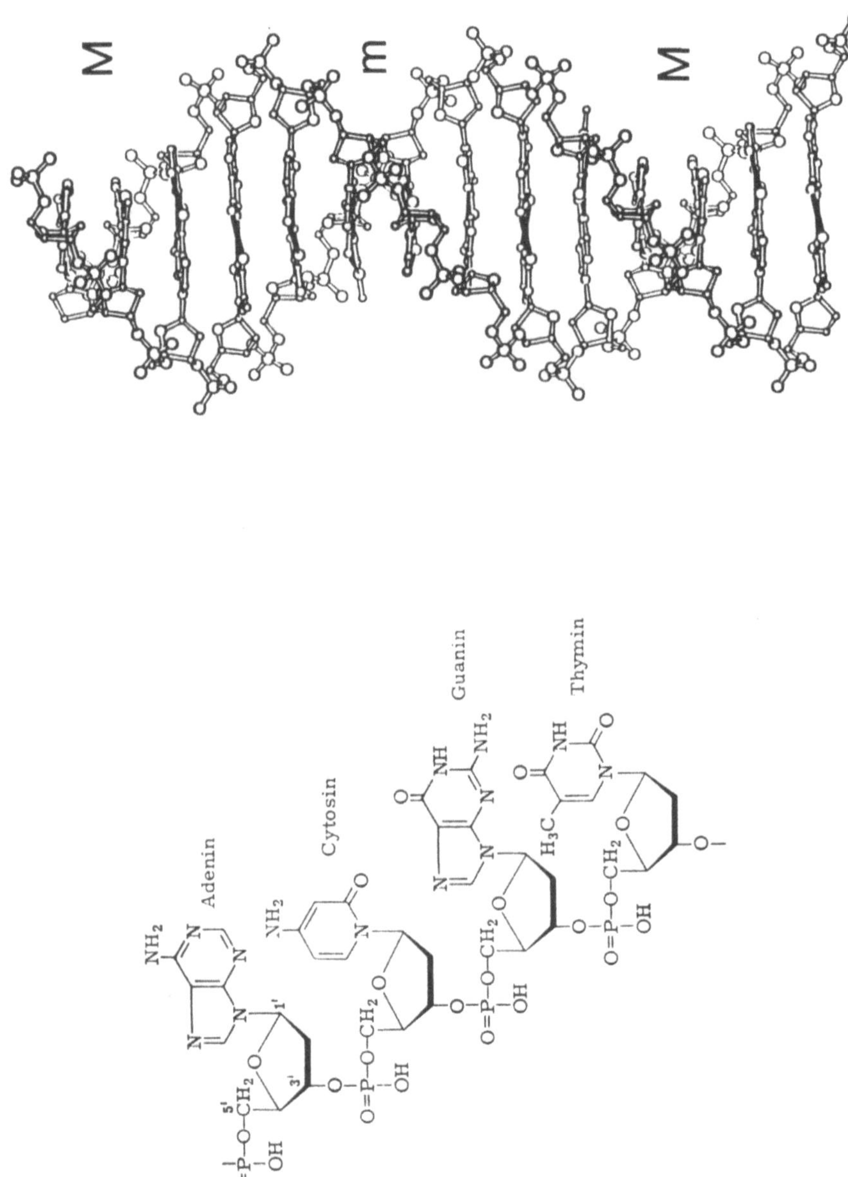

Figure 5-2   Chemical (left) and three-dimensional (right) structure of DNA. The helical repeat (pitch) of the double helix is 34 Å. Major and minor grooves are indicated by M and m, respectively.

5.2.4   Some Words about X-Ray Crystallography

The best method to determine the structures of molecules is X-ray crystallography. Its main disadvantage is that the respective molecule has to form crystals of about 0.5 mm in diameter. If such a crystal is illuminated by monochromatic X-rays, a diffraction pattern is obtained because the electromagnetic radiation interacts with the electron shells of the atoms in the crystal lattice which acts as a three-dimensional grid. The observed diffraction pattern is the Fourier transform of the electron distribution in the crystal. A Fourier back-transformation based on the diffraction pattern that contains several thousand individual reflections, permits us to obtain a picture of the electron density in the crystal. This electron density is then interpreted in terms of atomic structure of the respective molecule [4].

With small organic or inorganic molecules, one usually obtains a very strong diffraction pattern with diffraction extending to high resolution, i.e. 0.6 to 0.9 Å. This is sufficient to resolve individual atoms because the atom-to-atom bonding distances are in the range of 1.2 to 1.7 Å. With protein crystals, the obtainable resolution is usually lower because these crystals contain about 50% solvent which gives rise to structural fluctuations and consequently uncertainties in atom positions. This, in consequence, leads to a break-down of the high resolution diffraction data so that the maximum resolution is in the range of 1.5 to 2.5 Å, i.e. individual atoms cannot be resolved, but the envelope of atomic groups is defined. This in turn permits the interpretation of a protein structure in atomic detail, although the atoms are not individually "seen". Of the about 50% solvent in a protein crystal, oxygen atoms of water molecules can usually be located in the first hydration shell, but higher hydration shells are diffuse or not defined at all because water molecules are mobile and consequently do not contribute to the diffraction of the X-rays.

5.3 Three-dimensional Structure of FIS

FIS occurs as a homodimer and binds as such to the DNA double helix. The sequence of the 98 amino acids in the monomer is given in Figure 5-3. In the crystal structure, there is one FIS dimer in the asymmetric unit. Of both monomers, the first 23 N-terminal amino acids are so poorly defined in the electron density that they could not be located with cer-

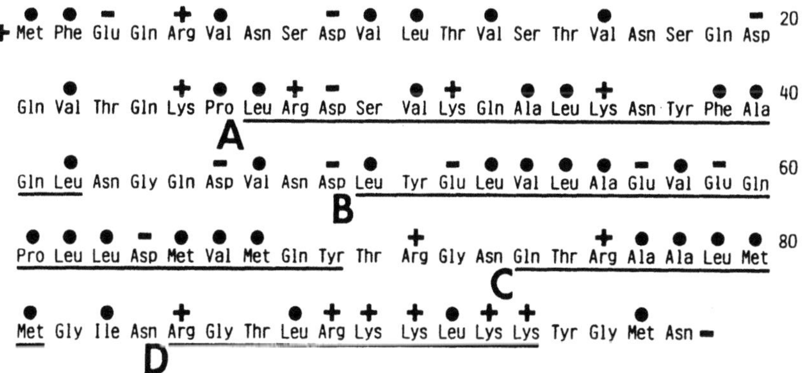

Figure 5-3   Amino acid sequence of FIS. α-Helices A, B, C, D are shown by lines; •, +, – indicate hydrophobic, positively and negatively charged amino acids respectively [5, 10].

tainty. From Pro 26 until Asn 98, the polypeptide chain is folded into four α-helices called A, B, C, D [5]. In the dimer, Figure 5-4, α-helices A and B form a hydrophobic core that stabilizes the FIS dimer, and helices C and D form the helix-turn-helix motif. The recognition helix D carries 6 positive charges (Figure 5-3) which is more than found in any other of the known helix-turn-helix motifs [3]. The reason for this accumulation of positive charges appears to be that FIS is not very specific and has to recognize and to bend DNA with a variety of different sequences. Consequently, ionic interactions with the negatively charged phosphate groups on DNA will predominate.

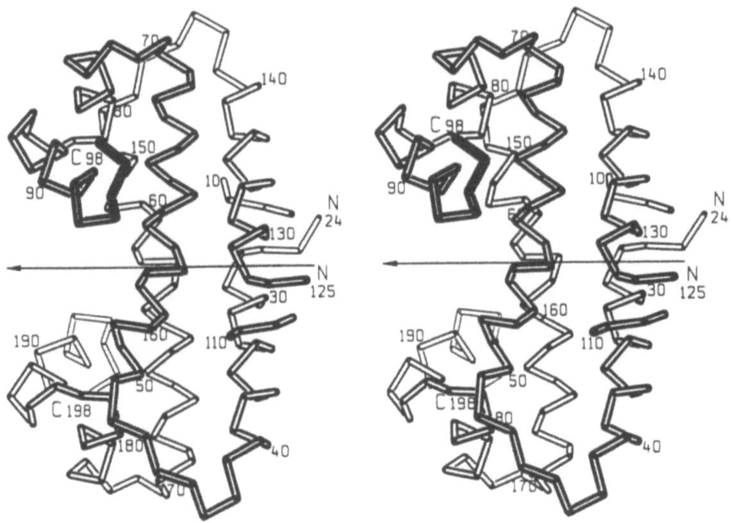

Figure 5-4   Stereo view showing the molecular structure of the FIS dimer. Only Cα-carbon atoms are presented, the N-terminus (Gln24) is indicated by N, the twofold axis relating the two monomers in the dimer is shown by an arrow. DNA recognition helices D, D' are to the left. In one of the monomers the amino acids are numbered 24 to 98, and in the other 125 to 198.

A particular feature of the FIS dimer is that the recognition helices in the helix-turn-helix motifs are separated not by 34 Å as required for binding to adjacent major grooves of a straight DNA double helix, but their separation is 5Å too short. If we assume that FIS does not change its structure significantly upon binding to DNA and that DNA is flexible, this means that DNA has to bend in order to bring its adjacent major grooves into contact with the recognition helices in FIS. This bending is not only obvious from the X-ray structure of FIS, but it has also been shown experimentally on the basis of gel retardation experiments.

Since attempts to cocrystallize FIS with DNA have so far been unsuccessful, we have done a model building study based on energy minimization calculations and computer graphics. In initial attempts, we contacted FIS with DNA that was bent smoothly as in the nucleosome core. This yielded a satisfactory first model, which was then subjected to a number of energy minimization cycles where the overall structure of the FIS dimer was kept rigid as found in the crystal structure, and the structure of the DNA double helix was allowed to change. This resulted in a model where the DNA is not bent smoothly around the FIS dimer, but it is straight in the center and around the twofold axis that passes through the FIS dimer and through the DNA, Figure 5-5 [6]. Further out, there are two bends symmetrically disposed about this twofold axis, and even further out the DNA double helix is straight again. This kind of bending was also found in the crystal structure of the complex formed between DNA and the catabolite activator protein (CAP [7]). However, the smooth bends found in the FIS-DNA complex are replaced in CAP-DNA by sharp kinks that occur at pyrimidine-purine sequence steps, which are known to be more flexible than other sequence steps.

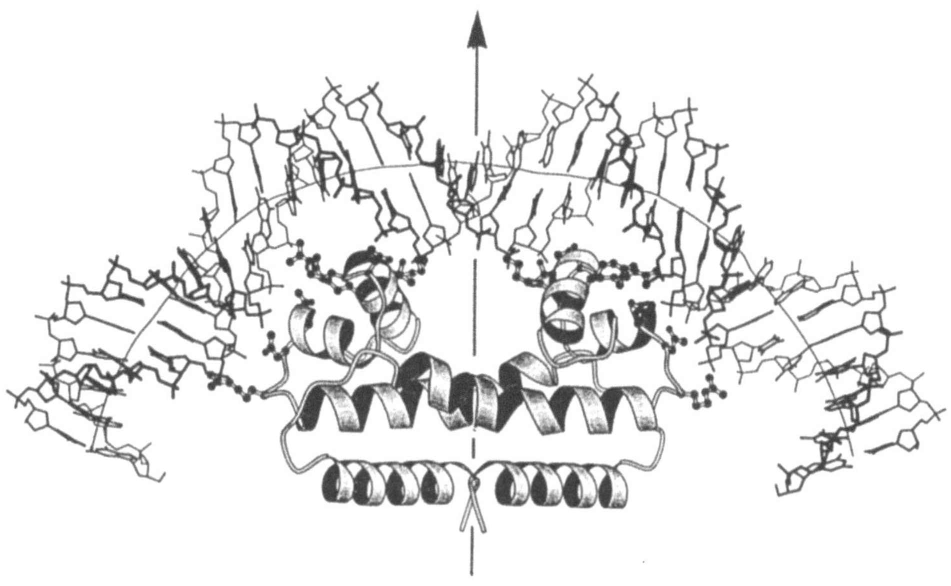

Figure 5-5   Computer-generated model of the FIS-DNA complex [6]. The twofold rotation axis that passes through the complex is indicated by an arrow.

How can this be rationalized on the basis of the DNA sequences that are recognized and bent by the FIS dimer? In Figure 5-6, DNA sequences are collected that are recognized by FIS. The bottom line of this table shows the "consensus" sequence, but the consensus is not very strict. The most conserved nucleotide is the one at the center of the consensus, A/T, which, however, is also not found in one of the DNA sequences, *fis I*. Another motif that is conserved are the two pyrimidine/purine steps at nucleotides −4,−3 and +3,+4, which are found in *all* of the sequences and represent the only really conserved feature. Since pyrimidine-purine steps are known to be flexible [8], it appears that FIS recognizes this part of the DNA, binds to it and stabilizes the bending by means of the ionic interactions between the positive charges on the recognition helix and the phosphate groups of the DNA double helix.

In two genetic engineering studies [9, 10], FIS was mutated in several positions to map its interactions with DNA and with the enzymes that are involved in the excision and inversion processes. These studies showed that mutations in the helix-turn-helix motifs interfere with binding of DNA, thereby abolishing the biological activity. On the other hand, if the 25 N-terminal amino acids that were not defined in the electron density and are disordered or mobile in the crystal structure were deleted, these FIS molecules still bound and bent DNA, but the action of the invertases was abolished. Consequently,

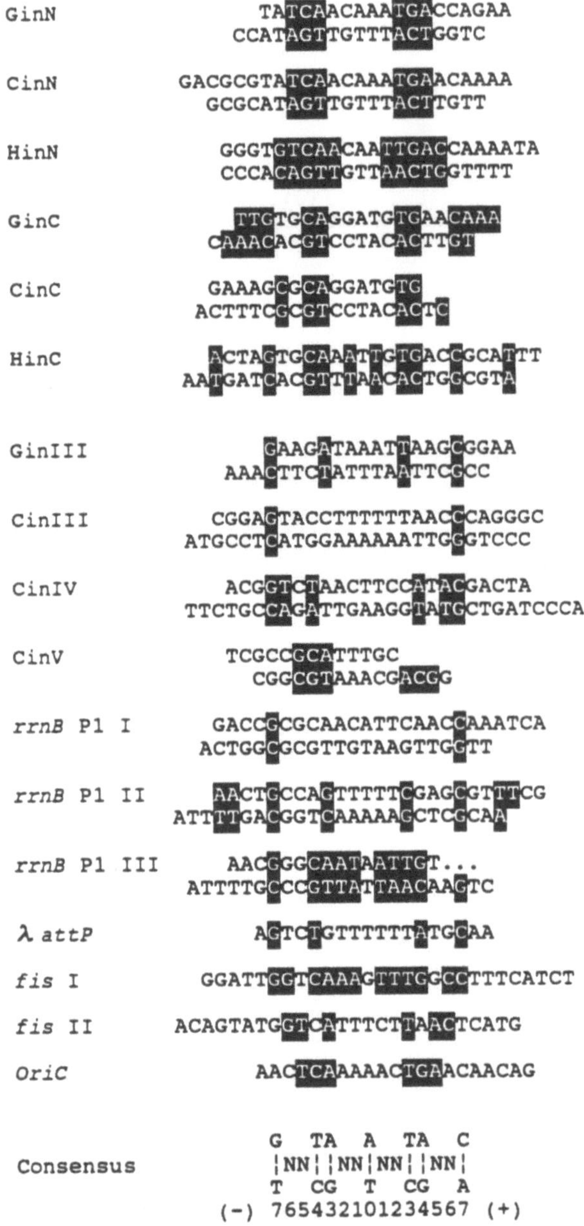

Figure 5-6   FIS binding sites on DNA. The DNA sequences are arranged according to consensus sequences; nucleotides in palindromic (i.e. twofold symmetric) arrangement are boxed.

FIS has two sites of interaction [11]: the helix-turn-helix motif interacts with the DNA and brings it into the proper configuration for the invertases to cut and religate the DNA, whereas the N-terminal tail interacts with these enzymes and brings them into proper orientation to interact with DNA.

5.4   Three-dimensional Structure of Tetracycline Repressor (TetR)

TetR consists of 207 amino acids and occurs also as a homodimer, with one monomer in the crystal asymmetric unit. In contrast to FIS, all the amino acids are well defined in the electron density obtained by X-ray analysis at 2.5 Å resolution [12]. The polypeptide is folded into 10 α-helices. The helix-turn-helix motif is found near the N-terminus and formed by α-helices 2 and 3. α-Helix 4 connects this motif with the core of the protein, α-helices 6 to 10, which are responsible for dimer stabilization and form a cavity that accommodates the inducer tetracycline, Figure 5-7.

We should keep in mind that the complex between TetR and tetracycline does *not* bind to DNA. The reason becomes clear from this crystal structure which shows that the separation between the two recognition helices 3 and 3' (the prime' refers to the "second" molecule in the homodimer) is 5 Å too wide to interact with adjacent major grooves which are 34 Å apart in the double helix of DNA. Since DNA cannot extend further without large structural deformations, there is just no interaction possible between DNA and TetR if complexed with tetracycline. The reason for this particular structure of TetR must be associated with the binding of tetracycline.

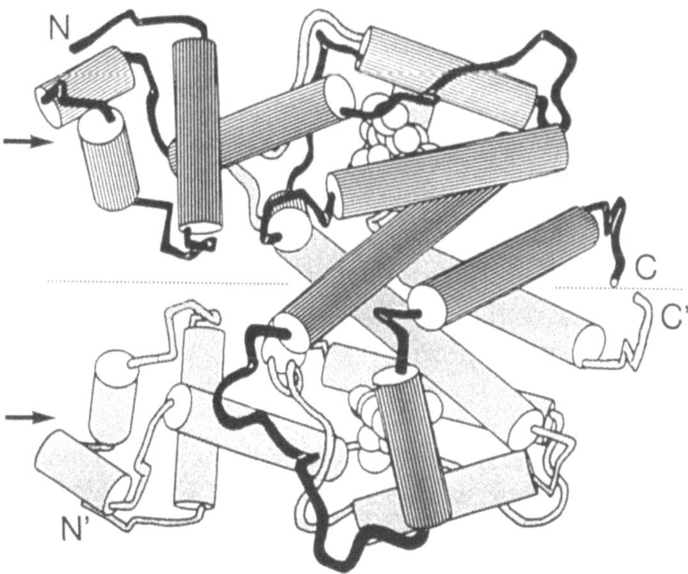

Figure 5-7   Three-dimensional structure of the tetracycline repressor (TetR) in complex with tetracycline. The dotted line indicates the twofold rotation axis that relates the two monomers in the TetR dimer. The N and C termini are marked (N, C for one monomer and N', C' for the other), and the atoms of the inducer tetracycline are shown as spheres. The arrows point at the helix-turn-helix motifs, α-helices 2,3 and 2',3'.

If we look at the binding site, tetracycline interacts with both monomers in the TetR homodimer in a very specific and complicated scheme that engages ionic, hydrogen bonding and hydrophobic interactions (see Figure 5-8). These interactions also include a divalent cation (Mg^{2+} in the crystal structure), which is required for proper binding of tetracycline to TetR. The binding involves amino acids in the core of the TetR homodimer and amino acids at the C-terminus of α-helices 4 and 6, which according to our present assumptions act as seesaw and lever. It appears that upon binding of tetracycline to the TetR dimer, the interaction with α-helix 4 triggers a conformational change in TetR that affects the spatial orientation of this α-helix. The conformational change widens the distance between the two helix-turn-helix motifs which are directly attached to the N-termini of α-helices 4 and 4' in the two monomers of the homodimer.

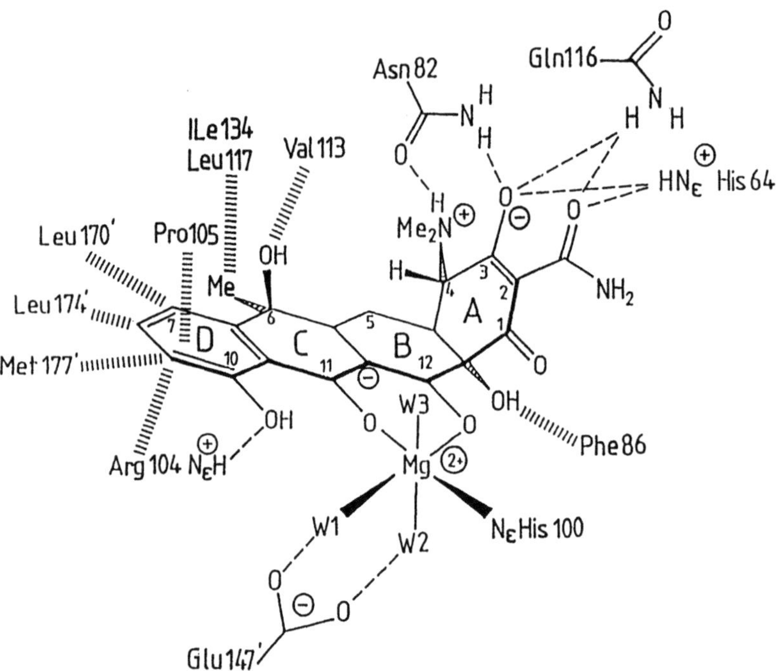

Figure 5-8   Schematic description of tetracycline-TetR interactions. Hydrogen bonds are indicated by dashed lines and hydrophobic (van der Waals) interactions by (llllllllll); W1, W2, and W3 are water molecules; amino acids with a dash belong to the "second" monomer in the TetR dimer.

Conversely, if TetR occurs without tetracycline or if tetracycline is removed from the complex with TetR, we assume that there is a conformational change that affects the seesaw-and-lever machinery and permits the helix-turn-helix motifs to approach each other at a distance of 34 Å that is perfect for interaction with straight DNA. In model building studies the orientations of the two α-helices 4 and 4' in the TetR homodimer were modified such that the separation between the helix-turn-helix motifs reduced to 34 Å. This TetR dimer was docked onto a straight DNA double helix with the specific sequence to which TetR binds,

5'-C T A T C A T T G A T A G-3'

•

3'-G A T A G T A A C T A T C-5'

Except for the outermost base-pairs, this sequence is palindromic, i.e. it has a twofold symmetry at the central T/A base pair (indicated by a dot), and this symmetry matches with the twofold symmetry of the TetR dimer. This, of course, facilitated the docking since the twofold axes of the DNA double helix and of the TetR dimer were superimposed, and there were only two degrees of freedom: the movement along and the rotation around this axis of DNA against TetR. This study showed that there is a good match in the interaction between the major groove of DNA and the recognition helices.

TetR has been characterized structurally and functionally by a number of genetic engineering experiments. Mutational studies on TetR and on the specific DNA sequence to which it binds have provided evidence that there are certain amino acids on α-helix 3 which interact specifically with the operator DNA. Other mutation experiments have indicated the amino acids that interact with tetracycline, and the results are consistent with the picture given in Figure 5-8. Moreover, a number of amino acids could be identified by mutational studies that are important for the induction of TetR, i.e. the conformational change that occurs on binding tetracycline. Most of these amino acids are positioned around the binding site for tetracycline, and some others are located between the two monomers in the TetR dimer, which supports the induction mechanism that we have proposed.

5.5   Conclusions

The two studies described in this paper shed light on the specific interactions between proteins and DNA. If DNA binds to FIS, it has to bend by about 90° because the recognition helices in the helix-turn-helix motifs of the FIS homodimer are 5 Å too close to match the 34 Å separation of adjacent major grooves in straight DNA. The bending occurs mainly at the two pyrimidine/purine steps which are 7 base pairs apart and the only conserved feature in all the known DNA sequences to which FIS binds. On the other hand, TetR binds to a specific DNA sequence in the absence of tetracycline because the separation of the helix-turn-helix motifs in the TetR homodimer matches the 34 Å between major grooves in straight DNA. If tetracycline is bound to TetR, this induces a conformational change that widens the separation between the recognition helices in the helix-turn-helix motifs by 5 Å so that recognition of DNA is no longer possible. As a consequence, the repressor releases DNA and transcription of the genes coding for the tetracycline resistance protein and for TetR can occur.

Acknowledgements

This work was supported by the Deutsche Forschungsgemeinschaft through the Sonderforschungsbereich 344 and the Leibniz-Programm, by the Fonds der Chemischen Industrie, and by the Bundesministerium für Forschung und Technologie (FKZ 05 5KEIAB5).

6 Function Based on Organization and Recognition: From "Complicated" Biological Systems to "Simple" Synthetic Systems

Steffen Denzinger, Achim Dittrich, Wolfgang Paulus, and Helmut Ringsdorf

6.1 Introduction: Function Based on Organization

Within the last few years, *Supramolecular Science* [1] has become the new title for describing the cumulative achievements at the interface between chemistry, physics, and biology. It bridges the gap between Life Science on one hand and Materials Science on the other.

What are functional supramolecular systems? One of the most perfect examples is a cell membrane [2]. Here, mother nature demonstrates the interplay of molecular self-organization and molecular recognition, assembling structures whose function is based on their organization. The glycocalix contributes to molecular recognition, lipids self-organize in double layers, the cytoskeleton proteins stabilize the cell and channel forming proteins provide access to the interior of the assembly. Thus, the scientific challenge to understand, build up, and mimic natural structures should focus no longer on single molecular performance but on that of molecular assemblies: the whole is certainly more than the sum of its parts in this case. Is this, however, a new way of thinking? Certainly not, since it was a basic insight of ancient philosophy in Asia and Europe: only the interaction between the parts creates the whole and its ability to function.

Supramolecular interaction was first postulated to be the source of molecular function as long ago as 1894 when Emil Fischer casually introduced his metaphor for selective interaction of molecules: the *lock and key principle* [3]. This idea has been a guiding light for one of the most stimulating fields in modern science, that is, for *Supramolecular Science*.

Today, the field has far surpassed its original focus on host-guest systems and is now located at the "borderline" between Life Science and Materials Science. It applies the principles of self-organization, regulation, replication, communication, and cooperativity, offering new routes for the design of materials in which organization precedes function. In this way, it has advanced to a promising area of applied science demonstrating that the "borderline" between scientific disciplines is in reality no borderline at all. Figure 6-1 represents this schematically.

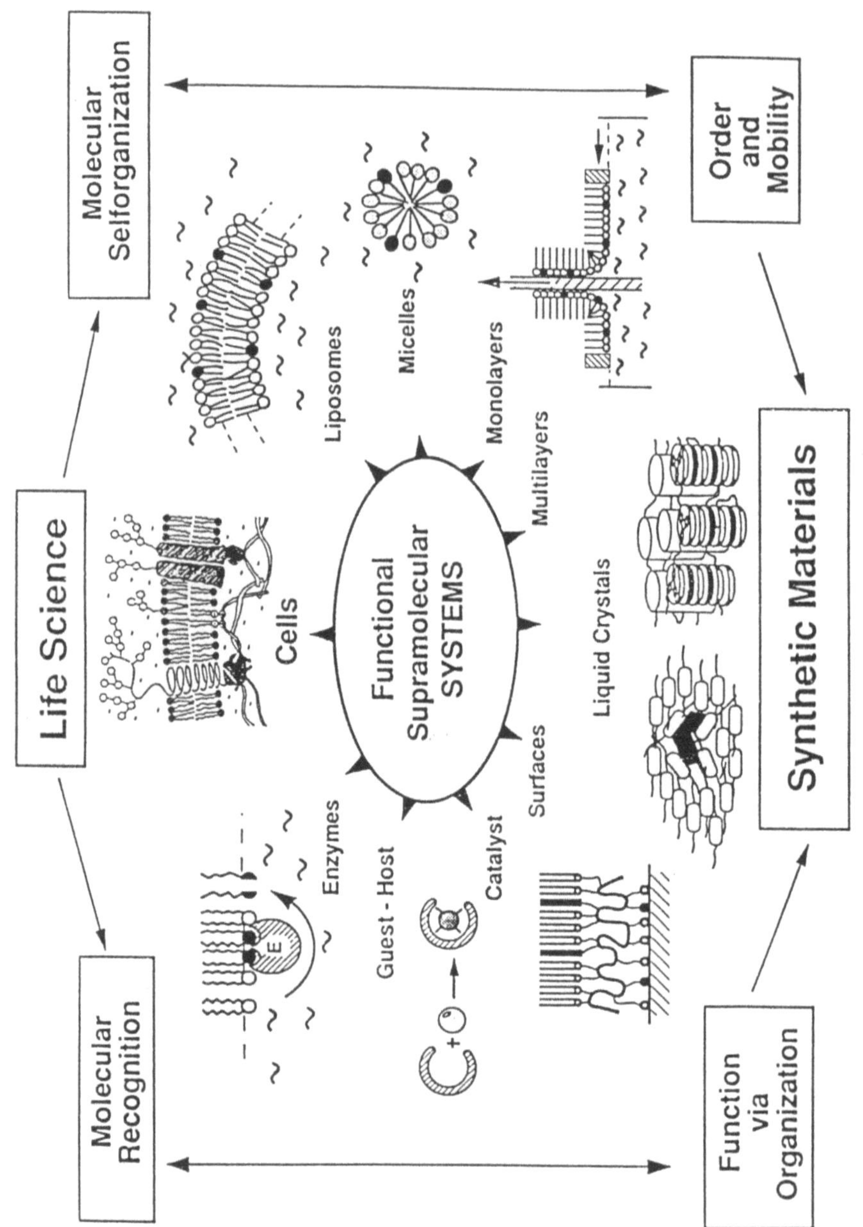

Figure 6-1 Functional supramolecular systems – a connecting link between Life Science and Materials Science.

The broad interest in supramolecular science also comes from the perspectives this field appears to offer [4]. The molecular evolution of life – as yet an unsolved problem – can be addressed in new light. Along the way to this ultimate intellectual goal, significant progress is expected e.g. in molecular biology, drug delivery systems, and in biophysics, where it should have a strong impact on pharmaceutics and medicine. Furthermore, the next decades will plunge us much deeper into what we already call the age of information. The growing need to increase information storage capacities and processing speed will finally force us to use to perfection the principles nature developed. We will learn to use nature's minute building blocks, molecules, as a source and carrier of information and to exploit their cooperativity and self-assembly to transmit information.

In the following pages, we want to give two examples designed to contribute to a better understanding of function based on molecular recognition and organization. One – the function of phospholipase A_2 at phospholipid monolayers – is a purely biological example where nature's skills in recognizing induced function are taken advantage of, the other – based on the interaction of barbituric acid amphiphile monolayers with triaminopyrimidine – is a purely "synthetic" approach to this problem. Figure 6-2 elucidates the principle for both examples discussed in this chapter.

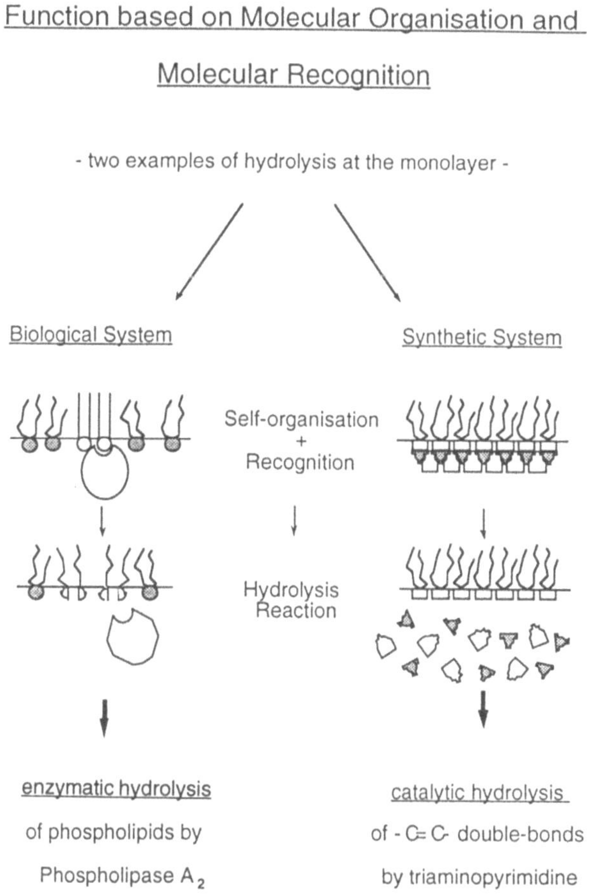

Figure 6-2   Molecular recognition and function in biological and synthetic systems.

6.2   Phospholipase A$_2$

6.2.1   Function and Properties of Phospholipase A$_2$

Phospholipase A$_2$ (PLA$_2$) constitutes a family of widely studied, ubiquitous, small, water-soluble lipolytic enzymes. PLA$_2$ is responsible for catalyzing hydrolysis of the 2-acyl ester bond of the 3-sn-glycerophospholipids membrane yielding an acyl fatty acid and the corresponding lysolipid (Figure 6-3). However, the exact mechanism is still unknown. The remarkable activation of this enzyme in response to organized interfacial phospholipid substrates [5] as opposed to disperse lipid monolayers, has made it an ideal tool for investigating membrane structure as well as for modeling enzymatic behaviour.

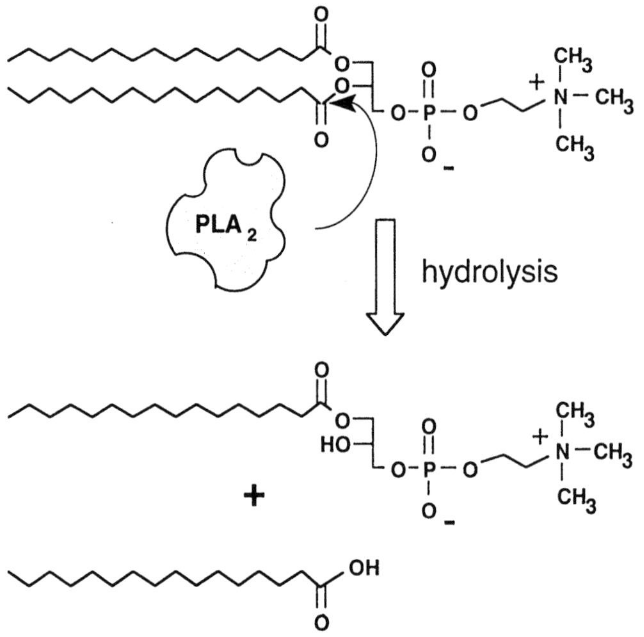

Figure 6-3   Cleavage reaction catalyzed by phospholipase A$_2$ (schematic): the enzyme hydrolysis of the ester bond in C-2 position of a lecithin, resulting in the corresponding lysolecithin and the free fatty acid.

Phospholipase A$_2$ activity depends strongly on the physical state of its substrate [6–9]. It increases many times when going from a homogeneous solution to an organized substrate such as micelles, liposomes or monolayers. The drastic enhancement of PLA$_2$ hydrolytic activity in the lipid phase transition region as well as a higher hydrolytic activity when confronted with gel-phase lipids as opposed to liquid-crystalline phases are important examples of how critical the physical state of the lipid substrate is to enzymatic activity.

6.2.2   Hydrolysis of Substrate Lipids of Phospholipase A$_2$ in Monolayers

In order to observe the hydrolytic action of phospholipase A$_2$ with the fluorescence microscope, two different fluorescence labels were needed. The monolayer was doped with a sulforhodamine-lipid (SR-DPPE = *N*-(Texas Red sulfonyl)dipalmitoyl L-α-phosphatidyl-ethanolamine), and the phospholipase with fluorescein isothiocyanate (FITC). In combination with a fluorescence microscope (Figure 6-4), these two labels make it possible to observe processes in the lipid monolayer and the behaviour of the enzyme separately [10]. By using a rhodamine filter in the fluorescence microscope, one only observes fluorescence from the monolayer doped with sulforhodamine, reflecting changes in the morphology of the lipid layer. On the other hand, using the fluorescein filter, the lipid layer becomes virtually transparent, and fluorescence is observed only from the fluorescein which is attached to the protein.

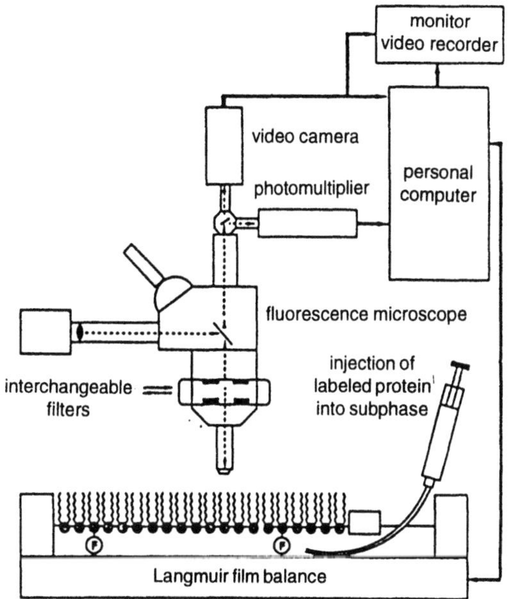

Figure 6-4 Schematic representation of a fluorescence film balance (combination of a Langmuir film balance with a fluorescent microscope: the fluorescence microscope is used to observe a monolayer at the gas-water interface. The processes taking place can be recorded by a video system. The fluorescence can be quantified by means of a photomultiplier. The interchangeable filter system can be used for selective observation of the monolayer and the protein by using different fluorescent dyes. An experiment of this type is outlined in Figure 6-5).

The lipid monolayer is compressed into the phase transition region. This leads to the formation of solid-analogous lipid domains in a liquid-analogous matrix (Figure 6-5A). Injection of the labelled phospholipase A_2 into the subphase (Figure 6-5B) is followed by specific recognition between the enzyme and its substrate. The attack by the protein takes place preferentially at the boundary between the solid-analogous and the liquid-analogous phase (Figure 6-5C) [11]. As hydrolysis progresses, the solid-analogous lipid domains are broken down, and the hydrolysis products accumulate in the monolayer (Figure 6-5D). After a certain time, the phospholipase A_2 starts to aggregate and to form domains of its own which have a specific shape (Figure 6-5E). This may indicate a change in the conformation of the protein due to interaction with the hydrolyzed monolayer. These processes can be observed directly under the fluorescence microscope [12], as will be illustrated by using L-α-dipalmitoylphosphatidylcholine (DPPC) and L-α-myristoylphosphatidylcholine (DMPC) monolayers as examples.

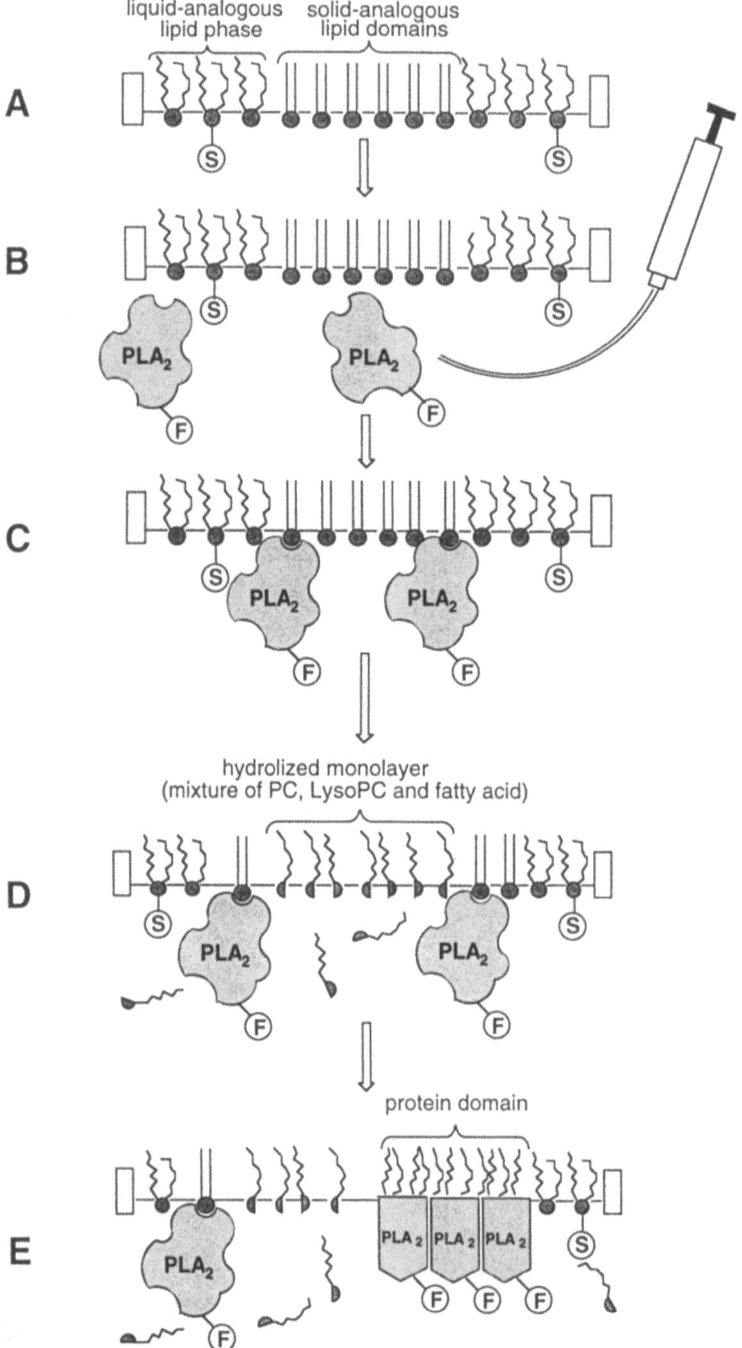

Figure 6-5   Hydrolysis of a lipid monolayer by phospholipase A₂ (schematic): (A) monolayer in the phase transition region with solid-analogous lipid domains in a liquid-analogous lipid matrix mixed with a sulforhodamine marker; (B) injection of the FITC-labelled phospholipase A₂; (C) specific recognition of the substrate lipids by the enzyme and preferential attack at the boundaries between liquid- and solid-analogous phase; (D) hydrolysis of the solid-analogous lipid domains and accumulation of the hydrolysis products in the monolayer; (E) aggregation of the enzyme; PC = phosphatidylcholine.

6.2.2.1   Hydrolysis of Solid-Analogous L-α-DPPC Domains

The enzyme is injected under a monolayer which is in the phase transition region. Viewed through the sulforhodamine filter, dark solid-analogous lipid domains which are typical for L-α-DPPC [13] are seen in the bright matrix of liquid-analogous lipid and sulforhodamine marker (see Figure 6-5A and the scheme in Figure 6-5A, B). With the fluorescein filter, the homogeneous fluorescence of the protein is seen under the monolayer (see Figure 6-6B). The first signs that the enzyme has recognized its substrate in the monolayer and has started to hydrolyze it appear after 10 minutes. First, areas of erosion are formed specifically at the concavities previously present in the lipid domains (see Figure 6-6C and the scheme in Figure 6-5C). The enzyme then starts to degrade the domains inwards from these points (Figure 6-6C, E). In consequence, there is local accumulation of the acid and lysolecithin, which may lead to autocatalytic activation of the phospholipase A_2 at this point [14]. This would explain why further hydrolysis from an initial point of attack is preferred. After about 60 minutes, only a few remnants of the lipid domains are left (Figure 6-6E, G, I, L).

Observation through the fluorescein filter reveals small bright dots after 30 minutes (Figure 6-6F), which appear dark through the sulforhodamine filter: the protein is starting to aggregate underneath the monolayer [12, 14a, 15]. These bright dots increase in number and size as the reaction progresses (Figure 6-6F, H, K) until, at its end (after about 60 minutes), they have a typical kidney shape (see Figure 6-6L). The monolayer now consists only of almost completely destroyed solid-analogous lipid domains and the protein domains in a matrix of liquid-analogous lipid (see Figure 6-6L, M).

It is noticeable that the domains of phospholipase A_2 are similar in shape to some of the lipid domains present at the start of the reaction (see Figure 6-6L, M, and Figure 6-5A). To examine whether there is a template effect, the hydrolysis experiment was repeated with L-α-DMPC which forms solid-analogous domains of different morphology.

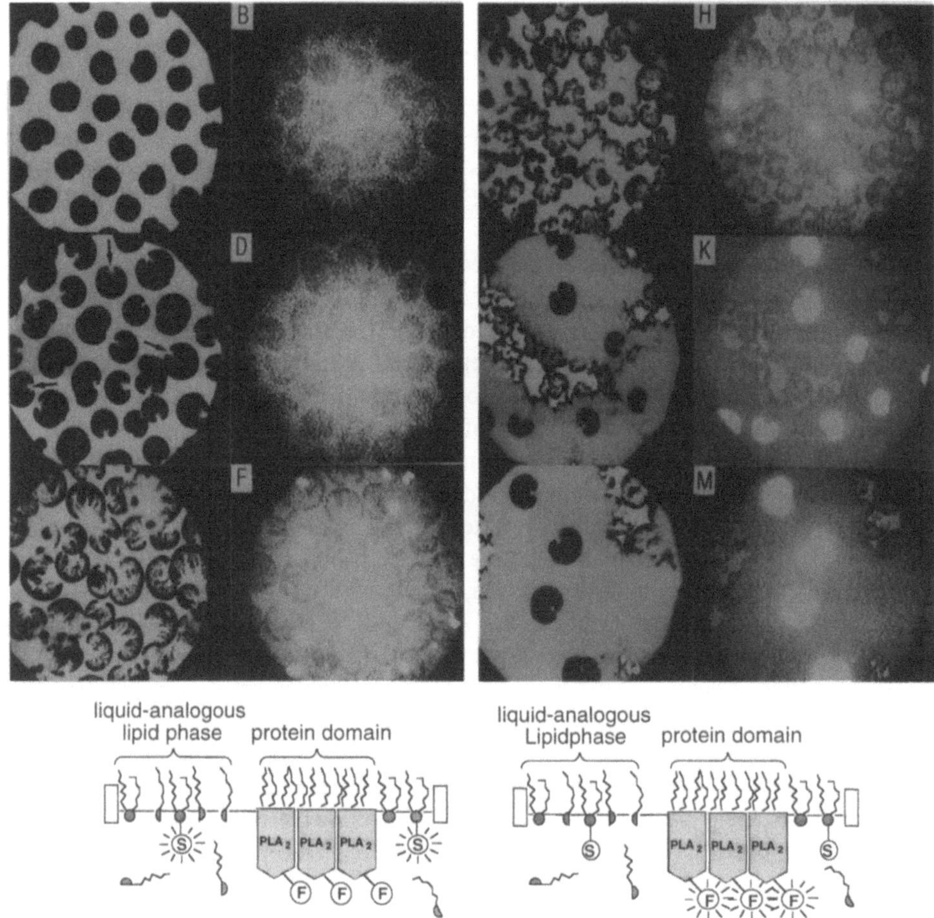

Figure 6-6   Hydrolysis of an L-α-DPPC monolayer by phospholipase A_2 observed by fluorescence microscopy; the monolayer is seen through the sulforhodamine filter (on the left) and the enzyme is seen through the fluorescein filter (on the right): (A) dark solid-analogous lipid domains in a bright lipid-analogous matrix; (B) homogeneous fluorescence of the enzyme in the subphase; (C) first signs of hydrolysis of the lipid domains (arrow); (D) homogeneous fluorescence of the enzyme in the subphase; (E), (F) progressive degradation of the lipid domains and first signs of the formation of enzyme aggregates below the monolayer (bright spots through the fluorescein filter (E), dark spots through the sulforhodamine filter (F)); (G)–(K) further hydrolysis of the lipid domains (G, I) with simultaneous growth of the protein domains (H, K); (L), (M) after almost complete degradation of the lipid domains, there are now only protein domains in a fluid matrix composed of the hydrolysis products and DPPC; these enzyme domains have a specific shape and appear dark through the sulforhodamine filter (L) and bright through the fluorescein filter (M); (A), (B) t = 0 min (immediately after enzyme injection); (C), (D) t = 10 min; (E), (F) t = 30 min; (G), (H) t = 40 min; (I), (K) t = 50 min; (L), (M) t = 60 min; π = 22 mN m^{-1}; T = 30 °C; 5 mm = 20 μm.

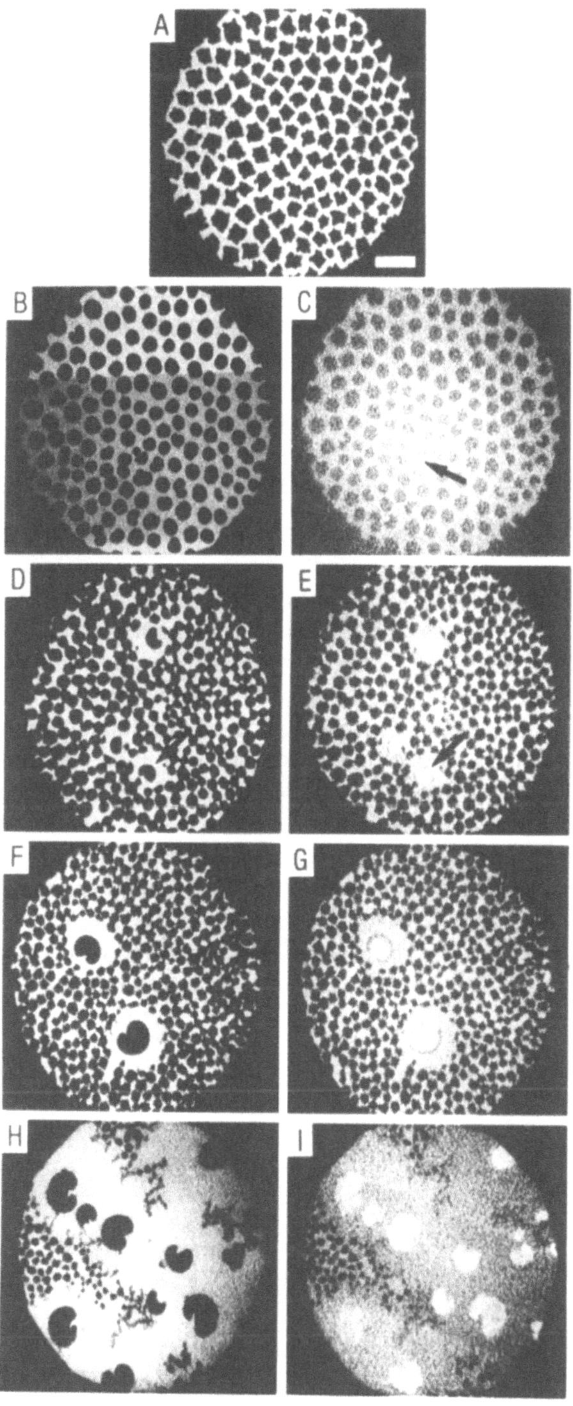

Figure 6-7   Hydrolysis of an L-α-DMPC monolayer by phospholipase A_2 observed by fluorescence microscopy; the monolayer is seen through the sulforhodamine filter and the enzyme is seen through the fluorescein filter: (A) dark solid-analogous lipid domains in a bright liquid-analogous matrix; (B), (C) first signs of hydrolysis of the lipid domains; small bright dots are seen at the corresponding sites through the fluorescein filter indicating that protein aggregation is starting (see arrows in B and C); (C)–(G) progressive hydrolysis leads to circular areas without solid-analogous lipid, with growing protein domains (arrow) in its centres; (H), (I) almost complete degradation of the lipid domains; only protein domains in a fluid matrix composed of the hydrolysis products and DMPC; these enzyme domains have specific shapes and appear dark through the sulforhodamine filter (H) and bright through the fluorescein filter (I); (A) t = 0 min (immediately after enzyme injection); (B), (C) t = 10 min; (D), (E) t = 20 min; (F), (G) t = 30 min; (H), (I) t = 50 min; $\pi = 22$ mN m^{-1}; T = 10 °C; bar = 20 μm.

6.2.2.2   Hydrolysis of Solid-Analogous L-α-DMPC Domains

The lipid domains in the phase transition region of a DMPC monolayer differ from those of DPPC in being small and star-shaped (see Figure 6-7A and Figure 6-6A). Five minutes after injection of the FITC-labelled phospholipase A_2, small concavities are evident in these domains in the sulforhodamine filter. With the fluorescein filter, bright dots are seen at the same locality, which is the first sign that protein aggregation is starting (see arrows in Figure 6-7B and 6-6C) [12, 14a, 15]. The resulting protein aggregates grow as the reaction progresses (Figure 6-6L, M). This shows that the shape of the protein domains does not depend on the original shape of the solid-analogous lipid domains.

However, observation of the progress of hydrolysis reveals some differences. Whereas individual domains of L-α-DPPC are specifically attacked and degraded, the enzyme produces large areas containing no solid-analogous lipid in the case of L-α-DMPC. The protein domains then appear in the middle of these areas (Figure 6-7D–G). It is not yet clear how these differences of degradation behaviour relate to the mechanism of hydrolysis.

6.2.2.3   Compression of a Hydrolyzed L-α-DPPC Monolayer

More information about the nature of the protein in the domains was obtained by changing the surface pressure in hydrolyzed L-α-DPPC monolayers. In the gas-analogous state ($\pi = 0$ mN m^{-1}), there is no effect on the shape of the protein domains. This indicates that they are very stable. Figure 6-8 shows the processes observed during compression, both schematically and as they appear under the fluorescence microscope. New solid-analogous lipid domains are formed in the fluid matrix (Figure 6-8B). This shows that even after complete hydrolysis of the original lipid domains, there is still uncleaved lecithin present in the monolayer. However, new lipid domains are produced not only in the fluid phase but also directly on the edges of the protein domains (which are bright when observed with the fluorescein filter). Aggregated PLA_2 is not active anymore.

The first evidence that the lysolecithin and fatty acid produced in the hydrolysis play an important role in the formation of the protein domains was gained from similar experiments with D-α–DPPC: although the enzyme is able to recognize this lipid, it cannot cleave it. Because the existence of hydrolysis products in the monolayer appears to be a necessary precondition for the formation of protein domains, investigations were carried out using mixed monolayers.

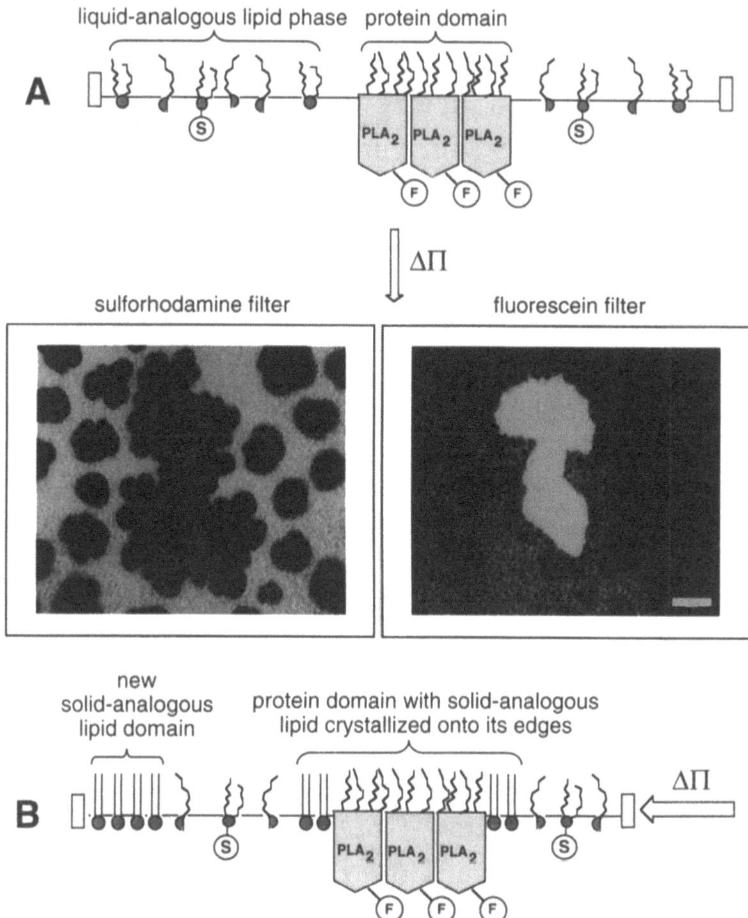

Figure 6-8   Compression of an L-α-DPPC monolayer after hydrolysis by phospholipase A_2: (A) schematic representation of a protein domain in a fluid matrix composed of hydrolysis products and DPPC; (B) compression leads to formation of new solid-analogous lipid domains, both in the fluid matrix and directly on the edges of the protein domains (appearance under the fluorescence microscope and schematic representation); 5 mm = 20μm.

6.2.3   Phase Behaviour of Mixed Lecithin, Lysolecithin, and Fatty Acid Mono-layers and their Interaction with Phospholipase A_2

The effect of the hydrolysis products was investigated separately from the hydrolysis process. This can be achieved by using mixed monolayers composed of lecithin and fatty acid in various ratios; such mixtures simulate the situation after a hydrolysis reaction. Assuming that only the cleavage products and not reactions during the hydrolysis (e.g. acylation [16]) induce domain formation and inhibition of PLA_2, protein domains should develop spontaneously whenever the composition of these mixed monolayers is suitable. This is demonstrated by the following example:

After spreading a 1:5:5-mixture of L-α-DPPC, L-α-LysoPPC, and palmitic acid, the liquid-analogous monolayer appears homogeneously bright under the fluorescence microscope (Figure 6-9). However, compression leads to phase separation and thus to formation of domains of unknown composition [17] (Figure 6-9B).

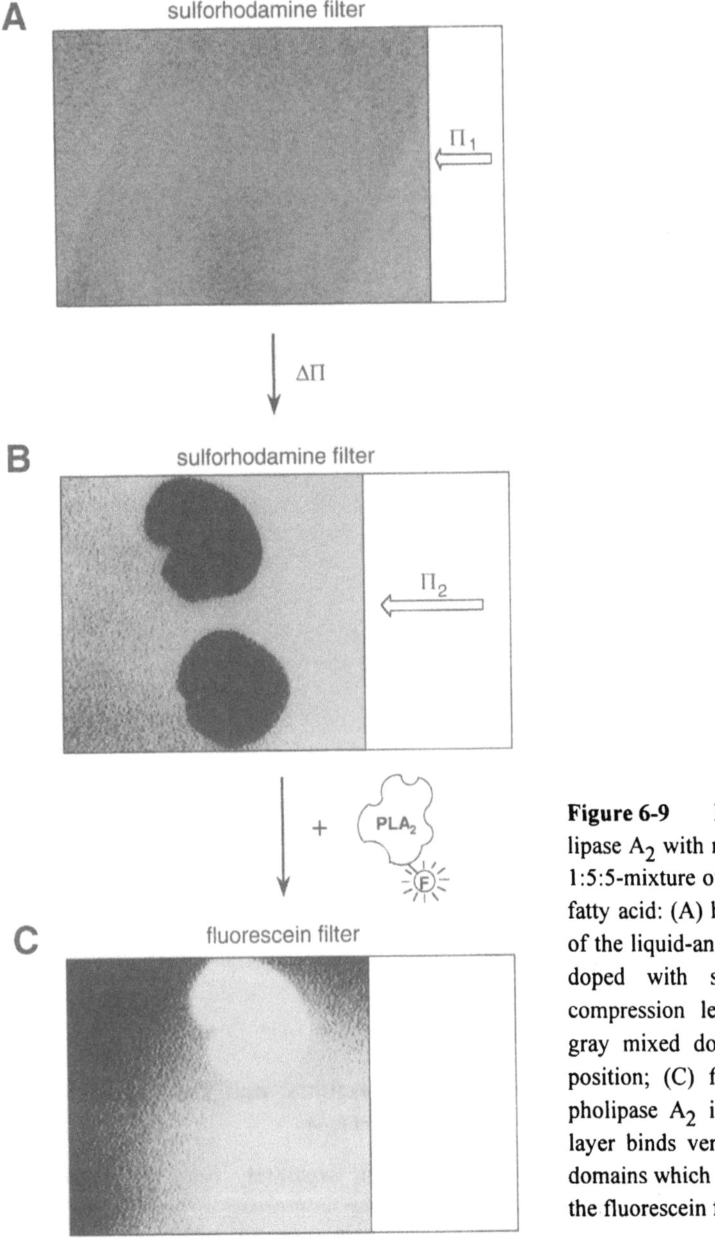

Figure 6-9 Interaction of phospholipase A_2 with monolayers composed of a 1:5:5-mixture of lecithin, lysolecithin, and fatty acid: (A) homogeneous fluorescence of the liquid-analogous mixed monolayer, doped with sulforhodamine-lipid; (B) compression leads to the formation of gray mixed domains of unknown composition; (C) fluorescein labelled phospholipase A_2 injected under this monolayer binds very quickly to these mixed domains which thus appear bright through the fluorescein filter. 5 mm = 20 μm.

These mixed domains are gray, unlike the black solid-analogous lipid domains described above; thus, these domains do not completely exclude the fluorescent dye.

Injection of the FITC-labelled PLA_2 under a mixed monolayer containing these gray domains, results in binding of the protein to the domains within two minutes without visible evidence of hydrolysis (Figure 6-9). This is much faster than the slow formation of the protein domains during hydrolysis with PLA_2. The mixed domains therefore represent a preferred environment for protein binding. The results are the same when the D form of DPPC is used in the mixing experiments. This is clear evidence that recognition of the head group of lecithin, but not its cleavage, is necessary for rapid protein aggregation.

A 1:1-mixture of lysolecithin and fatty acid, without DPPC, shows no domain formation by these hydrolysis products nor protein aggregation. Thus, uncleaved lecithin is necessary for inducing protein aggregation.

6.2.4   Conclusions

The experiments with fluorescein/anti-fluorescein antibody were attempts to use monolayers as systems to model specific recognition reactions at cell membranes. By using the relatively new method of fluorescence microscopy on monolayers, it was possible to observe other phenomena as well as the expected specific interaction. There was spontaneous formation of protein domains at the monolayer. The example of the enzymatic activity at the monolayer, that is the interaction of PLA_2 with lecithin domains, was originally based on ideas about the breaking down ("uncorking") of liposomes by PLA_2. In the monolayer experiments, it was possible for the first time to follow the progress of enzymatic reactions on organic substrates, the lipid domains, by direct visual observation. Once again, enzyme domains with a regular morphology were formed, which indicates inhibition of PLA_2 by the hydrolysis products. This might help to explain why the phospholipases which are widespread in the body do not destroy all the cell membranes.

There remains the interesting question of how functional supramolecular systems from living cells can be used as a basis for materials science and how material science can learn from life science in this context. There are, no doubt, very many biochemical processes which can be transferred to model membrane systems. Increasing numbers of examples are being provided by affinity chromatography and the relatively new area of biosensors.

6.3   The Complementary System of Barbituric Acid Lipids and 2,4,6-Triaminopyrimidine (TAP)

In order to examine the influence of preorientation on molecular recognition by spectroscopic methods more thoroughly, a purely synthetic lipid model system was synthesized for use at the air-water interface. In addition, the lipid system was designed with the intention that it should be capable of undergoing a substrate catalyzed reaction.

Our system is based on strand formation by hydrogen bonding in non-aqueous solvents as observed by J.-M. Lehn et al. and by G. M. Whitesides et al. [18]. It stimulates a recognition and orientation induced reaction at the air-water interface. This reaction is based on the interaction of barbituric acid lipids (1–4) with substrates such as 2,4,6-triaminopyrimidine (TAP), melamine, and urea, which can form complementary hydrogen bonds (see Figure 6-10).

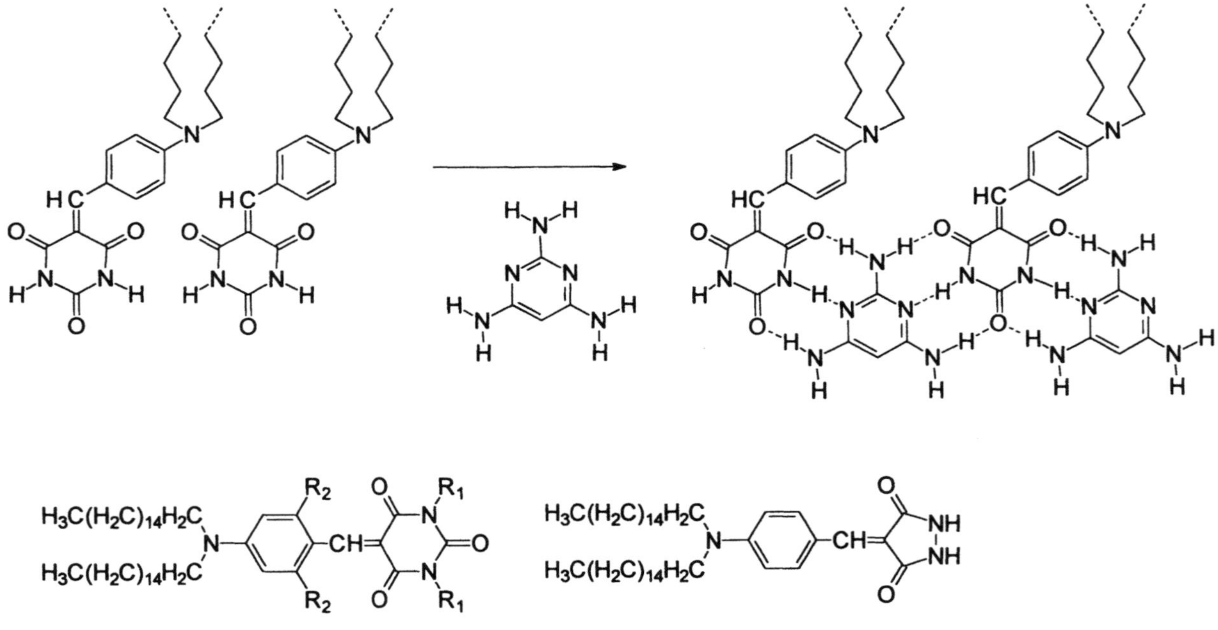

1 : R_1 = H; R_2 = H
2a : R_1 = CH$_3$; R_2 = H
2b : R_1 = CH$_2$CH$_3$; R_2 = H
2c : R_1 = CH(CH$_3$)$_2$; R_2 = H
3 : R_1 = H; R_2 = CH$_3$

Figure 6-10   Examined receptor lipids **1–4** and schematic picture of the strand formation of barbituric acid lipid **1** with 2,4,6-triaminopyrimidine at the air–water interface.

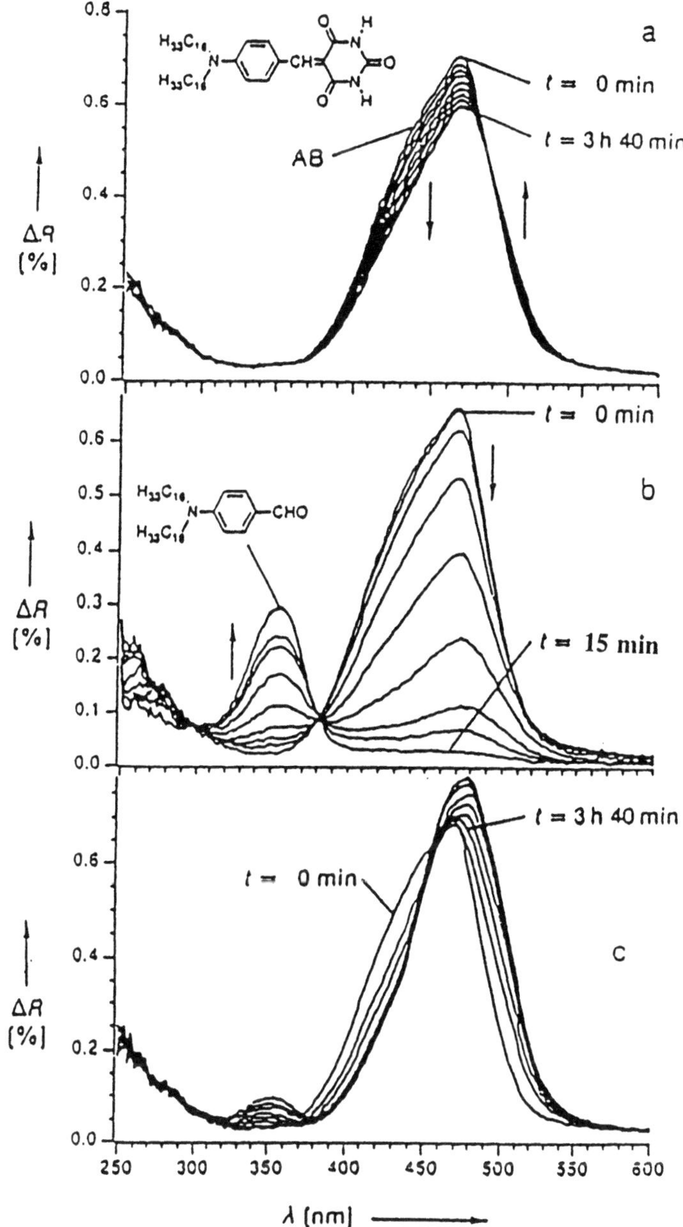

Figure 6-11   UV/Vis reflectance spectra of the barbituric acid lipids 1 at the air-water interface a) on an HCl containing 10^{-4} molar TAP-subphase (pH = 3); disappearance of the aggregation band at 430 nm; no hydrolysis; b) on a 10^{-4} molar TAP-subphase (pH = 6.5); fast recognition induced hydrolysis of barbituric acid lipid 1; c) on a 10^{-4} molar NaOH-subphase (pH = 10); slow basic hydrolysis of barbituric acid lipid 1.

The recognition reaction between the substrates and the barbituric acid lipids was determined by UV/Vis reflectance spectroscopy at the air-water interface. The detection of recognition is possible by the aggregation band which indicates when the interaction of the lipid chromophores disappears. The aggregation band cannot be found in solution. The phase of the monolayer plays an important role for the recognition reaction. Insertion of TAP into the monolayer can be observed most easily when the monolayer is in the fluid-analogous state which, for lipids **1–3**, happens at temperatures above 23 °C. All monolayer experiments shown here were carried out at 25 °C at the same area and the same pressure. One of these experiments is shown in Figure 6-11 [19].

Using TAP free water and a 10^{-4} molar HCl-subphase, the reflectance spectra do not change and the monolayer is stable over a period of at least 17 hours. In contrast, one can detect the disappearance of the aggregation band at 430 nm within four hours using TAP containing 10^{-4} molar HCl-subphase at pH = 3 (Figure 6-11a). This is a uniform and defined insertion process of TAP into the monolayer, as can be concluded from the three isobestic points at 360, 486 and 544 nm. While observing the insertion of TAP into a monolayer of **1** at pH = 6.5 which is the normal pH of a 10^{-4} molar TAP subphase, one can see a second process: a catalytic cleavage of the C=C double bond of the chromophore (Figure 6-11b). This leads to formation of the *p*-amino benzaldehyde ($\lambda_{max.} = 350$ nm) and of barbituric acid. Again, two isobestic points can be detected which indicate a uniform and defined process. The barbituric acid head of the lipid is already cleaved quantitatively after 15 minutes. Particularly astonishing is the fact that the hydrolysis on a 10^{-4} molar NaOH-subphase (pH = 10) has not even started at that time (Figure 6-11c).

The first step of a possible mechanism of this reaction which shows the characteristics of an enzymatic reaction, is shown in Figure 6-12.

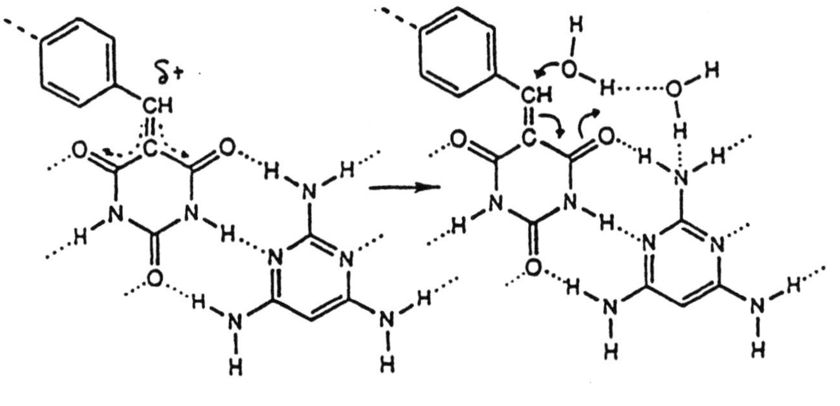

Figure 6-12   Addition of water to the activated C=C double bond of the chromophore of lipid **1**, polarization of the double bond, and nucleophilic attack of water.

In this mechanism, water which is situated in a hydrophilic surrounding and which therefore is very nucleophilic, attacks the double bond. The high activity of water can be explained by a recognition induced orientation: On the one hand, there is the activation of the C=C double bond by strand formation through intermolecular hydrogen bonds of the neighbouring carbonyl groups to TAP. On the other hand, hydrophobic pockets at the air-water interface are formed. This mechanism of a recognition and orientation induced C=C cleavage reaction (retro-Knoevenagel-reaction) is supported by several other experiments. When melamine ($pK_B = 8.9$) is used as a substrate instead of TAP ($pK_B = 7.2$), the speed of the insertion process is similar, but the hydrolysis process is approximately nine times slower. This can be attributed to the weaker hydrogen bonds that are formed with the less basic melamine, so that the activation of C=C double bonds is decreased. How important the formation of hydrogen bonds and, induced by this, the activation of the double bonds is, can be shown by the use of urea as substrate in the subphase (Figure 6-13).

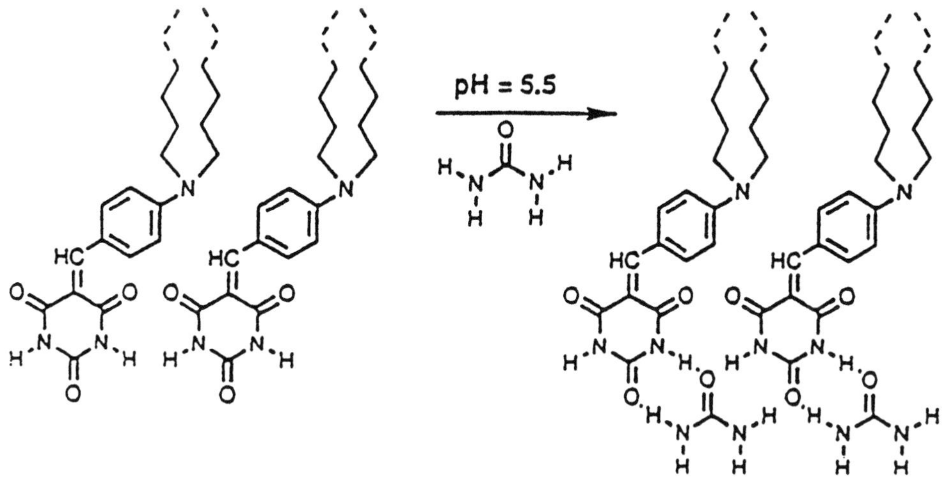

Figure 6-13   Scheme of the insertion of urea into a monolayer of lipid 1 and formation of a strand structure.

Figure 6-13 shows schematically the insertion of urea into the monolayer in order to form strands by hydrogen bonds. The reflectance spectra show a process similar to that reported in Figure 6-11a. The aggregation band disappears (again, one can detect three isobestic points). As there is no activation of the carbonyl groups in conjugation to the C=C double bond, no hydrolysis process can be detected.

In order to show that the enhanced concentration of hydroxy-ions due to the insertion of the base TAP is not – or at least is not the only – cause of this remarkable cleavage speed, monolayers of lipid 1 on subphases of water soluble organic bases, such as dimethylamine and trimethylamine, were examined. These organic bases are enriched at the surface and, therefore, there is an enhanced OH$^-$-concentration at the surface. They are orders of magnitude stronger than TAP ($pK_{b(dimethylamine)} = 3.29$; $pK_{b(trimethylamine)} = 4.26$; $pK_{b(TAP)} = 7.2$). Figure 6-14 shows the UV/Vis reflectance spectra of monolayers of 1 on these substrates. One can see that the reactions are both faster than with the even stronger base NaOH and the reaction with dimethylamine is, according to the higher basicity, faster than with trimethylamine. Nevertheless, even the faster one of these hydrolysis processes is still more than 20 times slower than with the weaker base TAP. This shows that there must be other factors in the mechanism which determine the reaction speed.

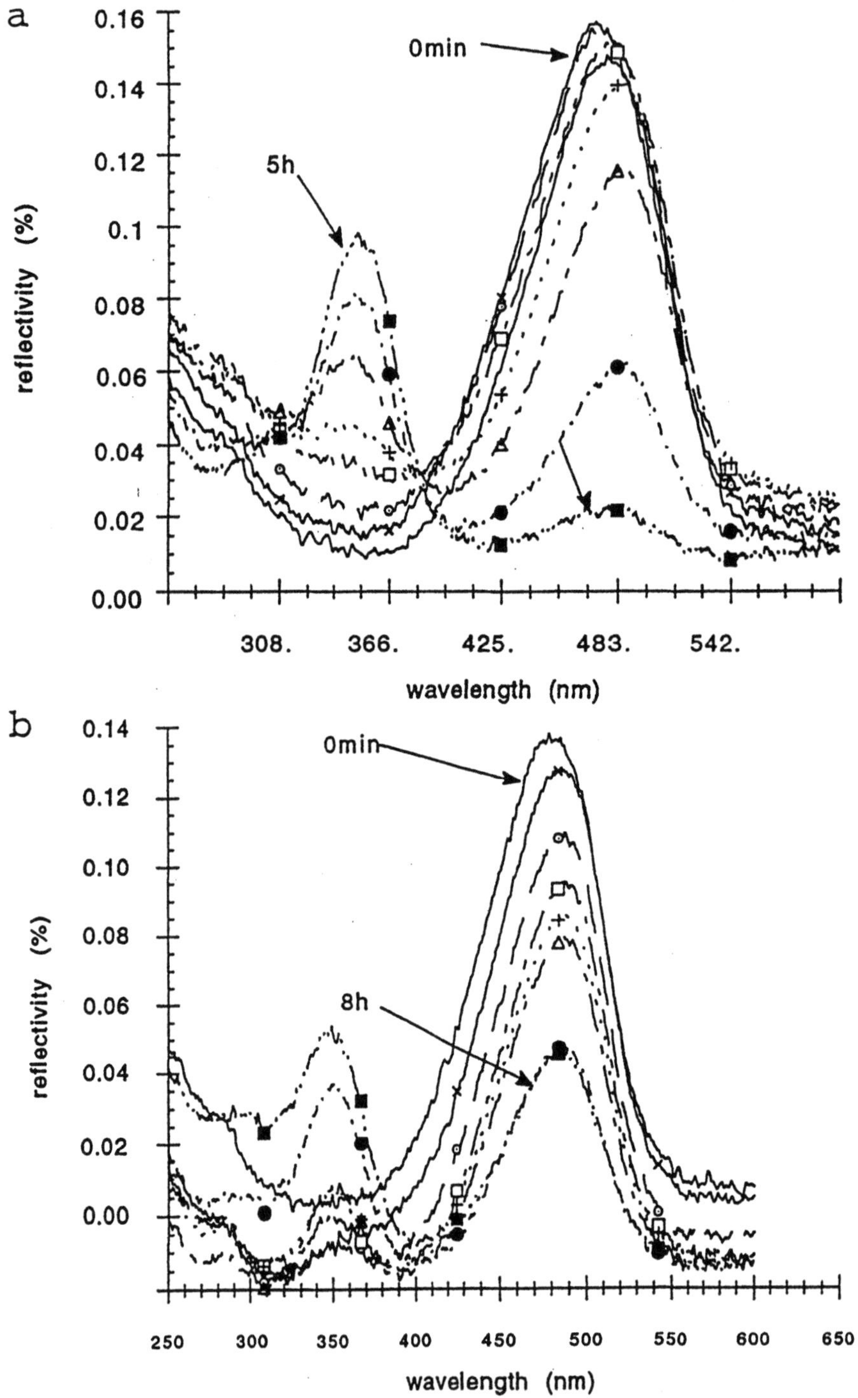

Figure 6-14   (a) UV/Vis reflectance spectra of a monolayer of **1** on a 10^{-4} molar dimethylamine subphase; the hydrolysis process is almost over after 5 hours; (b) UV/Vis reflectance spectra of a monolayer of **1** on a 10^{-4} molar trimethylamine subphase; there is only partial hydrolysis after 8 hours.

To investigate whether the proposed hydrophobic pocket plays a role in this reaction, it was blocked by introducing two methyl groups at the chromophore (**3**). The methyl groups are of about the same size as the water molecules in such a pocket. The UV/Vis reflectance spectra (Figure 6-15) show that even after nine hours only a small amount of the lipid is hydrolyzed under the same conditions as **1**. Hence, the hydrophobic pocket seems to play a crucial role in the reaction mechanism.

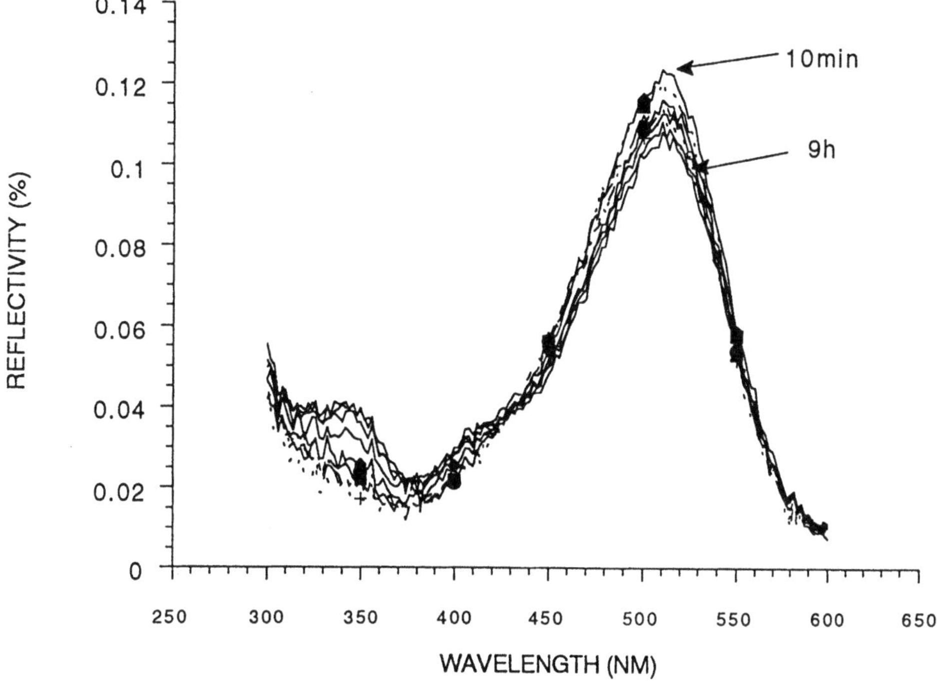

Figure 6-15   UV/Vis reflectance spectra of a monolayer of **3** on a 10^{-4} molar TAP subphase; a very small amount of hydrolysis only can be observed after 9 hours.

The fact that the insertion and cleavage reaction is caused by a strand like structure at the air-water interface, is demonstrated by conducting cospreading experiments [19] with various mixtures of barbituric acid lipid **1** and TAP. Figure 6-16 shows the UV/Vis spectra of the hydrolysis reaction of a 3:1-mixture (at pH = 5.5).

Three points have to be stressed: The speed of hydrolysis is considerably slower in a 3:1- than in a 1:1-mixture, the reaction stops after two thirds of the lipid are cleaved, and the aggregation band appears again. For the 2:1-mixture, the layer is cleaved completely but the reaction still proceeds much slower than for the 1:1-mixture. This means that TAP forms hydrogen bonds to two lipid molecules (which is again a proof of insertion); if that leads to strand formation, it reacts fast; if it merely leads to the formation of a 2:1-complex (2:1- and 3:1-mixture), it reacts very slowly as in that case only a one sided activation of the double bond of the barbituric acid chromophore is found. The fact that the aggregation band reforms in the case of a 3:1-mixture after the reaction is finished, is an additional hint that the destruction of the lipid aggregates is due only to the insertion of TAP. The result that for a 3:1-mixture not all the lipid molecules are cleaved, suggests that even in water there is an

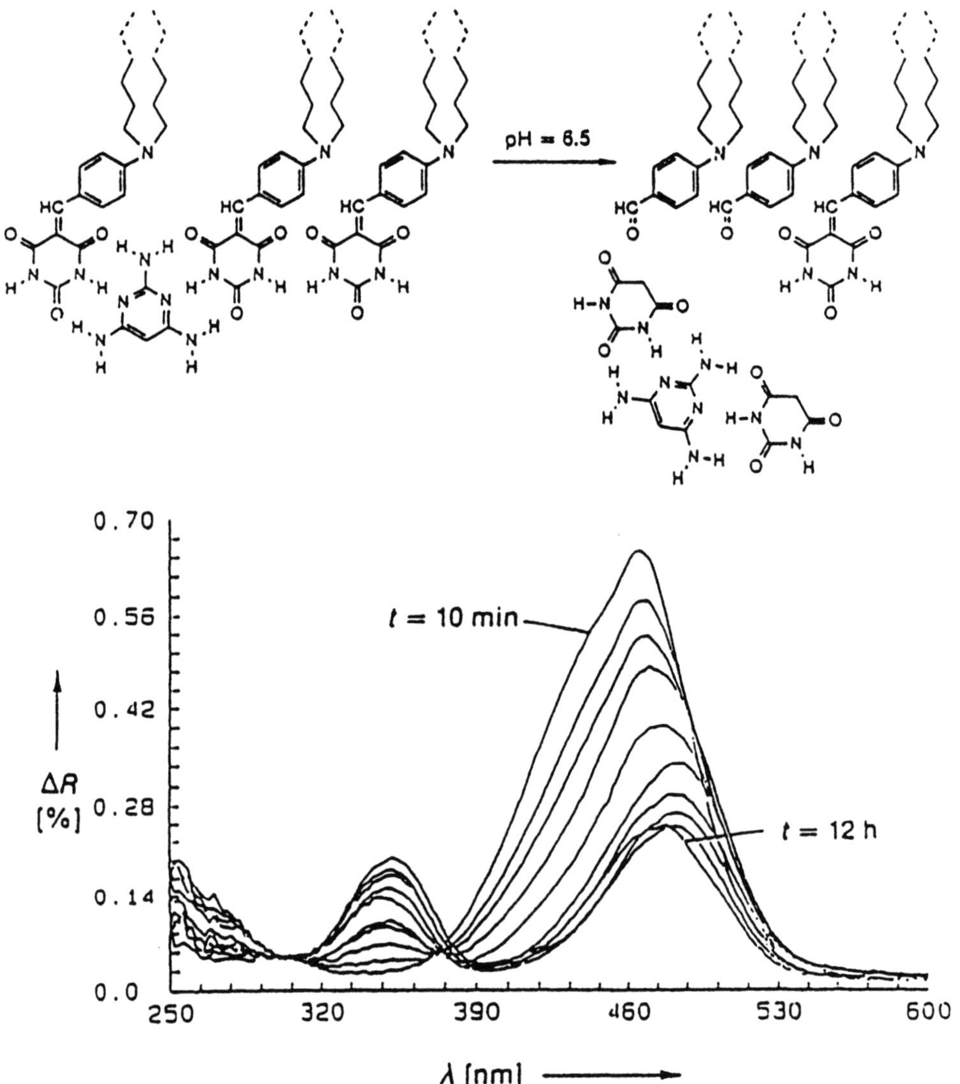

Figure 6-16   UV/Vis reflectance spectra of the partial hydrolysis of a 3:1-mixture of lipid **1** and TAP and schematic representation of a 2:1-complex of lipid **1** and TAP; reforming of the aggregation band after completion of hydrolysis (see Figure 6-11a).

aggregation of barbituric acid and TAP. This is indeed true. It has been proved by FTIR-ATR-spectroscopy that in water without any preorientation there is complexation of barbituric acid and TAP even in the presence of that strongly competing solvent [20]. Figure 6-17 shows the FTIR spectra of barbituric acid, TAP, and a 1:1-mixture of both. One can see that the spectrum of the 1:1-mixture is clearly not just a superposition of the spectra of the uncomplexed molecules. There are distinct shifts of the bands that show complexation. As none of the original bands of barbituric acid are found, one must assume that there is no uncomplexed barbituric acid.

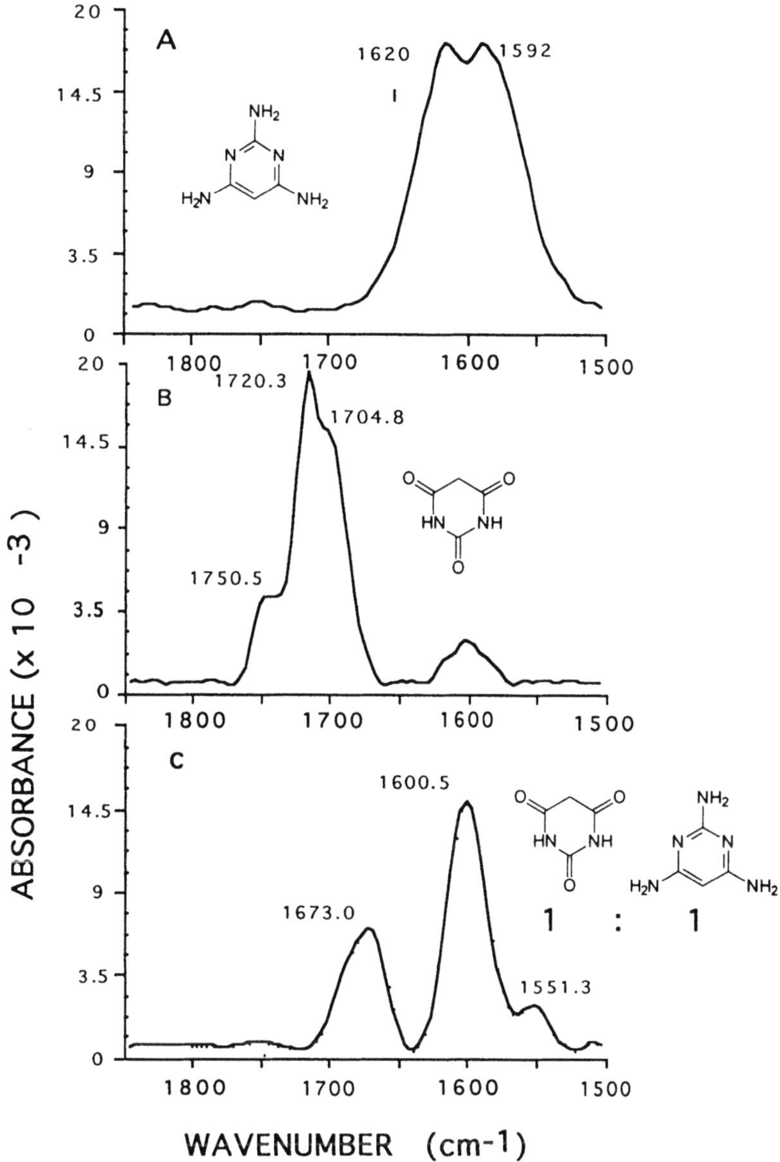

Figure 6-17   FT-IR-ATR-spectra of A) an aqueous solution of barbituric acid, B) an aqueous solution of TAP and C) an aqueous solution of a 1:1-mixture of barbituric acid and TAP.

How important a perfect fit is for the recognition reaction and the following hydrolysis, can be shown by variation of the receptor lipid (alkylation of the pyrimidine nitrogens; lipid **2a–2c**). The insertion process of TAP at pH = 3 into a monolayer of lipid **2a** is remarkably slower than the analogous interaction with lipid **1** (slow disappearance of the aggregation band).

When following the hydrolysis process of lipid **2a** on a TAP-containing subphase at pH = 6.5, one can find that the speed of hydrolysis is slowed down considerably (Figure 6-18). For lipids **2b–2c**, the hydrolysis is slowed down even further, but not prevented. For the lipids **2a** and **2b**, the aggregation band does not completely disappear during hydrolysis. For lipid **2c**, no aggregation band can be detected because the bulky head group probably prevents π-π-stacking.

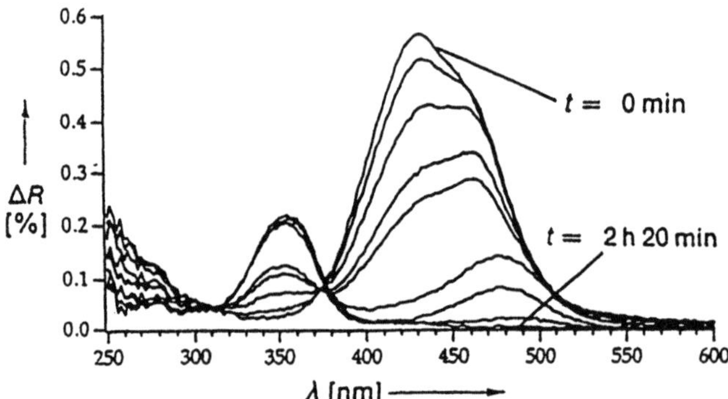

Figure 6-18   UV/Vis reflectance spectra of a monolayer of lipid **2a** on 10^{-4} molar TAP-subphase.

This experiment stresses the importance of the organization of the lipids at the air-water interface: None of the lipids of type **2** forms complexes with TAP in $CDCl_3$-solution (detected with ^{1}H NMR) [21], but they do so at the air-water interface, even though the system was transferred from a non-competing to a competing protic solvent. The preorientation of the recognition unit by the lipid monolayer is responsible for the stabilization of the weak complexes. As soon as the complex is formed, the TAP stays in the monolayer and the hydrolysis starts.

The principle of molecular recognition is underlined, when one examines the interaction of the pyrazolin-3,5-dion lipid **4** with TAP at the air-water interface. The aggregation band of the chromophore does not disappear because there is no fit for a strand structure, as the angles of the 5-membered ring structure are not compatible with the angles of the recognition unit, which is a six-membered ring. Due to this fact, no strand is formed and, as the proposed mechanism suggests, the monolayer is not cleaved at all.

7 Self-Organization of Molecules en Route to Crystal Formation

Ronit Popovitz-Biro, Isabelle Weissbuch, Jarek Majewski, Leslie Leiserowitz, and Meir Lahav

7.1 Introduction

Crystals, owing to their aesthetic shapes and symmetries have always attracted the attention of naturalists. They have played a paramount role in the advancement of the natural sciences in general including structural chemistry and biology. Nonetheless, the understanding and control of their formation from the molecular level is still at a rudimentary stage. Fundamental questions referring to the different pathways by which molecules self-assemble to form embryonic clusters which act as nuclei en route to crystal formation are still unanswered. In the absence of analytical tools which allow one to probe directly the structure and dynamics of formation of these clusters, we must rely still for a while on indirect methodologies.

The classical theory of crystal nucleation, inspired by the work of Gibbs on condensation of vapors into liquids, is based on the following scenario. When molecules start to aggregate to form a cluster, they create surfaces which separate them from the environment. This is associated with a positive free energy ΔG_S which is proportional to the square of its radius and thus destabilizes the nucleus. On the other hand, the interactions between the molecules within the bulk stabilize the cluster and the free energy of the bulk is therefore a negative term which is proportional to the volume of the cluster or to the cube of its radius. The total free energy is the difference between the energy of the bulk of the crystal and its surface [1].

$$\Delta G_V = \frac{4}{3}\Pi r^3 \Delta G_v \qquad\qquad \Delta G_S = 4\Pi r^2 \gamma$$

$$\Delta G = 4\Pi r^2 \gamma + \frac{4}{3}\Pi r^3 \Delta G_v \qquad \frac{d\,\Delta G}{d\,r} = 8\Pi r\gamma + 4\Pi r^2 \Delta G_v = 0 \qquad \text{at } r = r_c \qquad (7\text{-}1)$$

$$8\Pi r_c\gamma = -4\Pi r_c^2 \Delta G_v \qquad\qquad r_c = \frac{-2\gamma}{\Delta G_v}$$

ΔG_V = free energy of the bulk.

ΔG_S = free energy of the surface.

ΔG_v = free energy of the bulk per unit volume.

ΔG = overall free energy of the cluster.

γ = surface tension of the cluster.

r_c = $r_{critical}$.

The overall free energy depends upon the dimensions of the cluster, reaching a possible maximum at a size defined as the "critical size", therefore below the critical size, it has a tendency to disintegrate into smaller clusters or even isolated molecules. Below this size, the aggregate is metastable, whereas once it crosses the critical size its stability increases, and so it can grow and transform into a stable nucleus en route to crystal formation. This macroscopic mathematical treatment does not, however, take into account the structural and stereochemical aspects of the nucleus. The purpose of the present study is to approach this problem from the molecular level which takes into consideration the structure and morphology of crystals. Consequently, we follow a working hypothesis that on the onset of crystallization, close to critical size, the nuclei assume a structure akin to that of the mature crystal. This naive hypothesis permits the use of structural information stored in crystals for the design of auxiliary molecules which can, in a stereospecific manner, either inhibit or promote crystallization. We illustrate this approach with examples studied by us in recent years. These include the design of stereospecific inhibitors for the control of crystal polymorphism and for large-scale resolution of enantiomers by crystallization, as well as the design of crystallization promotors such as layers of amphiphilic molecules spread at the air-liquid interface. The structure of the latter were characterized by grazing incidence X-ray diffraction (GID) using X-rays from intense synchrotron source, cryo-transmission electron microscopy (TEM), and scanning force microscopy (SFM). Thus, it became possible to obtain direct insight into the structure of self-aggregates, which operate as template nuclei, inducing oriented growth of crystals at the interface.

7.2 Nucleation Inhibitors

The physical and chemical properties of solid materials depend not only on the structure of the molecules from which they are constructed but also on the way the molecules are arranged in these crystals. Molecules may generally crystallize in more than one crystal structure. When this phenomenon occurs, the system displays crystal polymorphism. So far, the existence of polymorphism for a given compound has been observed mainly by trial and error. In the present study we assume that supersaturated solutions of materials displaying crystal polymorphism contain molecular aggregates of structures resembling those of the various mature phases. At conditions close to equilibrium, however, only the nuclei corresponding to the thermodynamically stable phase will grow into crystals. In order to obtain the metastable polymorph, the crystallization of the stable one should be inhibited. A properly designed inhibitor should preferentially interact with precritical nuclei of the stable crystalline phase, and much less, if at all, with those of the desired metastable polymorph. Such a strong inhibition will lead to preferential precipitation of the thermodynamically metastable polymorph by virtue of kinetic control. This naive hypothesis is illustrated in Scheme 7-1.

Let us consider a supersaturated solution of 'A' molecules which self-aggregate to form embryonic clusters of different structure indicated by subscripts like $(\ )_\alpha$, $(\ )_\beta$, $(\ )_\gamma$ etc. For thermodynamic reasons only the crystalline phase $\{A\}_\alpha$ precipitates. However, in the presence of a "tailor-made" nucleation inhibitor for the nucleus $(\ )_\alpha$, we can prevent its transformation into the stable $\{A\}_\alpha$ form. If this inhibitor does not interfere with the structure of the clusters of the β form, under a kinetically controlled process, crystallization of the metastable $\{A\}_\beta$ polymorph will occur. Modelling and design of stereospecific nucleation inhibitors takes into consideration the packing arrangement of the different polymorphs.

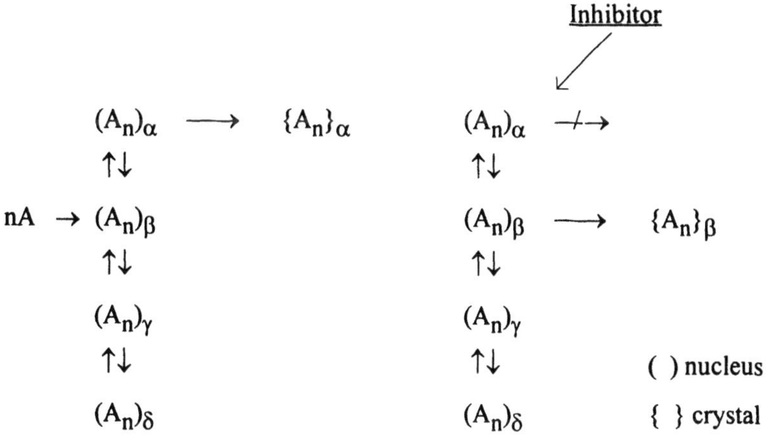

$$(A_n)_\alpha \longrightarrow \{A_n\}_\alpha \qquad (A_n)_\alpha \;\;\not\longrightarrow$$

Inhibitor

$$nA \rightarrow (A_n)_\beta \qquad\qquad\qquad (A_n)_\beta \longrightarrow \{A_n\}_\beta$$

$$(A_n)_\gamma \qquad\qquad\qquad (A_n)_\gamma$$

$$(A_n)_\delta \qquad\qquad\qquad (A_n)_\delta$$

( ) nucleus

{ } crystal

Scheme 7-1

This approach was successfully applied for the crystallization of a metastable polymorph of polar structure at the expense of more stable non-polar quasi-centrosymmetric form [2]. Figure 7-1 illustrates the packing arrangement of molecules in a polar and a non-polar crystal. The use of an inhibitor interferes with the growth of the non-polar crystal in the two opposite directions but can only inhibit growth in one direction in the case of the polar crystal. This approach has been instrumental for the control of crystal polymorphism of *N*-(2-acetamido-4-nitrophenyl)pyrrolidene (PAN), which exists in two distinct crystalline forms. Form α is non-polar quasi-centrosymmetric and is generally obtained as thick plates, either by slow cooling or slow evaporation.

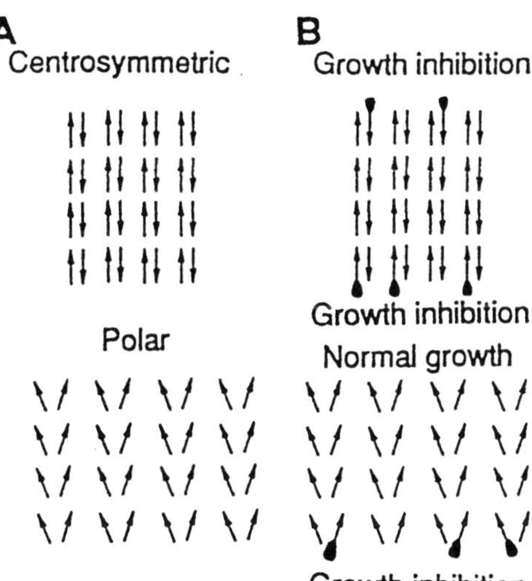

Figure 7-1   Packing arrangement of molecules in a non-polar and a polar crystal.

On the other hand, when aqueous solutions of PAN are cooled rapidly, a metastable β-form precipitates as dendritic needles. Upon standing in solution or heating to 183–186 °C, the needles transform into plates of the stable α-form. The packing arrangements of the two forms are shown in Figure 7-2 and Figure 7-3. In the α-form, the acetamido group of the two independent molecules are antiparallel and oriented towards both, the $+b$ and $-b$ directions, while in the polar β-form, the symmetry related molecules have the same component of the acetamido group oriented exclusively towards one direction along the b-axis, say $+b$. In both polymorphs the acetamido group also has components along the c-axis. On the basis of information from the two crystal structures, it became possible to model an efficient inhibitor for the nucleation of the stable polymorph, and to grow large single-crystals of the β-dimorph which displays optical second harmonic generation (SHG). The most efficient inhibitors of the α-form were derivatives of polyacrylic or polymetacrylic acid bearing units of PAN. Adding as little as 0.03% of this polymer inhibits the precipitation of the α-form and promotes the crystallization of the β-polymorph. The same approach has been applied recently for the induced crystallization of the metastable polar form of γ-glycine [3].

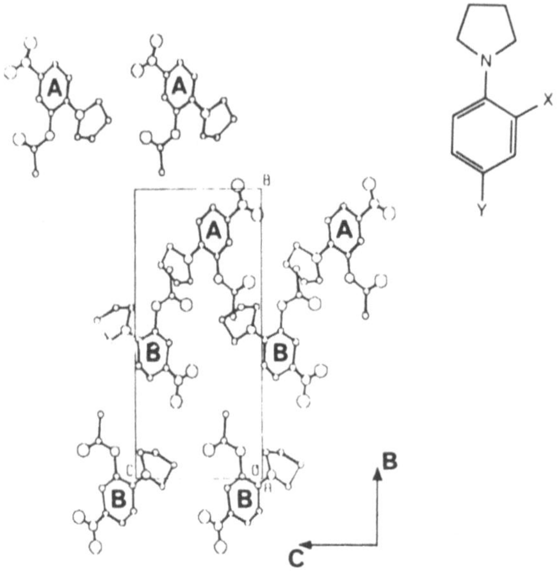

Figure 7-2   Packing arrangement of the stable α form of PAN, viewed down the a axis.

A different application encompasses large scale resolution of enantiomers by crystallization. There are many examples, where a racemate undergoes upon crystallization a process of spontaneous resolution. When this process occurs, all molecules of the R enantiomer segregate in one enantiomorph $\{\ \}_d$ whereas the molecules of the S enantiomer segregate in the second enantiomorph $\{\ \}_l$. Since the enantiomorphs are equi-energetic, they have equal probabilities to precipitate when crystallized in a non-chiral environment.

$$R \quad \neq \quad S$$
$$\downarrow \qquad \downarrow \qquad\qquad\qquad (7\text{–}2)$$
$$\{R\}_d \qquad \{S\}_l$$

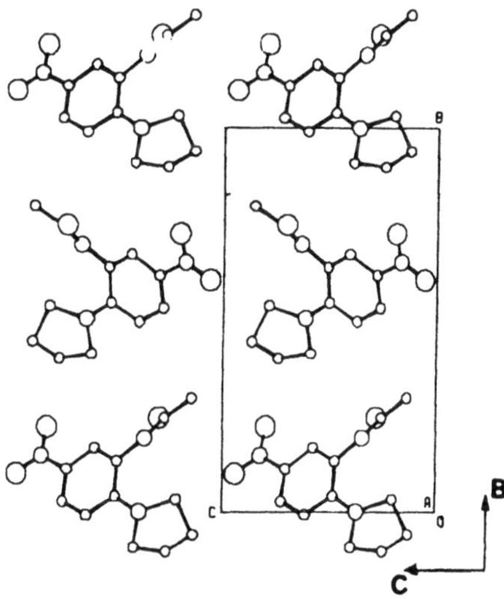

Figure 7-3   Packing arrangement of the metastable polar β form of PAN, viewed down the *a* axis.

If, however, we add to the solution, polyacrylic or polymetacrylic acid onto which resolved α-amino acids – such as resolved (R) lysine, glutamic acid etc. – have been grafted, this polymer will interact enantioselectively with the nuclei of (R) and will obstruct their transformation into mature crystals. On the other hand, these polymers will not interact with the nuclei of (S) and, therefore, they can continue to grow at a normal rate.

This method appears to be very general and thus a large variety of racemates, including a dozen of α-amino acids, have been resolved via this route [4]. For example, racemic histidine which precipitates at room temperature as a racemate was resolved by chiral polymeric inhibitors which were designed to inhibit the growth of the stable racemate and one of the enantiomorphs [5]. The chiral polymeric inhibitors were polyacrylic or polymetacrylic acid onto which resolved *p*-aminophenyl alanine had been grafted. Crystallization in the presence of such inhibitors resulted in precipitation of the unaffected enantiomorph in almost quantitative yield.

7.3   Crystal Nucleation Promotors

The energy barrier for crystal nucleation is always lowered when the nucleus interacts with foreign surfaces. This process can happen at various levels of specificity that may range from non-specific adsorption to oriented growth. One can envisage oriented crystal nucleation by designed nucleation promotors that match the structure of a specific crystal plane and thus may result in epitaxial crystallization. This approach has been applied for several systems such as induced crystallization of α-glycine [6, 7], *p*-hydroxy benzoic acid [8], silver propionate [9] and ice [10] by the use of designed promotors. Here we illustrate the concept with the system of α-glycine and ice.

The crystalline structure of the monoclinic α-glycine is composed of hydrogen bonded layers in which the molecules are related in the *ac* plane by translation symmetry only (Figure 7-4). The area per molecule within such a layer is *ac* sin β = 25.9 Å^2. The layers are interlinked by N–H⋯O bonds via centres of inversion to form centrosymmetric bilayers. Since each layer is chiral, the replacement of the C–H group of an α-glycine molecule, which emerges from the (010) crystal surface, by an alkyl group for example, would generate a chiral α-amino acid surface of R-configuration. By virtue of the glide or inversion symmetry elements of the crystal, a layer of glycine molecules at the opposite crystal surface may be replaced by a chiral layer of (S) α-amino acids. The hydrophobic optically pure α-amino acids valine, leucine, norleucine, isoleucine, phenylalanine and α-amino octanoic acid, all pack in hydrogen bonded layer structures similar to that of α-glycine. Surface tension measurements of aqueous solutions of these amphiphilic molecules, both pure as well as saturated in glycine, indicate positive surface accumulation which implies that these molecules aggregate at the solution surface. According to our hypothesis, the molecules at the surface may form ordered 2D aggregates which are stabilized by the hydrogen bonds between their amino acid groups. Proving this proposed lateral ordering involved the design of structured surfaces which may act as templates for an epitaxial-like crystallization of 3D crystals such as glycine at the air-solution interface [6].

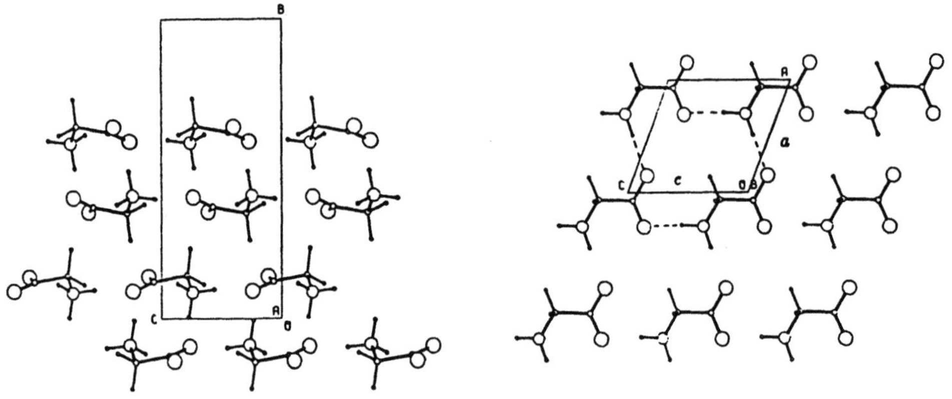

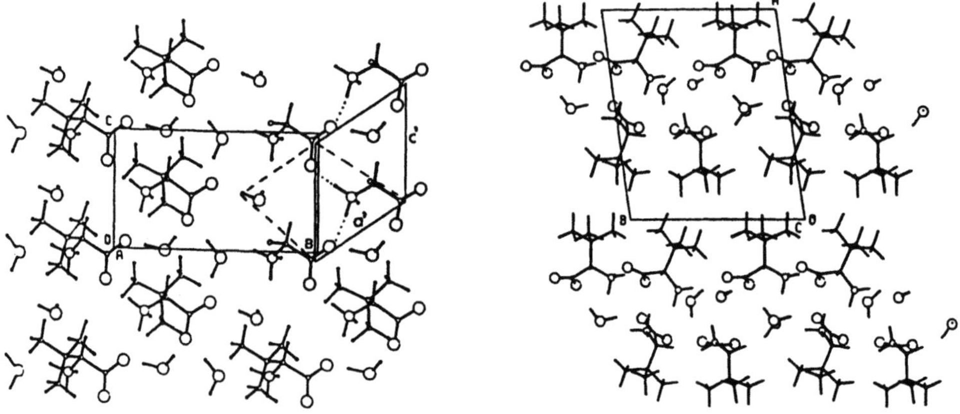

Figure 7-4   Packing arrangement of α-glycine, viewed along the *a* and along the *b* axis.

Figure 7-5   Packing arrangement of (S)-*tert*-butyl glycine monohydrate, viewed along the *b* and the *c* axis.

With this idea in mind, we investigated the oriented crystallization of glycine in the presence of low concentrations of optically pure hydrophobic α-amino acids. Two types of hydrophobic α-amino acids were used: one class such as leucine and valine forms crystalline layer structures similar to that of glycine, while the other class was selected so that the hydrophobic moieties are too bulky to allow formation of a layer structure akin to that of glycine, for example *tert*-butyl glycine. Indeed the packing arrangements of compounds such as *tert*-butyl glycine and neopentyl glycine show the formation of a hydrogen bonded layer that occludes solvated water, since separation distances between neighbouring molecules imposed by the bulky hydrophobic groups are too big (Figure 7-5).

As expected, the R α-amino octanoic acid induced a very fast nucleation of glycine crystals at the air-solution interface, oriented so that their (010) faces were exposed to air, whereas the S enantiomer induced the crystallization of glycine crystals oriented so that their ($0\bar{1}0$) faces were exposed to the air. In contrast, addition of α-amino acids with the bulky hydrocarbon side chains did not induce crystallization at the solution surface despite the high surface accumulation of the additive, as determined from surface tension measurements. The results are in agreement with a lattice mismatch between the proposed 2D packing of the bulky α-amino acids at the water surface and the structure of layers of α–glycine in its crystal [6].

7.4   Oriented Crystallization of α-Glycine under α-Amino Acid Monolayers

In order to obtain direct information on the structure of molecular aggregates formed at the air-solution interface, we extended our studies to the induced crystallization of glycine by insoluble amphiphilic α-amino acids. These surfactant molecules, in contrast to the short chain α-amino acids, were characterized by surface pressure/area isotherms. Recently, with the advent of intense, highly collimated and monochromatic X-ray beams from synchrotron sources, it became possible (in cooperation with K. Kjaer and J. Als-Nielsen) to probe the structure of the two-dimensional crystalline domains formed by these amphiphiles at the air-solution interface [11]. Crystallization experiments of glycine underneath amphiphilic molecules with a surface area per molecule of ~ 25 Å^2, close to that of glycine at the {010} face, yield results akin to those of the corresponding hydrophobic α-amino acids [7]. The same behaviour was observed for both, the compressed and the non-compressed monolayers. Thus, for example, monolayers of palmitoyl-R-lysine and stearyl-R-glutamate yield complete oriented crystallization of glycine. On the other hand, optically pure α-amino acid monolayers which bear a fluorocarbon chain as the hydrophobic part and which have a surface area per molecule of 28.5 Å^2, induce crystallization of glycine with both (010) and ($0\bar{1}0$) orientations. An α-amino acid monolayer with a bulky steroid as the hydrophobic part of the molecule, with a surface area per molecule of 38 Å^2, does not induce crystallization of glycine. The structure of two of these monolayers was determined by grazing incidence X-ray diffraction (GID) measurements [12, 13].

Figure 7-6 depicts the diffraction pattern obtained from GID measurements performed on a monolayer of palmitoyl-R-lysine. The monolayer proved to be a 2D powder with a "coherence length" (in other words, a perfect crystalline domain size) of about 500 Å and unit cell dimensions $a = 5.03$, $b = 5.46$ Å, and $\gamma = 117.8°$, implying a molecular area of 24.3 Å^2. The molecular layer thickness was determined from X-ray reflectivity measurements and compares well with the molecular model deduced from the GID data and X-ray powder diffraction data of 3D crystalline palmitoyl-R-lysine. It was deduced that the hydrophobic chain of the monolayer molecule is tilted by 30° from the vertical, and such an orientation is favourable for hydrogen bonding between neighbouring amide groups (Figure 7-7) [12]. On this basis, a model was proposed (Figure 7-7) in which the packing arrangement of the α-amino acid head groups is very similar to that of an *ac* layer of glycine in its own crystal structure. We propose that this similarity in the packing arrangement is responsible for the epitaxial crystallization of glycine at the air-solution interface.

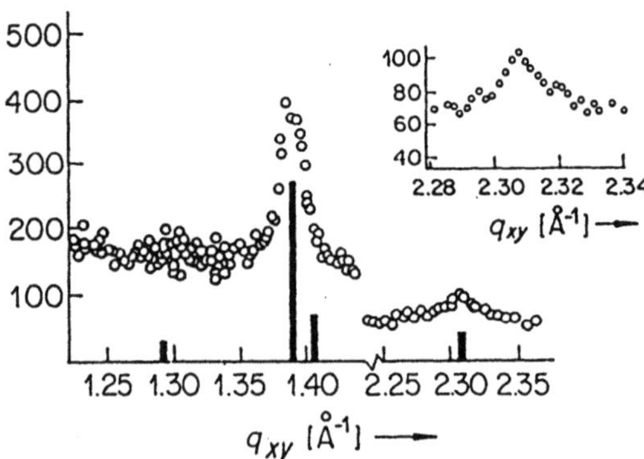

Figure 7-6   GID diffraction pattern of the palmitoyl-R-lysine monolayer.

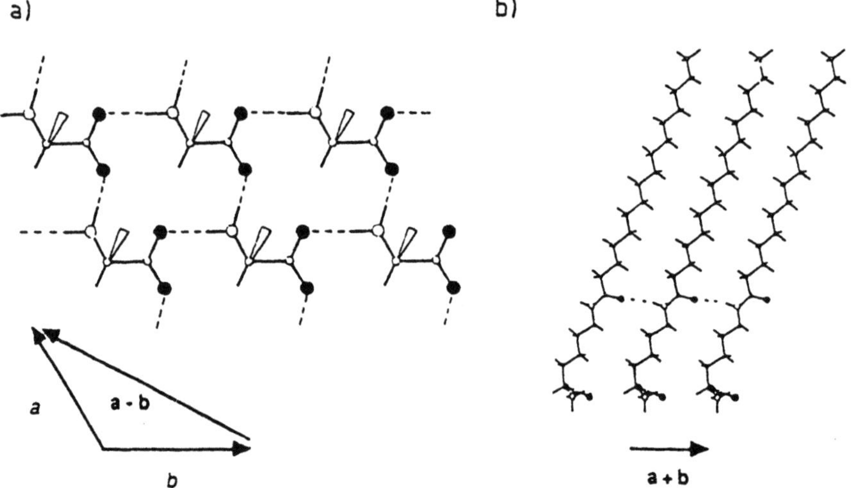

Figure 7-7   Views of the packing arrangement model of palmitoyl-R-lysine over pure water. (a) Perpendicular to the monolayer plane, the hydrophobic chains are represented by a wedge. The NH···O=C hydrogen bonds are designated by dashed lines. (b) View along the *a–b* axis showing molecules in a row parallel to the *a+b* direction.

In contrast to the structure of palmitoyl-R-lysine, GID data for the perfluorinated α-amino acid $F_3C(CF_2)_9(CH_2)_2OCOCH_2CH(NH_3^+)CO_2^-$ (PFA) in the compressed state yielded a cell which differs significantly from that of α-glycine. Recently, we obtained GID patterns from a monolayer of docosanoyl-lysine on water in the uncompressed state, and the packing arrangement of the α-amino acid head groups are very close to that of glycine at the (010) and (0$\bar{1}$0) faces [14].

7.5 Induced Nucleation of Ice by Monolayers of Aliphatic Alcohols

Recently, we focused our efforts on investigating the self-aggregation of amphiphilic alcohols which organize the molecules of water at the interface and induce freezing of supercooled water into ice.

Pure water can be supercooled to temperatures of $-20\,°C$ to $-40\,°C$. Therefore, the induction or inhibition of freezing of ice, in particular through the role of auxiliaries such as membranes or proteins, has far reaching ramifications for the living and non-living world. Water soluble alcohols generally reduce the freezing-point of water and thus act as anti-freeze substances. On the basis of the above-mentioned studies on self-organization of amphiphilic molecules, we anticipated that amphiphilic alcohols can also self-organize into structured self-aggregates at the air-water interface. Moreover, since the surface area per molecule of such alcohols is $18–20\,Å^2$, which is close to that of a molecule of water in the ab plane of ice, we expected that the OH groups of the monolayer may organize the water molecules underneath, into ice-like structures and thus induce freezing of the supercooled water [10, 15].

The stable form of ice (ice I) crystallized under atmospheric pressure, is hexagonal, with space group $P6_3/mmc$ and axes $a = b = 4.5\,Å$, $c = 7.3\,Å$, $\gamma = 120°$. The crystal structure may be described, in terms of the oxygen atoms, as composed of $0.9\,Å$ thick bilayers

(a)

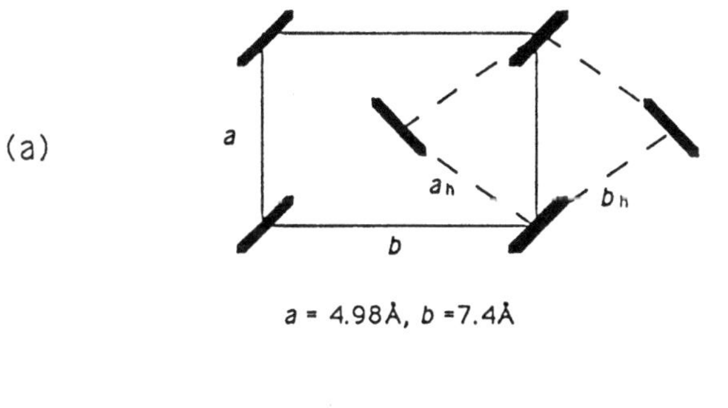

(b)

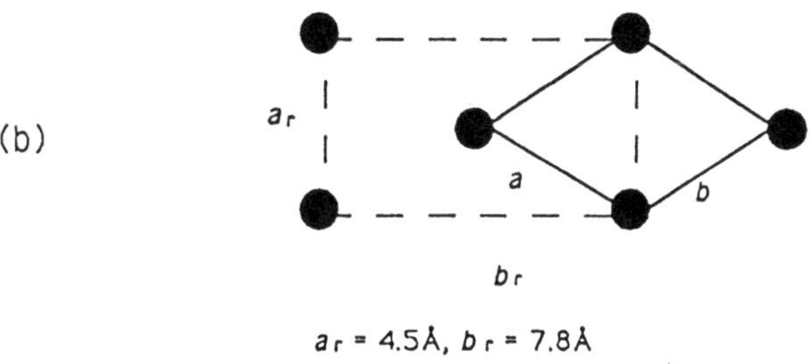

Figure 7-8   (a) Schematic representation of the orthogonal packing O⊥ of hydrocarbon chains of an alcohol monolayer in a rectangular cell (a, b), viewed along the chain axis. The distorted hexagonal representation (a_h, b_h) is also depicted. (b) Schematic representation of hexagonal ice in terms of the hexagonal cell (a, b) and the c-centred rectangular cell (a_r, b_r).

parallel to the *ab* plane, and separated by 2.75 Å along the hexagonal *c* axis. The axial length of 4.5 Å in the *ab* plane is the separation distance between two water molecules, both hydrogen bonded to the same water molecule within the bilayer. Each oxygen atom participates in three O–H⋯O hydrogen bonds within the bilayer and a fourth O–H⋯O bond of 2.75 Å interlinking the neighbouring bilayers along the c-axis. The area per molecule of water within an *ab* layer is *ab* sin 120° = 17.5 Å^2. When amphiphilic alcohols of the general formula $C_nH_{2n+1}OH$ were spread over water, they aggregated into crystalline domains whose structure could be directly investigated by GID. The packing arrangement of the alcohol groups could be correlated with that of the *ab* plane in hexagonal ice (Figure 7-8).

The efficiency of these alcohol monolayers as ice nucleators was evaluated by measuring the threshold freezing temperatures of supercooled water drops (10 µl) covered by the monolayers and compared to drops covered by monolayers of the corresponding carboxylic acids, for which we do not expect structural match.

Figure 7-9 shows the freezing temperatures of supercooled water drops covered by monolayers of alcohols $C_nH_{2n+1}OH$ and carboxylic acids $C_nH_{2n+1}CO_2H$ *vs* the number of carbon atoms in the chain. This figure indicates that the amphiphilic alcohols induce ice nucleation at much higher temperatures and more reproducibly than the corresponding acids. Furthermore, the freezing-points depend on the length of the aliphatic chain, and it is remarkable that there is a pronounced odd/even effect. The freezing-point curve for the *n* odd series increases with chain length, reaching a value of –1 °C for *n* = 31, whereas in the *n* even series a plateau at –8 °C is reached for *n* = 22. Freezing temperatures obtained for drops covered by monolayers of the analogous aliphatic carboxylic acids were scattered in the range of –12 °C to –18 °C.

In order to explain the difference in behaviour of these systems, we studied the structure of the deposited amphiphilic alcohols on water [16] before and during ice freezing [17]. From these studies we could conclude that the lattice match between the monolayer and ice and the degree of crystallinity of the monolayer clusters, are among the important parameters that determine the efficiency of ice nucleation. Monolayer films of long amphiphiles n > 22, with many methylene-methylene interactions (about 2 kcal mol^{-1} for

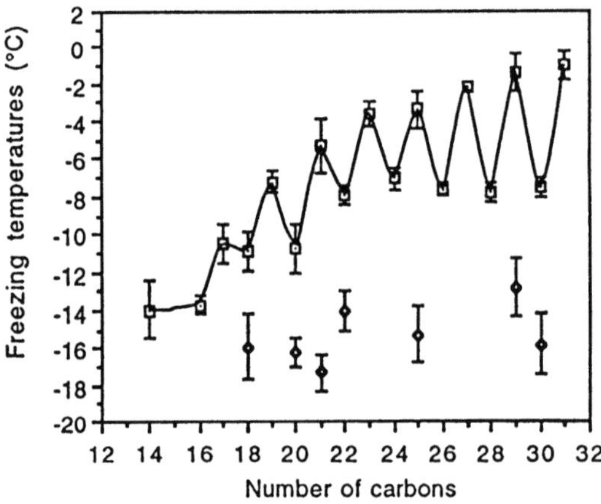

Figure 7-9   Freezing temperatures of supercooled water drops covered by monolayers of aliphatic alcohols $C_nH_{2n+1}OH$ (□) and carboxylic acids $C_{n-1}H_{2n-1}CO_2H$ (◊) *vs* the number of carbon atoms in the chain.

each methylene) have a tendency to be more ordered and crystalline than the alcohols bearing shorter hydrocarbon chains. Indeed, GID measurements on the series $C_nH_{2n+1}OH$, $n = 16, 20, 23, 30, 31$ indicate that the extent of the lateral order (i.e. crystalline coherence length) of these spontaneously formed two-dimensional (2D) crystalline clusters decreases with shorter chain length. In addition, the overall degree of crystallinity for the alcohols of longer chains ($n = 23, 30, 31$) is higher compared with the shorter alcohols ($n = 16, 20$).

From the IR measurements performed on the spread films on water, one can observe the antisymmetric and symmetric CH_2 stretching vibrations which are conformation-sensitive and may be empirically correlated with order (i.e. the trans-gauche character of the hydrocarbon chain). Systematic studies on this series of materials indicate that gradual increase of the chain length is associated with a blue shift of $1.0-1.5$ cm^{-1} and a decrease of $2-4$ cm^{-1} in band width, in both the antisymmetric and symmetric stretching bands, indicating that the chains became more ordered. In conclusion, the monolayers of shorter chains are less crystalline than the longer ones, they assume a higher tilt angle, and therefore, the structural match with that of ice decreases gradually.

The explanation of the difference in the induced freezing-point by hydrocarbon alcohols bearing an odd or even number of carbon atoms is more complex. GID measurements of monolayers with $n = 30$ and $n = 31$ on water resulted in similar packing arrangements. However, this analysis does not provide information regarding the exact orientation of the OH groups at the water surface. From the results of the ice nucleation experiments, we may conclude that the interatomic hydrocarbon arrangements of the two monolayers with $n = 30$ and $n = 31$ are almost the same, but the two structures are not superimposable in terms of the arrangements of the head group moieties CH_2CH_2OH. The orientation of the latter with respect to the water surface is not the same, which may manifest itself in differences in ice nucleation.

In order to clarify the role played by the orientation of the OH head groups at the air-water interface on nucleation efficiency, we prepared amphiphilic alcohols bearing functional groups, such as amides and esters, along the aliphatic chain. In these systems the packing arrangement is strongly influenced by the hydrogen bonds between the amide groups interlinking the chains, which fix their packing arrangement and molecular tilt in a way that influences the orientation of the OH group at the interface. This class of materials can be divided into two groups according to the difference in temperature at which they nucleate ice, which is about 4 °C. The structural difference between these two groups is specified by the parity of the number of carbon atoms in the fragment interlinking the polar group to the alcohol head group. The packing arrangement of two such ester alcohols $C_{19}H_{39}CO_2C_mH_{2m}OH$ ($m = 9, 10$) as determined by GID measurements and corroborated by atom-atom potential energy calculations, is shown in Figure 7-10. From the packing arrangement of the systems, we notice that the OH groups of the alcohols with $m = 9$ and $m = 10$ assume different orientations relative to the water surface. In the monolayer with $m = 9$ the C–OH bond is almost perpendicular to the water surface so that the O–H bond and the two lone-pair lobes of the hydroxyl oxygen atom are equally well exposed to water. In the monolayer $m = 10$, the C–OH bond forms an angle of 20° to the water leaving two possibilities; either the O–H bond points into the water, or one oxygen lone-pair lobe and the OH group are equally well exposed to the water. Therefore, a correlation exists between the orientation of the hydroxy groups including their lone-pairs, with the freezing-point of supercooled water.

In order to establish which layer of ice is nucleated under the alcohol monolayer, X-ray powder diffraction measurements of ice were carried out under various alcohol monolayers.

$C_{19}H_{39}CO_2(CH_2)_mOH$

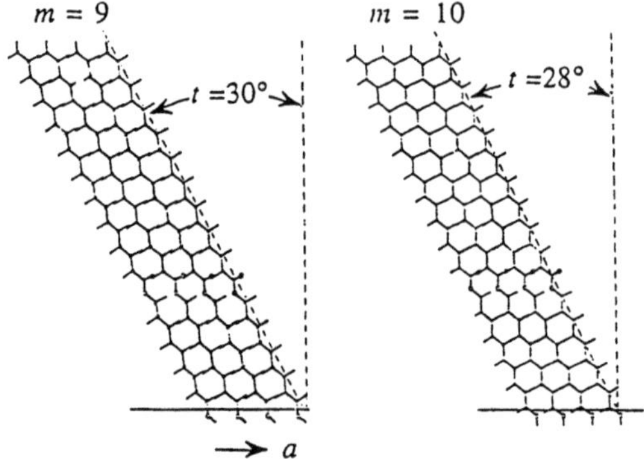

Figure 7-10   Two-dimensional structures of monolayers of ester alcohols $C_{19}H_{39}CO_2C_mH_{2m}OH$, ($m$ = 9, 10), viewed along the b axis.

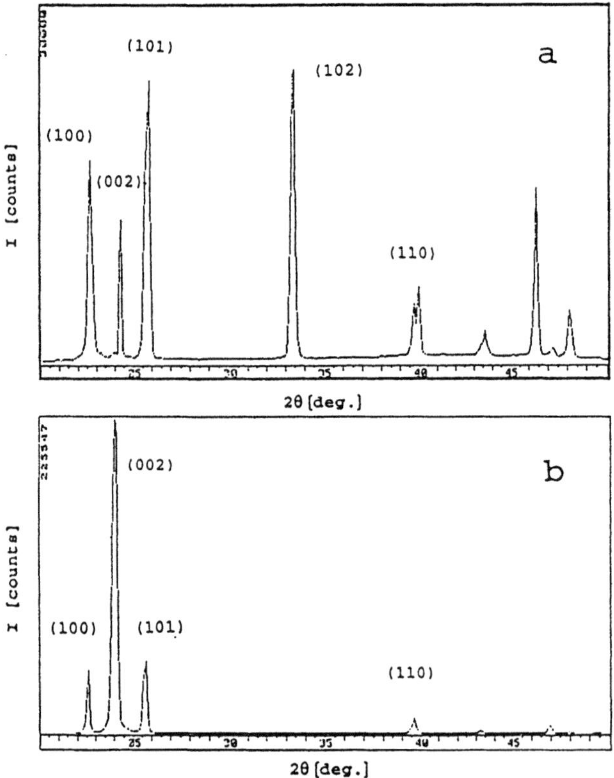

Figure 7-11   X-ray powder diffractograms (CuK$_\alpha$ radiation) of (a) ice frozen from pure water and (b) ice nucleated under a monolayer of $C_{31}H_{63}OH$ at 70% surface coverage.

Figure 7-12   (⟶) GID measurements made on a monolayer of $C_{31}H_{63}OH$ over pure water cooled to freezing-point. (a) The two Bragg peaks {1,1} and {0,2} of the monolayer on water at 4 °C.

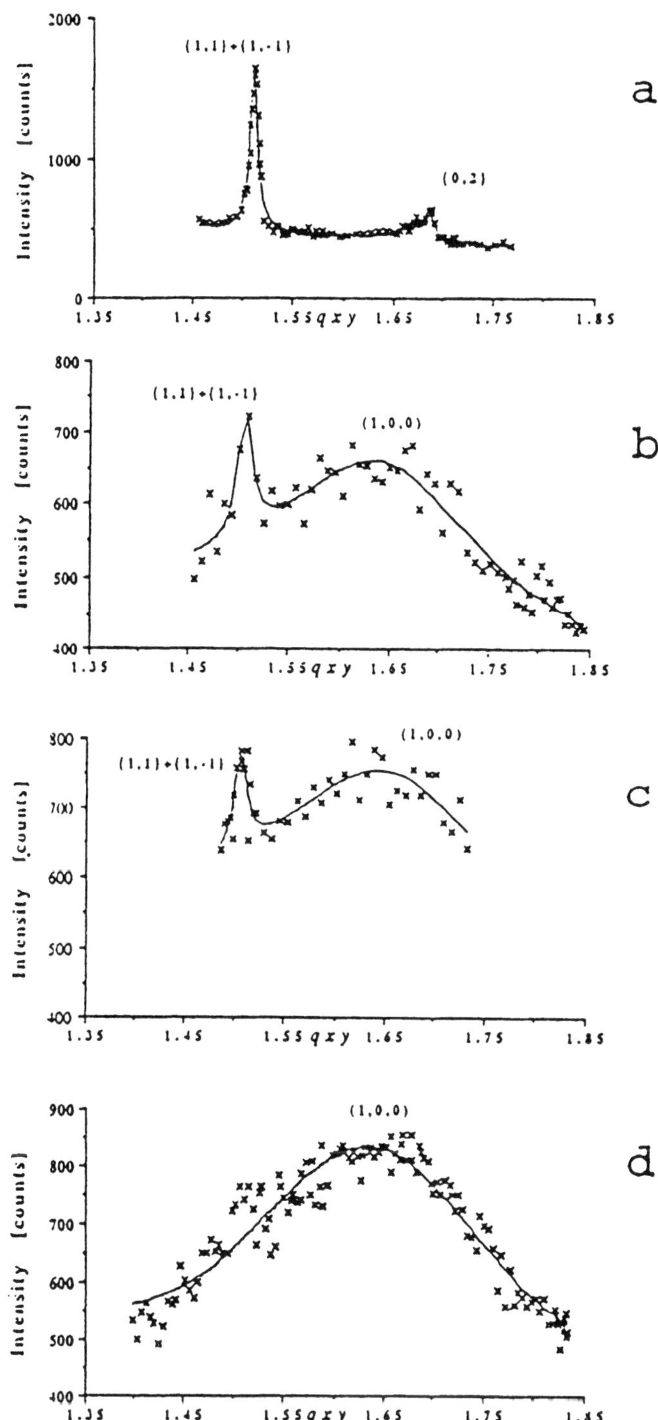

(b) First stage of ice crystal nucleation under the monolayer at 0 °C, the {1,1} peak of the monolayer and the (1,0,0) peak of ice are visible. (c) as (b), after about 15 minutes. (d) as (c), after a further 15 minutes, only the (1,0,0) peak of ice is visible.

Diffractograms obtained from ice nucleated under a monolayer of the alcohol $C_{31}H_{63}OH$ indicated a preferred orientation of the ice crystals with the c-axis perpendicular to the plane of the sample. This is in contrast to ice nucleated in the absence of the monolayer, or under a perfluorinated monolayer where no lattice match between the monolayer and ice exists, Figure 7-11.

Further independent evidence for the epitaxial relation between the structure of $C_{31}H_{63}OH$ monolayers and hexagonal ice could be obtained from cryo-transmission electron microscopy (TEM) studies [18]. The diffraction pattern of a thin layer of water, covered with a monolayer and cooled slowly, showed the typical {1,0,0} reflections of a single-crystal of hexagonal ice, four {1,1} reflections, and two {0,2} reflections of a single-crystallite of $C_{31}H_{63}OH$ in an epitaxial relationship.

In another series of experiments, GID measurements were performed on a monolayer of $C_{31}H_{63}OH$ on the surface of water during and after ice formation [17]. Here, again, one could demonstrate that the ice crystals were oriented with their {0,0,1} face attached to the monolayer and that the monolayer maintained its 2D crystalline integrity on nucleating ice as well as after melting the ice back to water (Figure 7-12). Furthermore, from the width of the (1,0,0) ice diffraction peak, a lateral coherence length of approximately 25 Å was estimated. Such a low coherence length is indicative of multiple nucleation sites separated by about 50 Å. Consequently, the upper threshold value for the diameter of the critical nucleus of ice under the monolayer can be estimated to be say maximally 25 Å, which would correspond to an "ice-like" hemisphere containing about 70 molecules of water.

7.6 Conclusions

Mother Nature has elaborated methodologies under physiological conditions for the construction of complex functional materials such as teeth, bones, shells etc.. There is still a long way for man to go until he will be able to mimic the preparation of similar composites from inception. Progress in this direction requires a deeper understanding of the dynamics of self-organization of molecules into complex clusters or solids. At the present time, studies along these lines are focused towards understanding self-organization of simple systems such as thin molecular films or mono- and dicomponent crystallites. Once the rules for design of such organized assemblies are elaborated, they will presumably be applied in the construction of more complex functional solids with desired optical, electric, and magnetic properties. The realization of this dream will, of course, take some time. However, with the advent of new technologies in recent years, new sophisticated analytical tools emerged, such as intense coherent X-rays from synchrotron sources, new microscopic tools such as STM (Scanning Tunnelling Microscopy) and AFM (Atomic Force Microscopy), to mention but a few. These tools, together with more advanced computational methods, will furnish a more accurate understanding of the van der Waals and electrostatic forces between molecules which are responsible for the constitution of surfaces or materials on the molecular level. Indeed, by applying some of these methods it became possible to monitor the growth of self-aggregates of several amphiphiles at the air-water interface [19].

Acknowledgements

This work was carried out with the financial support of the Minerva Foundation, Munich, Germany and the Israel Academy for Science and Humanities. We thank also J. Als-Nielsen and K. Kjaer for the fruitful collaboration in GID measurements.

8 Towards Molecular Devices

J. Fraser Stoddart and Natalie M. Rowley

Abstract

The challenge that confronts scientists and engineers fascinated by the prospect of fabricating molecular devices by a "bottom up" approach – that is, starting from atoms and molecules – calls for radical departures from current attitudes and activities in many areas of science and technology. Whilst current sophisticated electronic technologies are being questioned, the delicately perfected art and unquestionably elegant science of chemical synthesis must now be prepared to undergo one of the most profound conceptual changes it has ever witnessed. With their anticipated reliance on organized molecular assemblies, man-made molecular information processing systems are almost certainly going to emerge as we learn from nature how to self-assemble and self-replicate at a molecular level and how to self-organize at a supramolecular level.

8.1 Introduction

The living world is made up of molecular compounds that interact physically and react chemically with each other, often in rather specific and selective ways. The phenomenon of molecular recognition has evolved around the principles of self-organization, self-assembly, and self-synthesis to the extent that participating molecules must not only be blessed with a reasonably precise chemical form, but that they must also usually perform a prescribed biological function. One of the most exciting and potentially rewarding challenges that the chemist faces today, is to devise and realize wholly synthetic systems [1] that function like their biological counterparts, by accepting, storing, processing, transferring and disseminating information at a molecular level. The manufacture and manipulation of materials on effectively a nanometer scale raises [2] the prospect of being able to perform physical feats with microscopic devices, reminiscent of the automatic control units comprising the nervous system and the brain. The fundamental investigation of the chemical science that will enable the future development of mechanoelectrical and photoelectrical communication systems and devices is a challenging field of research – one that could be referred to as *molecular cybernetics* [3].

The question is how to construct materials that express their properties at a molecular level, and how to gain access to these properties. While the problems associated with the successful development of electronic [4a] and photonic [4b] molecular device systems are numerous, the most pressing fundamental ones are those of synthesis and fabrication. What is required is the development of a modular chemical approach in which molecular-size fragments are incorporated into a molecular assembly in a highly controlled and totally precise manner. As stated back in 1984 by Carter [5], *"this challenge will probably only be mastered when we have learned the principles of self-organization and self-synthesis from the biological world and have applied them more broadly to both organic and inorganic chemistry"*.

The answers will probably emerge from the new fields of supramolecular [6] and host-guest chemistry [7] that have been developing with such vigour ever since Pedersen announced [8] his discovery [9] of the so-called crown ethers in 1967. This contemporary area of chemistry has focused the attention of the synthetic chemist on the nature of the non-covalent bond as never before. In particular, template effects [10] have been observed during the course of many chemical reactions that yield complexing macrocycles and molecular receptors as products, often associated with remarkable efficiencies and selectivities. Alkali metal and alkaline earth metal cations play an important role as non-covalently bound templates during the preparation of many oxygen- and nitrogen-containing macrocyclic compounds. The template effect not only speeds up rates of reaction but it also leads to much higher yields of products, i.e. the evidence for templating is both kinetic and thermodynamic. Organic cations can also act as non-covalently bound templates during crown ether syntheses. In the case of transition metals, covalent (coordinative) bonds are invariably formed during templated syntheses of nitrogen-, sulphur- and phosphorous-containing macrocycles. These covalently-bound templates are usually much more difficult to release, by demetallation, from within the macrocyclic compound.

The relevant background information is now available, within the research fields identified above, to enable the evolution of a completely new synthetic methodology for the modular construction of large, discrete and ordered molecular assemblies from prefabricated molecular components [11]. This approach, it could be argued, should appeal to the self-assembly aspects of the template effect involving not covalently bonded species, but weak non-covalently bonded components that do not implicate, in the beginning at least, relatively strong coordinative interactions with (transition) metal ions. The bond-forming processes should then occur easily within the ordered environment of the non-covalent templating forces in such a way as to retain the order originally imposed by the weak interactions. Finally, the molecular fragments, which are organized in this manner, should subsequently be unable to dissociate from each other as a consequence of simple mechanical constraints after the reactions are complete. The usual host-guest relationship involves rapid association between host and guest to form a 1:1-complex. In this situation, dissociation of the complex into its molecular components occurs, albeit less rapidly, i.e. a reversible process ensues. If, however, the guest is trapped mechanically around the associated host, then dissociation is prevented and the process becomes an irreversible one. There are several ways of achieving this end. One is the formation of molecules which have been called rotaxanes. The name rotaxane is derived from the Latin words *rota* meaning "wheel" and *axis* meaning "axle". In chemical terms, this type of molecule contains a linear component (the axle) encircled by a macrocyclic component (the wheel). In order to prevent the wheel slipping off the axle, the linear component must be terminated at both ends by large stoppers, i.e. blocking groups. Although rotaxanes are already well-documented in the literature [12, 13], they have rarely been regarded as much more than academic curiosities. Generally speaking, they have been prepared in relatively low yields by using the concept of statistical threading or by directed synthesis through temporary covalent bonding to a central core, which is subsequently dissected out to release the rotaxane. Examples of the use of non-covalent bonding to direct the syntheses of rotaxanes were rare until recently. There now follows a description of how non-covalent bonding can be used in their synthesis, via templates, as well as how to produce high structural order that is incorporated automatically into these new molecular assemblies.

8.2   Molecular Shuttles

During self-assembly processes, appropriate precursors form often a complicated product in a single step, replacing a process that would normally require numerous synthetic steps. The self-assembly of rotaxanes has recently been approached in a number of different ways. The *threading* approach [14–17] (**a** in Scheme 8-1) relies on statistical or non-covalent bonding directed association of a rod-like component with a bead-like macrocyclic component of the emerging rotaxane. This 1:1-complex is then trapped mechanically by the covalent attachment of blocking groups (stoppers) which are of sufficient size to prevent the dissociation of the macrocycle from the newly created dumbbell component. In the *clipping* approach [14, 15, 18–23] (**b** in Scheme 8-1), the bead-like macrocyclic component is assembled around the preformed dumbbell component to form the [2]rotaxane [*].

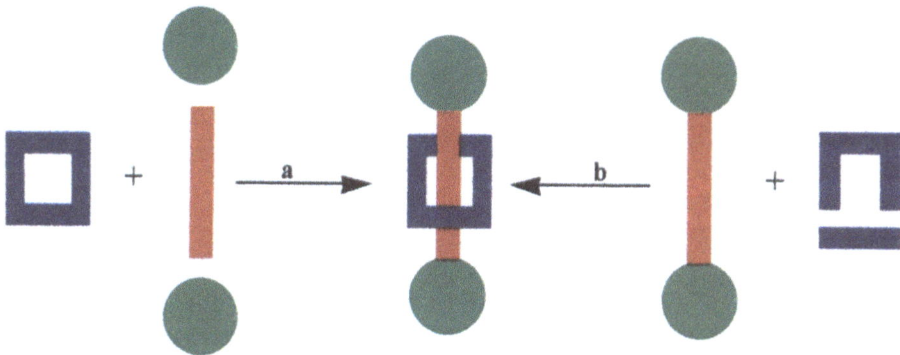

Scheme 8-1   Assembling a [2]rotaxane by (**a**) threading and (**b**) clipping.

The clipping procedure has been used to self-assemble a so-called molecular shuttle [15, 18, 24, 25] (Figure 8-1). It is a [2]rotaxane in which the dumbbell component is constructed from a polyether thread containing *two* π-electron rich hydroquinone rings (stations), terminated by triisopropylsilyl stoppers. The dumbbell component is encircled by a π-electron deficient bead which is a tetracationic cyclophane containing two bipyridinium units bridged by two *p*-xylyl spacers. The bead becomes localized at one of the two degenerate stations as a result of non-covalent bonding interactions with the stations. ^{1}H NMR spectroscopic studies [15, 18, 24, 25] have shown that, at room temperature, the bead moves back and forth between these stations at a rate of ca. 500 times a second. Although both degenerate positions (with the bead at either station) have equal energy, there is a barrier of ca. 13 kcal mol^{-1} between the two states.

---

[*]   A [n]rotaxane consists of n–1 'wheels' and 1 'axle'.

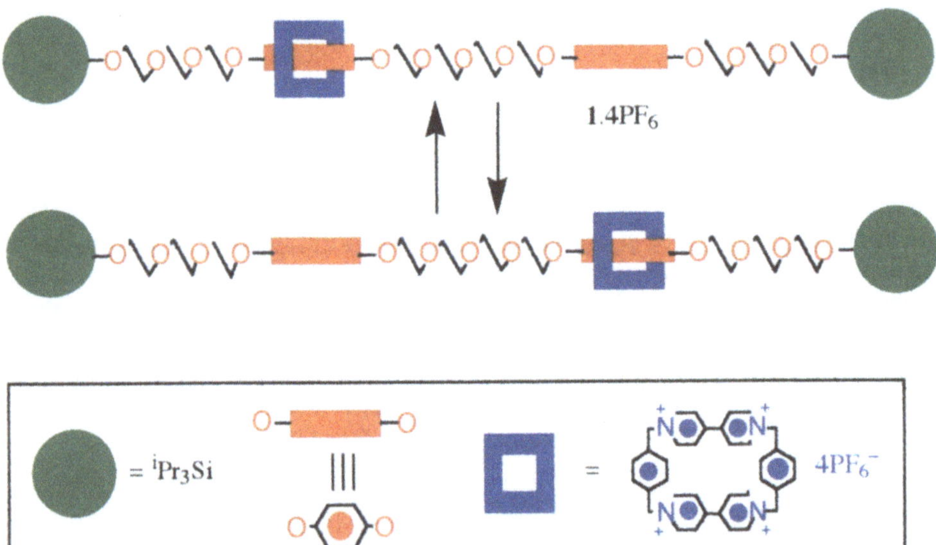

Figure 8-1   A molecular shuttle (**1**.4PF$_6$).

More recently, an "intelligent" self-assembly approach [26] was devised as an alternative route to forming [2]rotaxanes. This approach is based upon the mutual recognition between π-electron deficient 4,4'-bipyridinium dications and π-electron rich hydroquinol rings contained within a macrocyclic polyether. This self-assembly route is highly selective in that it requires *two* 4,4'-bipyridinium dications to be present in the [2]rotaxane in order that

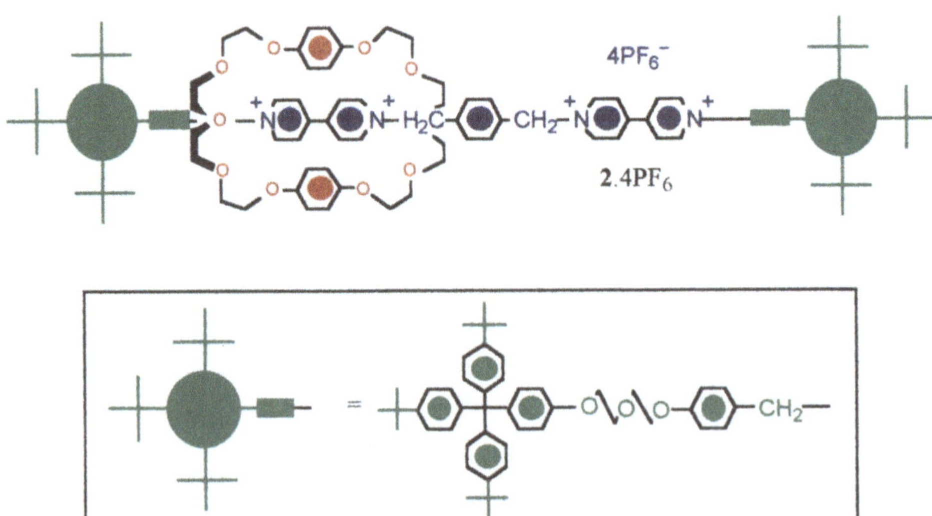

Figure 8-2   A molecular shuttle comprising a dumbbell with π-electron deficient stations and a π-electron rich bead (**2**.4PF$_6$).

one bisparaphenylene-34-crown-10 ring can be incorporated. Hence, the molecular shuttle **2.4PF$_6$** (Figure 8-2) formed by this route can be considered to be the reverse of the molecular shuttle **1.4PF$_6$** described in Figure 8-1, in that it contains a dumbbell with π-electron deficient stations and a π-electron rich bead, i.e. it takes the form of a dumbbell-shaped component containing two π-electron deficient 4,4'-bipyridinium units in the thread terminated by tris(4-*tert*-butylphenyl)methyl stoppers, and is encircled by a bisparaphenylene-34-crown-10 macrocyclic bead containing two hydroquinone rings. ^{1}H NMR spectroscopic studies [26] have shown that, at room temperature, the bead moves back and forth between stations at a rate of 300 000 times a second, and that there is a barrier of ca. 10 kcal mol^{-1} between the two states.

Molecular structures of this kind can be considered as possible first steps towards molecular-scale information processing devices. They are prototypes for molecular assemblies that would be able to "receive, store, transfer and transmit information in a highly controllable manner" [24]. In the second type of molecular shuttle, the stoppers incorporating tris(4-*tert*-butylphenyl)methyl groups are too large to permit passage of the crown ether over them and allow interaction with the 4,4'-bipyridinium dications. It was therefore reasoned that, by judicious adjustment of the size of the stoppers, it should be possible to arrive at the situation where the size complementarity between the crown ether and the stoppers was such that *slippage* [27, 28] of the macrocycle over them would be possible at elevated temperatures (**c** in Scheme 8-2). The complexation of the 4,4'-bipyridinium dication within the dumbbell-shaped component by the bisparaphenylene-34-crown-10 would then provide a *thermodynamic* trap for the crown ether, raising the activation energy for the extrusion process relative to that for slippage. Thus, the [2]rotaxanes formed should be stable at room temperature.

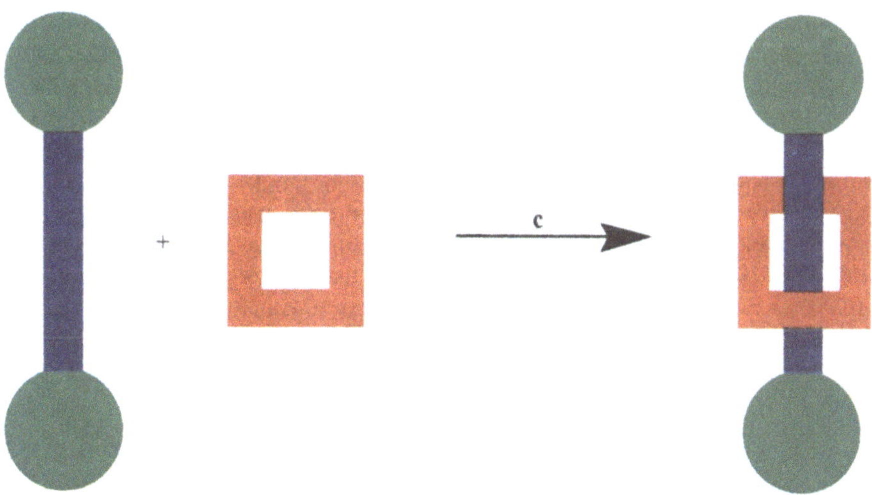

Scheme 8-2   Assembling a [2]rotaxane by (**c**) slippage.

The slippage route did indeed permit the successful self-assembly, in good yields, of [2]rotaxanes, and thus demonstrates the preparative utility of the slippage method as a third route to rotaxane formation [27, 28]. A range of 4,4'-bipyridinium dications (Figure 8-3), where the size of the stoppers was systematically varied, was synthesized. The success of

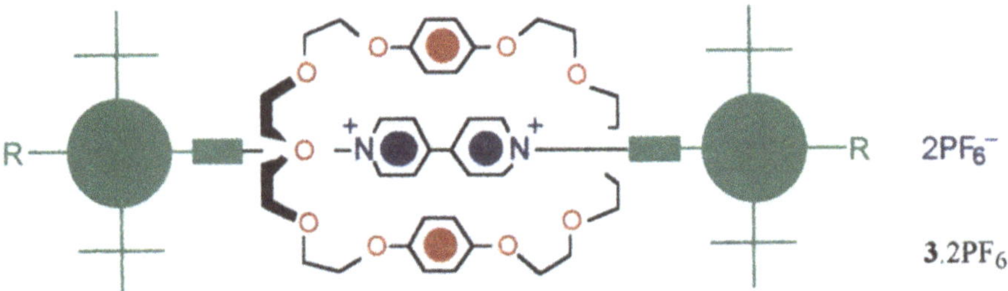

Figure 8-3   Examples of [2]rotaxanes (**3.2PF$_6$**) formed by the slippage of a bead over varying sizes of stoppers.

the slippage method in the thermally-promoted self-assembly of a number of room temperature stable [2]rotaxanes **3.2PF$_6$** containing *one* 4,4'-bipyridinium dication in the dumbbell encircled by *one* crown ether, led to the belief, that, by employing larger dumbbells with n dicationic binding sites in slippage experiments, it should be possible to self-assemble, thermally, [n]rotaxanes, along with the corresponding [n–1], [n–2] etc. rotaxanes. Indeed, it has been shown [28] that it is possible to self-assemble [3]rotaxanes, e.g. **5.4PF$_6$**, along with [2]rotaxanes, e.g. **4.4PF$_6$**, by this route (Figure 8-4).

Although the size complementarity has been exploited previously [29] in the statistical synthesis of [2]rotaxanes, the yields of the [2]rotaxanes obtained in these cases were very poor (<1.5%). The addition of a thermodynamic trap, in the form of non-covalent bonding interactions between the two components of a [2]rotaxane, not only enhances the overall yield obtained during the slippage process, but it also increases the inherent stability and information content of the resulting structures. These features, together with the sheer synthetic simplicity of the experimental approach, recommend slippage as a viable and alternative synthetic procedure for the construction of larger [n]rotaxanes and polyrotaxanes. The self-assembly of the [2]rotaxane **4.4PF$_6$** and the [3]rotaxane **5.4PF$_6$** [28] (Figure 8-4) in good yields by slippage illustrates the potential of this method in the synthesis of oligo- and poly-rotaxanes. The principle of size-complementarity between the rings and stoppers of [n]rotaxanes, in combination with stabilising non-covalent bonding interactions between the dumbbell and ring components, opens up yet another avenue for the construction of novel architectures and new advanced materials.

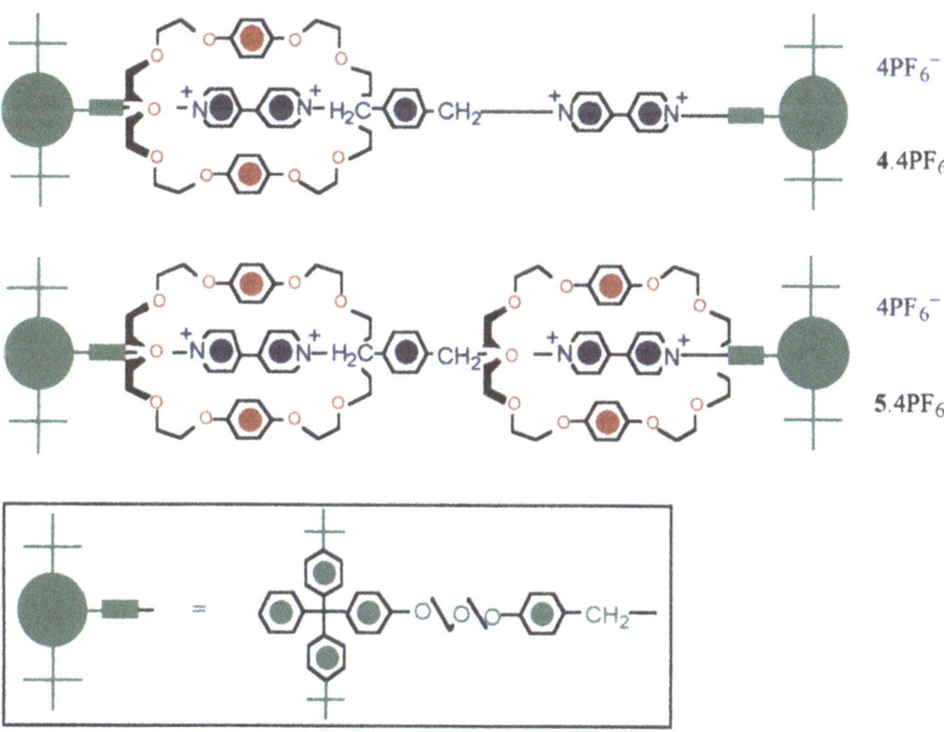

Figure 8-4   A mixture of a [2]rotaxane (**4**.4PF$_6$) and a [3]rotaxane (**5**.4PF$_6$) formed by the slippage route.

In **1**.4PF$_6$, **2**.4PF$_6$ and **4**.4PF$_6$, movement of the bead between stations is a random process. The next logical step is to devise a "desymmetrized" system in which the two stations represent distinguishable states and in which movement between these states is controllable. Controllable movement of the bead might be achieved chemically, electrochemically or photochemically. "A controllable shuttle would be a first step towards building molecular systems into which one can read information, process it, transfer it from one part to another, then write it back out again. The possibility exists – but of course it is very much in the future – of being able to build up systems of this type in the direction of a molecular-scale computer" [24].

8.3   Towards the Electrochemical Control of Molecular Shuttles

In an attempt to design a molecular shuttle which may be controlled electrochemically, it was decided to replace one of the two degenerate hydroquinone rings in **1**.4PF$_6$ by a unit with a lower π-donating ability, i.e. of higher oxidation potential. It was thought that this act would permit control of the positioning of the tetracationic macrocycle within the molecule by electrochemical means. A suitable replacement for one of the hydroquinone rings was thought to be a *p*-xylyl residue on account of its considerably lower π-donating ability.

Exercising control in such a system would require the preferential occupation of the more π-electron rich hydroquinone site by the tetracationic macrocycle. Oxidation of this site to its corresponding radical cation should then lead to the preferential occupation of the less π-electron rich *p*-xylyl site, thus effecting control of the shuttling process. Thus, a molecular shuttle was synthesized [21], using the principles of self-assembly, which contained one hydroquinone unit and one *p*-xylyl group (Figure 8-5).

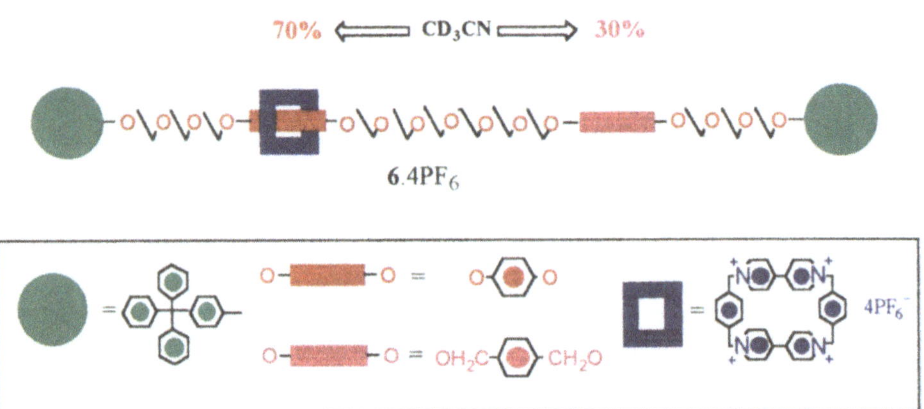

Figure 8-5   A desymmetrized molecular shuttle (**6**.4PF$_6$) incorporating a *p*-xylyl and a hydroquinone station.

However, the ^{1}H NMR spectrum of **6**.4PF$_6$ (Figure 8-5) at 233 K in CD$_3$CN revealed [21] that the macrocycle distributed itself between the two sites, with 70% occupancy at the hydroquinone rings and 30% at the *p*-xylyl residues. Thus, the self-assembly of **6**.4PF$_6$ has demonstrated the relative ease of construction of an apparently complex system with potential switching action, although the position of the site occupancy equilibrium in this case is not as satisfactory as it might be and other problems preclude electrochemical control of the switching action. Nevertheless, it has been demonstrated that controllable molecular shuttles should be attainable using the synthetic strategy of self-assembly if the question of occupancy can be addressed satisfactorily.

Recognition of the shortcomings of the first generation of a desymmetrized molecular shuttle **6**.4PF$_6$ – namely the finely balanced nature of the translational isomerism operating within the system and the inherent inconvenience associated with the very high oxidation potential of the *p*-xylyl residue – prompted a reversal of the design logic of molecular shuttles. Thus, it was decided that the desymmetrisation of the molecular shuttle had to be achieved by replacement of a hydroquinone ring by a more π-electron rich unit. Ideally, the π-electron donor chosen to replace the hydroquinone unit should possess an oxidation potential of less than +1.0 V, form a stable radical cation, be non-basic, and have a small steric demand. A 2,3,5-trisubstituted indole residue was identified as promising to satisfy most of the above criteria. It was hoped that the highly π-electron rich nature of the indole nucleus would force preferential occupation of this site within the [2]rotaxane by the tetracationic macrocycle. Furthermore, oxidation of the indole unit to its corresponding radical cation should result in the transfer of the macrocycle to the hydroquinone site. Thus, it was decided to synthesize [22], using self-assembly, the [2]rotaxane **7**.4PF$_6$ containing one indole unit and one hydroquinone unit (Figure 8-6).

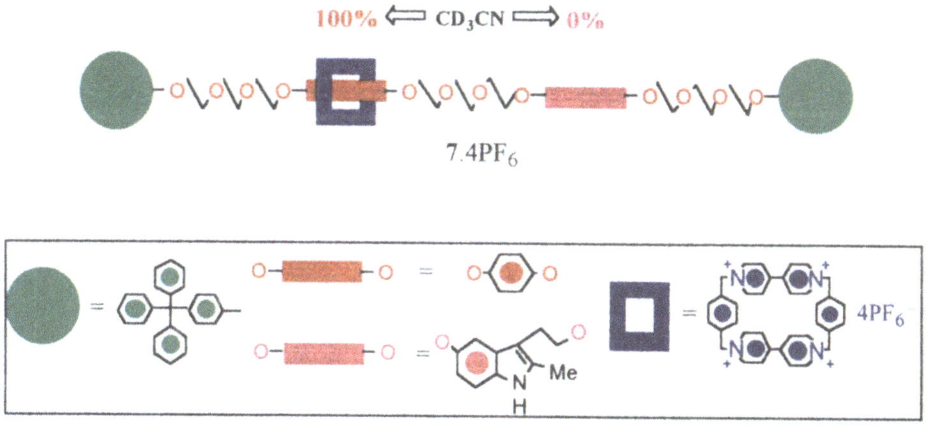

Figure 8-6   A desymmetrised molecular shuttle (7.4PF$_6$) incorporating an indole and a hydroquinone station.

In order to establish the basis for an electrochemically-driven switching action, the macrocyclic tetracation must occupy the more π-electron rich indole unit exclusively. Although ^{1}H NMR spectroscopy of 7.4PF$_6$ (Figure 8-6) at 233 K in CD$_3$CN revealed [22] that the macrocycle was indeed located exclusively at one site, it was the hydroquinone site which was the occupied one! This situation is undoubtedly a reflection of the conflict between steric and electronic demands within the system. Although the indole unit is much more π-electron rich, its inclusion within the cavity of the tetracationic macrocycle is more sterically demanding, and so the hydroquinone ring is preferentially included within the tetracation. The preferential inclusion of the less π-electron rich hydroquinone ring within the tetracationic cyclophane indicated that a more fundamental redesign of the molecular shuttle was necessary in order to achieve the goal of a controllable molecular shuttle.

In view of these shortcomings in the properties of the second generation molecular shuttle 7.4PF$_6$ and questionable stability of the indole radical cation, it was decided to replace the indole nucleus with a sterically less demanding residue. Tetrathiafulvalene (TTF) immediately became an extremely attractive candidate to fulfil this function. Not only are TTF and hydroquinone residues of comparable bulk but TTF also displays highly reversible redox behaviour at low potentials. Thus, a molecular shuttle containing two hydroquinone sites and one TTF site was synthesized [23] in the form of the [2]rotaxane 8.4PF$_6$ (Figure 8-7). Preliminary ^{1}H NMR spectroscopic studies of 8.4PF$_6$ (Figure 8-7) in CD$_3$COCD$_3$ and CD$_3$SOCD$_3$ at 298 K showed [23] that in CD$_3$COCD$_3$ the tetracationic cyclophane occupied exclusively the two equivalent hydroquinone rings, whilst, in CD$_3$SOCD$_3$, the macrocycle was located predominantly on the TTF residue. UV spectroscopic investigations of 8.4PF$_6$ in acetone supported the ^{1}H NMR spectroscopic evidence, i.e. a charge transfer band (λ_{max} = 752 nm) was present in DMSO, but not in acetone. This charge transfer band corresponds to occupation of the TTF residue by the tetracationic cyclophane. The preferential occupation of the TTF site in DMSO bodes well for the prospect of electrochemical switching of the macrocycle position within the [2]rotaxane. Currently, the temperature dependence of the NMR spectra of 8.4PF$_6$ in a range of deuterated solvents of different polarities is being investigated. These results, along with attempts to influence the dynamic behaviour of 8.4PF$_6$ by electrochemical means, will be discussed subsequently in a full paper.

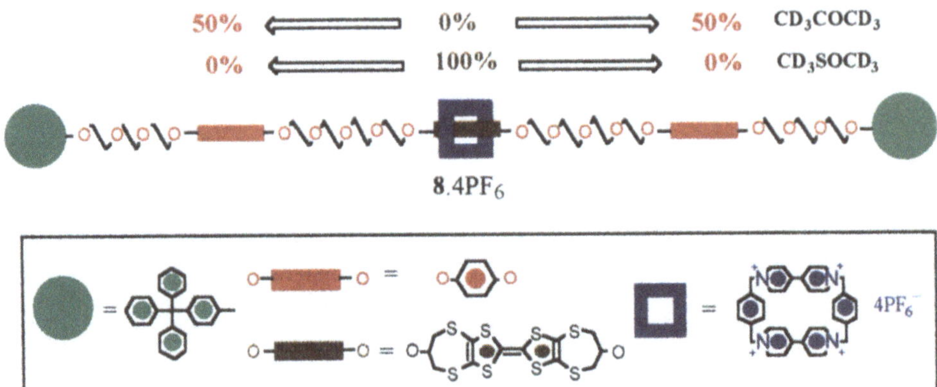

Figure 8-7   A desymmetrized molecular shuttle (**8**.4PF$_6$) incorporating a tetrathiafulvalene (TTF) and two hydroquinone stations.

These studies led to the realization that it would be desirable to increase the number of recognition subunits suitable for the self-assembly of these complicated molecular structures in order to enhance the likelihood of preparing systems having the sought-after switching functions. Thus, it was decided [17] to prepare a [2]rotaxane **9**.4PF$_6$ incorporating two

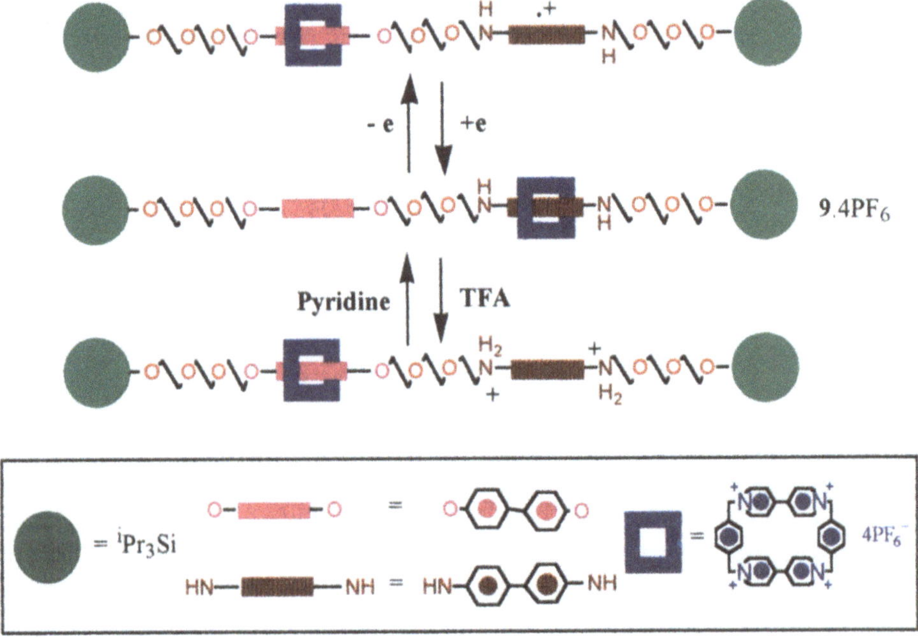

Figure 8-8   A chemically and electrochemically switchable molecular shuttle (**9**.4PF$_6$).

different π-electron donor stations derived from a biphenol and a benzidine unit (Figure 8-8). Electrochemical stimuli or proton concentration can be used to impart control on the positioning of the bead at the stations on the thread of the dumbbell component.

^{1}H NMR spectroscopic studies at 229 K in CD_3CN on **9**.4PF$_6$ (Figure 8-8) show that the bead occupies preferentially the benzidine station. Upon protonation (or oxidation) [17] of the benzidine residue, the rotaxane switches to a different 'state' in which the bead interacts only with the biphenol station. Switching is completely reversible since the original molecular 'state' can be recovered by the corresponding deprotonation (or reduction) process.

8.4   Towards the Photochemical Control of Molecular Shuttles

The concept of a molecular shuttle, which may be addressed photochemically, has led [20] to the identification of the use of tetraarylporphyrins as groups which could serve the dual purpose of stoppers and of photochemically-active functions in a [2]rotaxane with molecular switching possibilities. Self-assembly was used to produce one and two station [2]rotaxanes – **10**.4PF$_6$ and **11**.4PF$_6$ – with one or two recognition sites, respectively. In both [2]rotaxanes, tetraarylporphyrin groups act as the stoppers and hydroquinone rings act as the recognition sites (Figure 8-9).

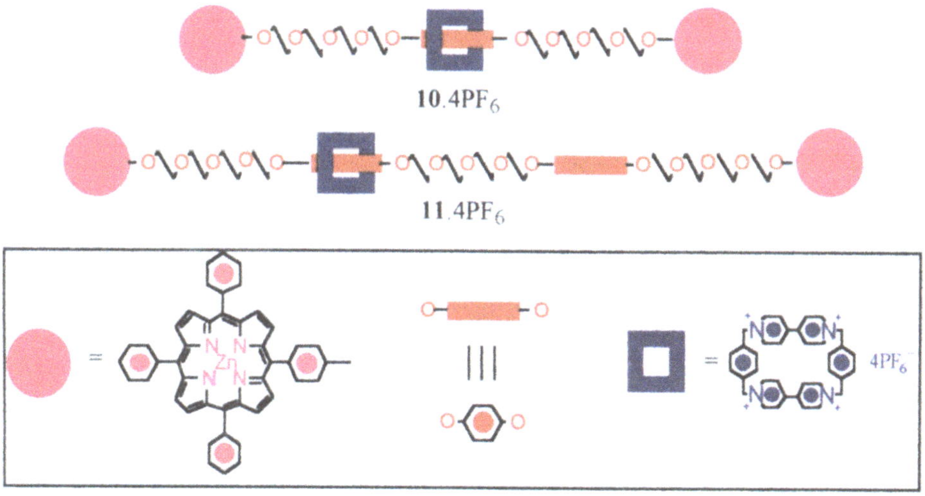

Figure 8-9   Examples of porphyrin-stoppered [2]rotaxanes (**10**.4PF$_6$ and **11**.4PF$_6$).

^{1}H NMR spectroscopic studies showed [20] that degenerate shuttling occurred in the two station system **11**.4PF$_6$ (Figure 8-9) at a rate of 25 times per second at 263 K, corresponding to an energy barrier of ca. 14 kcal mol^{-1}. Photochemical studies have yet to be carried out on these systems. They will be presented subsequently in the form of a full paper.

The control of energy and/or charge distribution within molecules is of fundamental importance in the development of molecular electronic information processing systems. A molecular shuttle offers a unique opportunity to study both inter- and intra-component photochemically-induced electron transfer in a system in which two non-covalently linked components are constrained within one system. In contrast to previously studied systems, a molecular shuttle enables a photochemically-active π-electron donor and a π-electron acceptor to be held in close proximity without the possibility of solvent molecules intervening between them. In order to develop a molecular shuttle which may be addressed photochemically, it will be necessary to understand the manner in which energy or charge is distributed within the molecular shuttle following photon absorption.

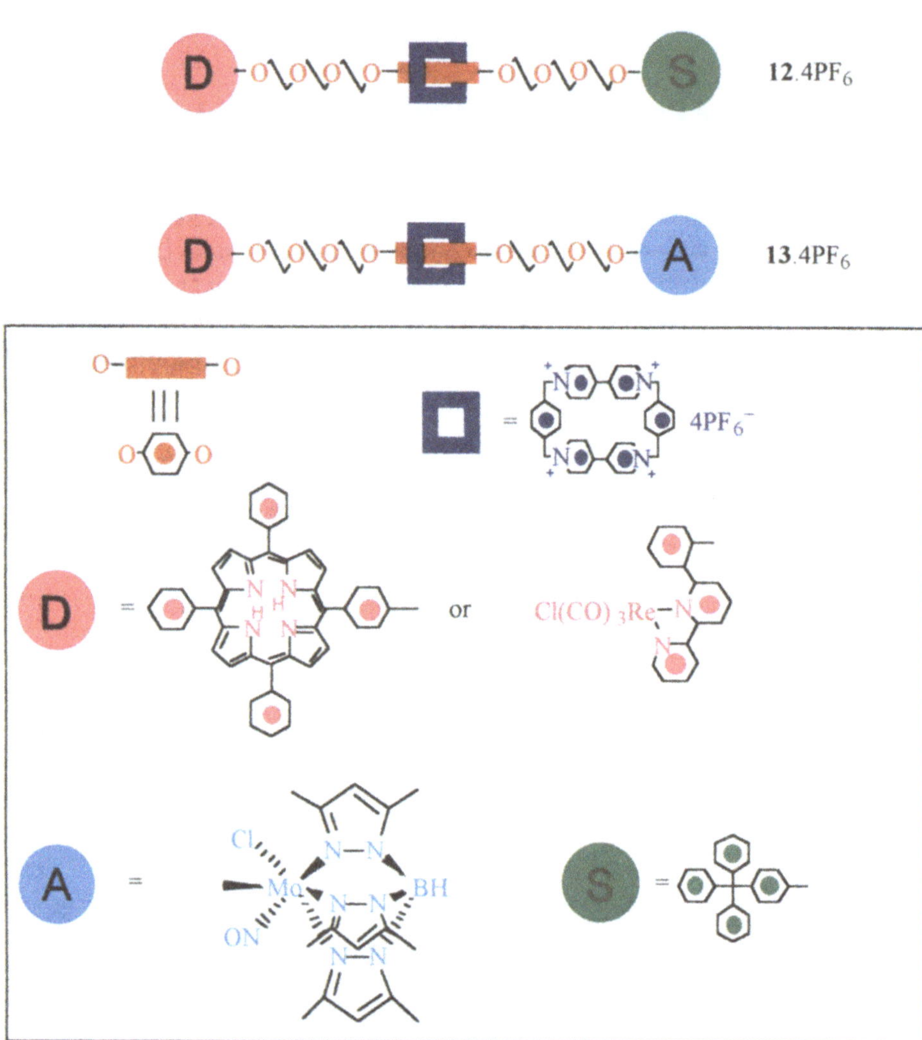

Figure 8-10   Examples of [2]rotaxanes for preliminary photochemical investigation (**12**.4PF$_6$ and **13**.4PF$_6$).

Research is now in progress to produce some simple second generation "one-station" rotaxanes for preliminary photochemical investigation (Figure 8-10). The [2]rotaxane **12**.4PF$_6$ is designed to enable study of photochemically-induced intercomponent electron transfer from the photochemically active porphyrin or Re(bpy)(CO)$_3$Cl π-electron donor to the tetracationic cyclophane bead. The [2]rotaxane **13**.4PF$_6$ will facilitate an investigation of the competition between photoinduced intercomponent electron transfer (from the π-electron donor to the bead, and from the bead to the π-electron acceptor) and photoinduced intracomponent electron transfer (from the π-electron donor to the -MoL*(NO)Cl π-electron acceptor).

A different approach to molecular switches has also been investigated. It has involved [30] the forming of a pseudorotaxane, *i.e.* an unstoppered [2]rotaxane **14**.4Cl comprised of a thread, incorporating a naphthalene unit as a station and a tetracationic cyclophane bead (Figure 8-11). When anthracene carboxylic acid is irradiated with light in aqueous solution, in the presence of triethanolamine (TEA), present as a sacrificial electron donor, it undergoes photoinduced electron transfer, reducing the tetracationic cyclophane bead component of **14**.4Cl to form probably the bisradical cation [30]. This bisradical cation form of the bead no longer binds the naphthalene substrate, and so the thread is ejected from the cavity of the bead. This decomplexation can be monitored by the disappearance of the charge-transfer band and the emergence of fluorescence from the naphthalene residue. The process can be reversed by bubbling oxygen into the initially oxygen-free system in the absence of light.

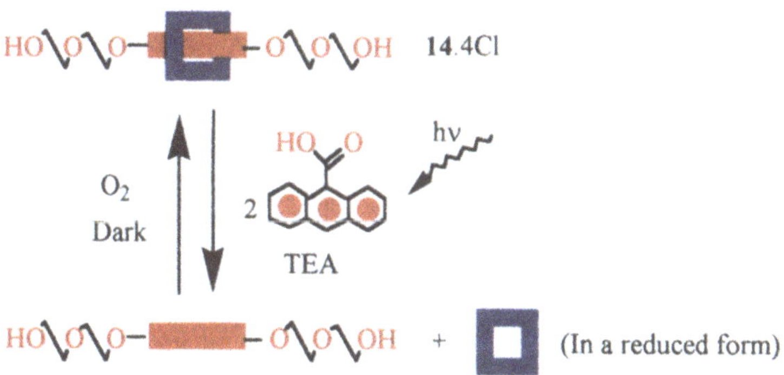

Figure 8-11   A photochemically-driven molecular machine (**14.4Cl**).

8.5 Conclusions

In this article, we have outlined a general approach to establishing the concept of self-assembly in the making of [2]rotaxanes. It is dependent on rather precise stereoelectronic matching of the various different types of non-covalent bonding interactions between appropriate molecular components. The syntheses rely upon the use of template-directed methods. Not only does this mean that the synthetic methods are efficient but it also dictates that the order responsible for the production of the 1:1-complexes leading to rotaxane formation "lives on" in the products after they have been made. The fact that the separate components of these molecular assemblies interact so intimately and precisely with each other conjures up the prospect of molecules that have machine-like properties [31] to them, in so far as the components move with respect to each other in a related and interdependent manner. It is for this reason that we can look forward to constructing molecular structures in a modular fashion and then studying their mechanical properties.

[2]Rotaxanes are currently being constructed with intramolecular electrochemical gradients that cause the constituent units to undergo reduction at distinct potential values. This property is a consequence of the distinctly different environments of each reducible subunit. These differences can easily be extended and utilized to build up a *vectorial electron shuttle*, i.e. a molecular device capable of efficient unidirectional electron transport through identical redox sites. The electrochemical gradient driving the unidirectional electron flow could be created by the gradual variation in the environments to which the identical redox sites are exposed. This type of molecular electronic device could find important applications in artificial photosynthetic schemes or electrocatalysis.

[2]Rotaxanes may also be regarded as prototypes for the construction of photochemical molecular devices, i.e. ordered assemblies of molecular components designed to achieve specific photoinduced functions, such as energy or electron migration [4b]. Hence, research is now in progress to synthesize, *via* self-assembly, some [2]rotaxanes which exhibit photochemical control of the shuttling process. The various applications of such devices include the antenna effect, i.e., the absorption of light energy by several components and migration of the resulting electronic energy to a single, specific component, and the developments in the field of molecular electronics, e.g. the recently described molecular shift register [32].

It is therefore important to establish how supramolecular structures can be self-assembled in a spontaneous manner so that information can be written into them, stored in them, processed in them, transferred between them, and eventually read back out of them, i.e. how the molecular-scale computer will start to become a reality. The late George Pimentel predicted [33] in 1985 that it is no longer a matter of *whether* there will be man-made molecular-scale computers but *when* they will come into existence and *who* will be leading in their development. He went on to add, "The *when* question will be answered on the basis of fundamental research in chemistry; the *who* question will depend on which countries commit the required resources and creativity to the search."

9 Non-Covalent Interactions Between Aromatic Molecules [1]

Christopher Hunter

Abstract

Studies of π-π interactions in covalently-linked porphyrin dimers have been used to develop a simple but powerful model for understanding and computing non-covalent interactions between aromatic molecules. The model has been tested on experimental data from protein X-ray crystal structures and by the successful design and synthesis of a molecular receptor for p-benzoquinone. Applications of the model to the organization of molecules in the liquid crystalline phase and sequence-dependent DNA structure show that ideas derived from simple chemical systems can be useful for understanding more complex molecular assemblies.

9.1 Introduction

In this contribution, we will focus on one type of non-covalent interaction, the π-π interaction [2], in a range of different systems. We will see how studies of simple chemical systems can be used to derive general principles which are useful in developing an understanding of more complex biological systems. π-π interactions play an important role in a range of molecular recognition and self-assembly phenomena. Some examples are listed below.

1. The organization of aromatic molecules in the solid state and hence materials properties [3].
2. Sequence-dependent DNA structure and recognition is determined by the base-stacking interactions [4].
3. The three-dimensional structures of proteins [5].
4. Host-guest chemistry and recognition of drugs by biological receptors [6].

9.2 π-π Interactions in a Simple Chemical System

We began to study π-π interactions in simple systems such as the cofacial covalently-linked porphyrin dimer shown below.

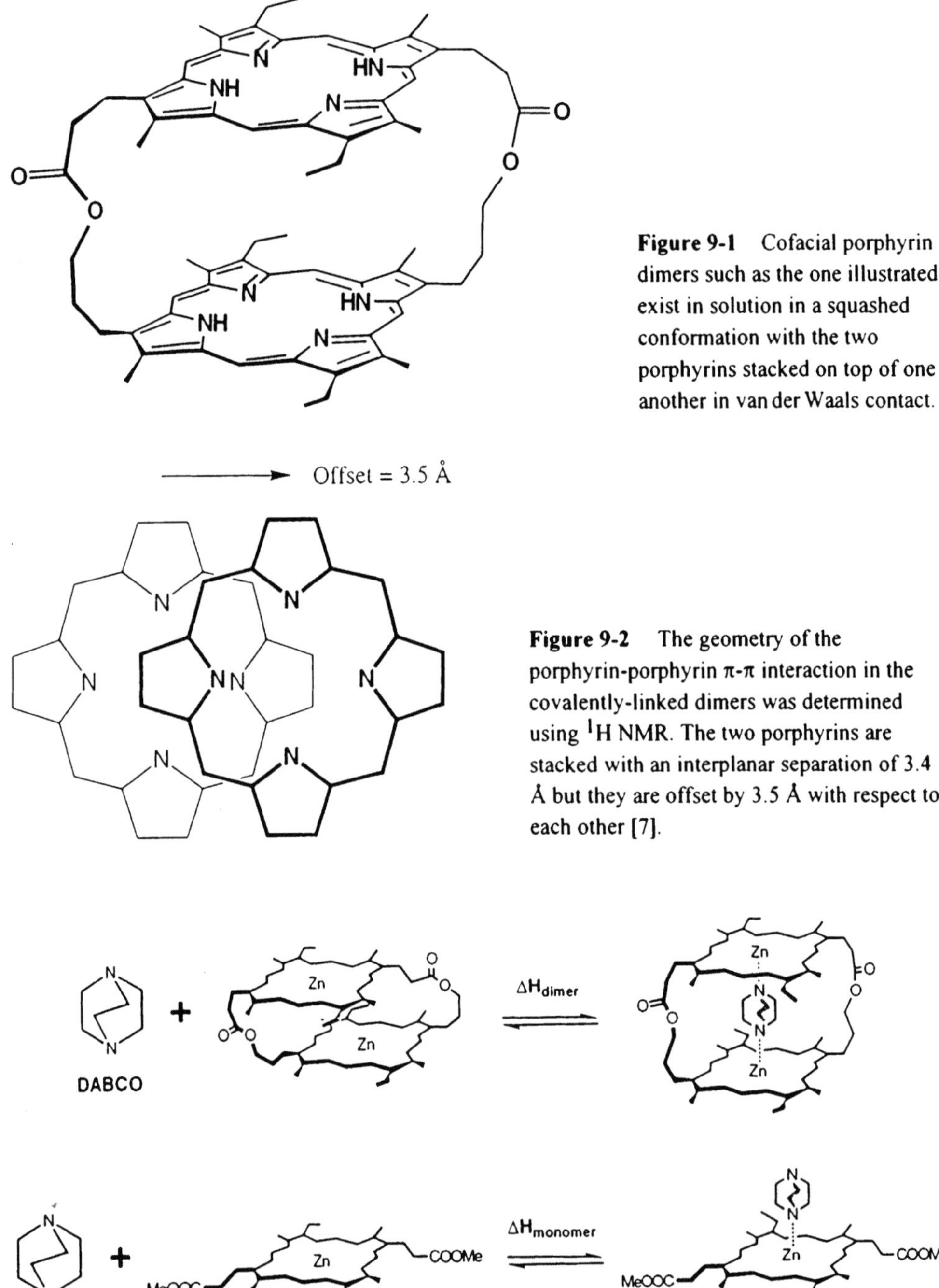

Figure 9-1   Cofacial porphyrin dimers such as the one illustrated exist in solution in a squashed conformation with the two porphyrins stacked on top of one another in van der Waals contact.

Figure 9-2   The geometry of the porphyrin-porphyrin π-π interaction in the covalently-linked dimers was determined using ^{1}H NMR. The two porphyrins are stacked with an interplanar separation of 3.4 Å but they are offset by 3.5 Å with respect to each other [7].

Figure 9-3   Metallation of the porphyrin did not affect the geometry of the stacking interaction and provided a method for probing the magnitude of the interaction. Coordination of a bifunctional ligand such as DABCO forced the two porphyrins apart so that each nitrogen could coordinate to the zinc centres simultaneously as shown. By comparing the intrinsic zinc-nitrogen interaction energy found in a monomeric porphyrin ($\Delta H_{monomer}$) with that found for the dimeric porphyrin (ΔH_{dimer}), we estimated the porphyrin-porphyrin π-π interaction to be 48 ± 10 kJ mol^{-1} in CHCl$_3$ [8].

At this point, we felt in a position to think about the nature of the interaction between the two porphyrins. Many different terms are used to describe non-covalent interactions between π-systems: π-π interactions, π-stacking, electron donor acceptor interactions, π-acid π-base interactions, charge transfer interactions, and this has led to much confusion.

There are several contributions to any non-covalent interaction, some or all of which may be important in any given system. The important interactions for aromatic molecules are

1. Van der Waals interactions which define the size and shape specificity of the interaction. For aromatic molecules, these interactions are maximized when the molecules are stacked on top of one another with maximum overlap of the two π-systems. This is clearly not the case for our porphyrin system (see Figure 9-2).

2. In solution, desolvation of the interacting surfaces can make a significant contribution to the observed interaction. For example, in polar solvents, the solvophobic effect can increase the interaction, but in solvents which compete for the recognition sites (e.g. H-bonding in water), the interaction can be reduced. In the case of the porphyrin-porphyrin interaction discussed above, solvophobic interactions are expected to be very small in chloroform [9]. Even if they were important, they would tend to force maximal overlap of the porphyrin π-systems which is not observed.

3. Electrostatic interactions between the static molecular charge distributions. This is the last important class of interactions (induction and charge transfer may be ignored for reasons discussed elsewhere [10]), and it must therefore be these interactions which cause the observed geometry for the porphyrin system shown in Figure 9-2.

Electrostatic calculations based on simple charge distributions which use charges placed at the nuclei of the atoms, the kind of monopole distribution used by molecular mechanics force-fields, yielded results which did not compare at all well with the experimental results. We therefore decided to resort to a more complex charge distribution which more accurately represented the electron distribution found in aromatic molecules [11].

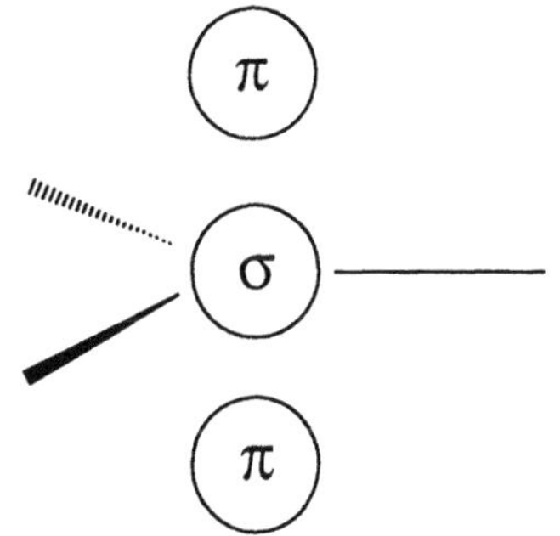

Figure 9-4   For each atom in the π-system we used three charges, one σ-charge and two π-charges placed 0.47 Å above and below the plane of the nuclei. This charge distribution was used to recalculate the electrostatic interaction between the two porphyrins and gave quite different results.

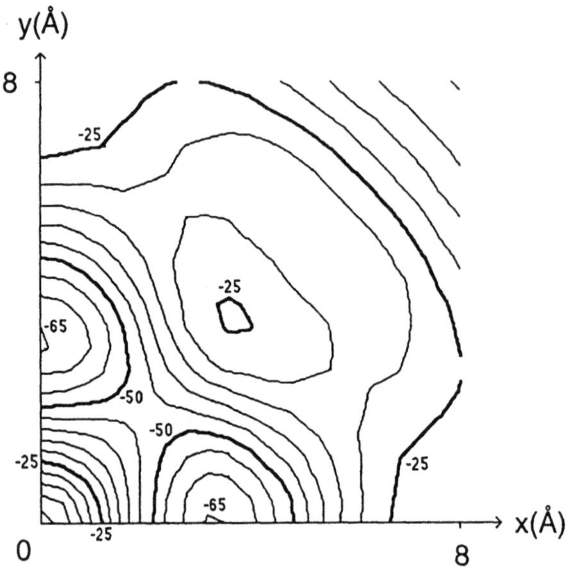

Figure 9-5   The π-π interaction between two porphyrins calculated using the model shown in Figure 9-4 (van der Waals + electrostatic interaction). The porphyrins are fixed in a parallel orientation 3.4 Å apart and the energy is plotted as a function of the displacement of the centre of one porphyrin relative to the other. There is a well-defined energy minimum at a lateral displacement of the two porphyrins of 3.5 Å. This is precisely the experimental geometry shown in Figure 9-2. The magnitude of the interaction at the minimum is 65 kJ mol^{-1}, which is close to the 48 ± 10 kJ mol^{-1} found experimentally (the calculation does not include desolvation which will reduce the calculated value somewhat). The displaced or offset geometry in this system is caused by the strong repulsive electrostatic interactions between the two closely approaching π-clouds when the porphyrins are in a face-to-face geometry with maximal π-overlap. The preferred orientation shown in Figure 9-2 shows very little overlap of the π-systems which reduces this repulsion significantly.

9.3   A General Model for π-π Interactions

Thus we had developed a simple model which accounted for our experimental observations on the porphyrin-porphyrin π-π interaction. We then attempted to extend these ideas to produce some general guidelines which could be used to predict the behaviour of molecular systems in which these interactions are important [11].

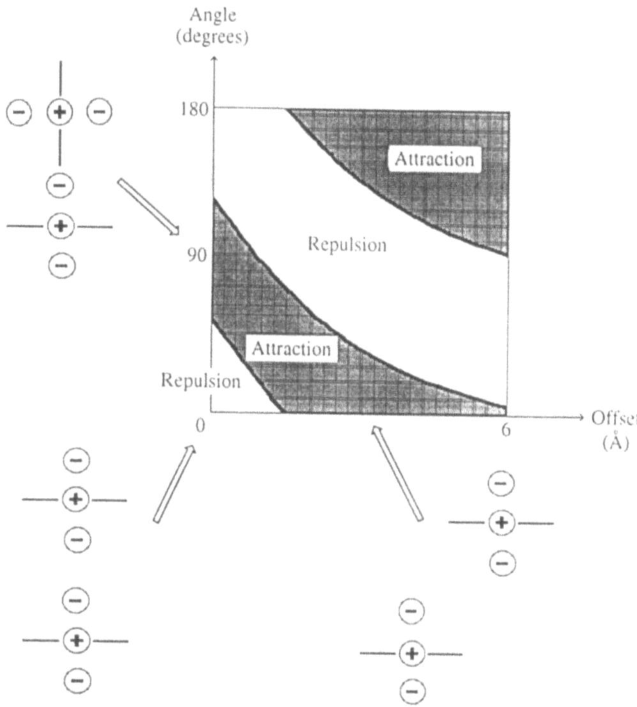

Figure 9-6   We investigated the electrostatic interaction between two idealized π-atoms (simple three centre charge distributions such as shown in Figure 9-3) as a function of their relative orientation. This figure again shows that face-to-face geometries are unfavourable due to repulsion between the two closely approaching π-electron charges. However, in the edge-to-face or offset stacked orientations illustrated, attractive interactions between the σ-charges and the π-electron charges dominate. Thus face-to-face stacking is much less favourable than one might expect on the basis of van der Waals and solvophobic effects. The electrostatically favourable offset stacked geometry is the one we saw for the porphyrin-porphyrin interaction earlier.

This figure provides a general framework for understanding interactions between non-polarized π-systems. Polarization of π-systems by heteroatoms alters this picture and will be addressed later.

9.4   Experimental Evidence for the Model

9.4.1 π-π Interactions in Proteins

The high resolution X-ray crystal structures of proteins represent a vast database of geometrical preferences of non-covalent interactions between various functional groups. In collaboration with Janet Thornton and Jus Singh, we investigated the interactions between aromatic amino acid side-chains [12].

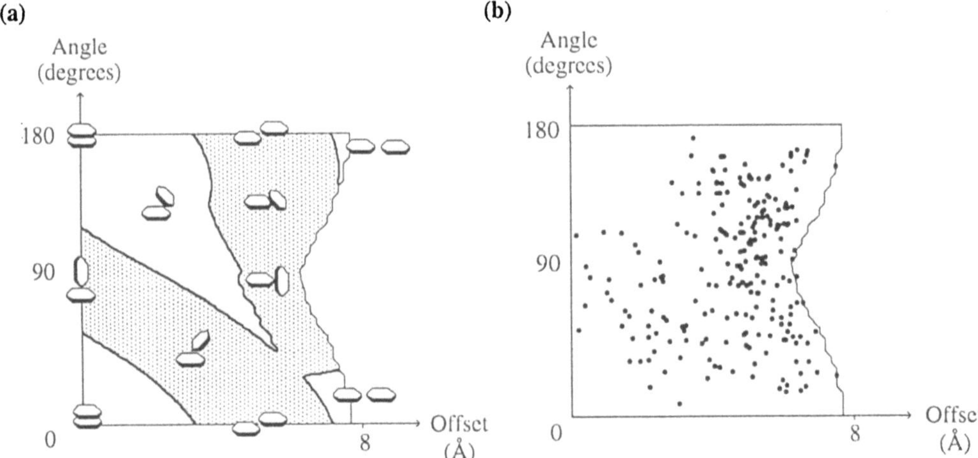

Figure 9-7   (a) shows the electrostatic interaction between two benzene molecules as a function of their relative orientation. The geometries are illustrated for clarity. Attractive regions are shaded and repulsive regions are not. The plot clearly resembles that for the two idealized π-atoms (Figure 9-6): face-to-face stacked orientations are unfavourable; edge-to-face orientations are favourable, and offset stacked orientations are favourable.

(b) shows the experimental geometries for the 220 different phenylalanine-phenylalanine side-chain side-chain interactions which were found in the Singh-Thornton database. There is an excellent match with the calculation: repulsive orientations (in particular the face-to-face geometries) are not observed, and the scatter is concentrated in the attractive zone.

9.4.2   π-π Interactions as an Element in Host Design

For an organic chemist, the real test of any model such as this is "Can I use it to design a new molecule or reaction?" We therefore decided to synthesize a host for a small molecule, *p*-benzoquinone [13].

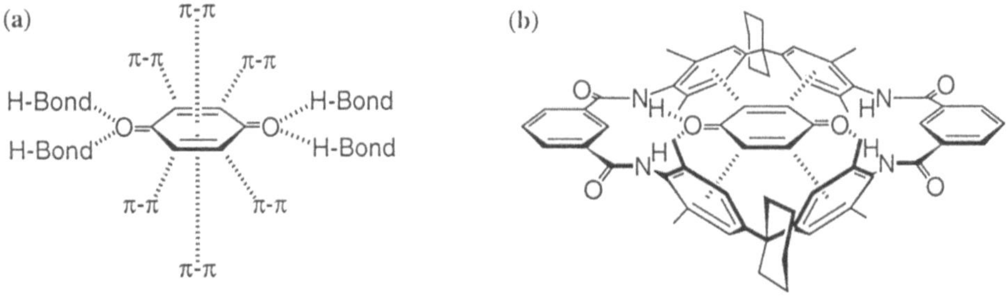

Figure 9-8   (a) illustrates the approach to host design. The recognition sites that we chose to use were the 4 H-bonds and 4 edge-to-face π-π interactions with the edge of the quinone ring.

(b) shows the complex formed between the host and *p*-benzoquinone. Evidence for this structure comes from changes in the ^{1}H NMR chemical shifts on complexation which show the presence of the 4 H-bonds as well as the edge-to-face alignment of the π-systems. This does not prove that the edge-to-face π-π interactions are attractive but is consistent with the ideas presented above. Moreover, if the quinone hydrogens are replaced by fluorines, chlorines, methyl groups or fused aromatic rings where the edge-to-face π-π interactions are not possible, no complexation can be detected at all.

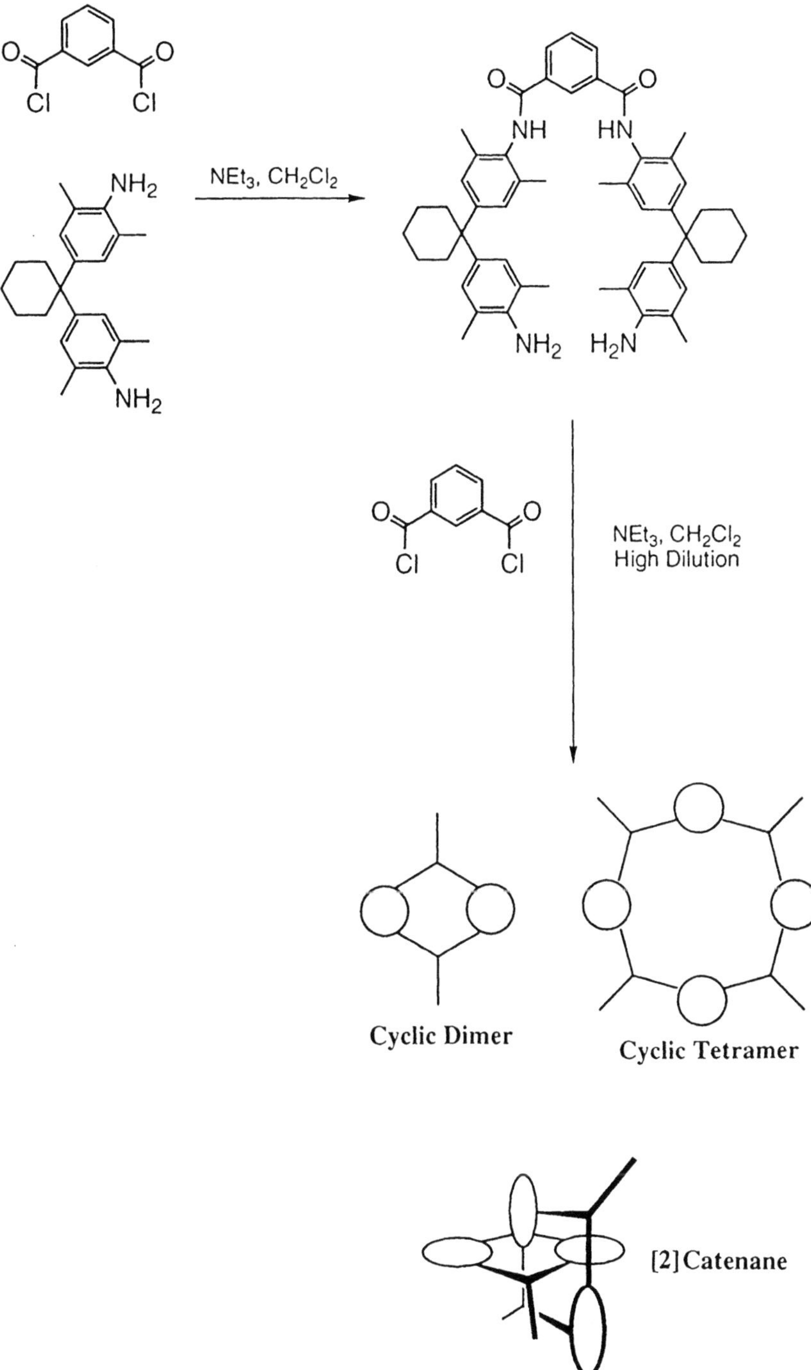

Figure 9-9   The synthesis of the quinone host. The final macrocyclization reaction produced a cyclic dimer (the quinone host), a cyclic tetramer and a [2]catenane comprised of two interlocked cyclic dimers. The catenane is formed in a remarkably high yield (34%) considering that the reaction is performed under high dilution conditions [14].

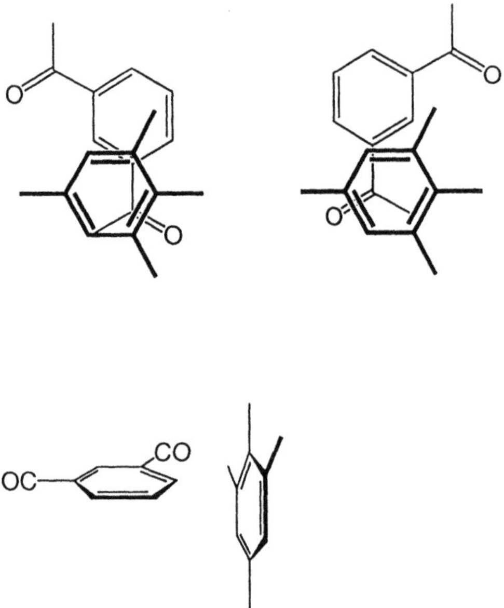

Figure 9-10 A clue as to the intermolecular interactions which lead to this efficient templating process comes from the final structure of the [2]catenane which was determined by two-dimensional NMR. There are several H-bonds and π-π interactions between the two macrocycles which comprise the catenane. The geometries of the π-π interactions are again edge-to-face and offset stacked as shown above.

9.5 The Effect of Polarization by Heteroatoms

The model as discussed so far refers to non-polarized π-systems such as aromatic hydrocarbons. But most systems of interest contain heteroatoms which can significantly perturb the π-electron distribution and hence the π-π interaction. In order to understand these effects, we divide the total electrostatic interaction into three parts [4].

1. πσ-πσ interactions. These are the interactions associated with the out-of-plane π-electron density if no polarization were present, i.e. the kind of interaction we have already discussed.
2. Atom-atom interactions. These are the interactions between the net partial atomic charges.
3. Atom-πσ interactions. This is the cross-term of the other two interactions, i.e. the interaction between the partial atomic charges on one molecule and the out-of-plane π-electron density on the other.

To illustrate this approach, we will look at a simple system related to the porphyrins we saw earlier. Phthalocyanins attached to long alkyl chains form discotic mesophases, but the organization of the molecules in the columns depends on the chemical nature of the substituents [15].

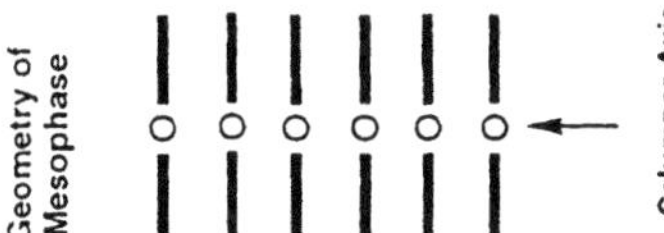

←———

Figure 9-11   (a) This octa-alkyl phthalocyanin forms a discotic mesophase in which the molecules are tilted by 45° with respect to the columnar axis as shown.

(b) This octa-alkoxy phthalocyanin forms a discotic mesophase in which the molecules are aligned at 90° to the columnar axis as shown.

(c) The minimum energy orientation for an octa-alkyl phthalocyanin dimer calculated using the model described above for π-π interactions. This geometry is very similar to that found for the porphyrin dimer in Figure 9-2 and is dominated by the $\pi\sigma$-$\pi\sigma$ interactions which are minimized when the overlap of the π-systems is lowered to reduce π-electron repulsion. The staggered arrangement of the phthalocyanins is what causes the tilting in the mesophase.

(d) The minimum energy orientation for an octa-alkoxy phthalocyanin dimer calculated using the model described above for π-π interactions. The molecules are not offset but twisted, so that in the mesophase, they are aligned at 90° to the columnar axis. This geometry is dominated by atom-$\pi\sigma$ interactions. The oxygen substituents donate electron density into the neighbouring six-membered rings giving them a net negative charge. The negative partial atomic charges will have unfavourable atom-$\pi\sigma$ interactions with the π-electrons of the other molecule, so the system adopts the geometry which minimizes overlap of these regions with the π-systems. In the geometry shown in Figure 9-11(c), two of these negatively charged rings would lie over the phthalocyanin core of the other molecule which would give rise to a very large repulsive interaction. In the preferred geometry in Figure 9-11(d), there are attractive $\pi\sigma$-$\pi\sigma$ offset interactions between these electron-rich rings.

9.6 π-π Interactions in a Complex Biological System

At this point, we felt that the model could be usefully applied to more complex systems where experimental evidence is more difficult to acquire but where π-π interactions play an important role. The three-dimensional structure and properties of DNA depend critically on the sequence of the aromatic bases in a manner that is not well-understood [16, 17]. The key link between DNA sequence and three-dimensional structure is the π-π interactions between the stacked aromatic bases: the preferred orientation of the stacking interaction between two base-pairs juxtaposed in a sequence defines the conformation of that step in the helix. We have applied the π-π interaction model above to elucidate the chemical basis for sequence-dependent effects on DNA structure [4].

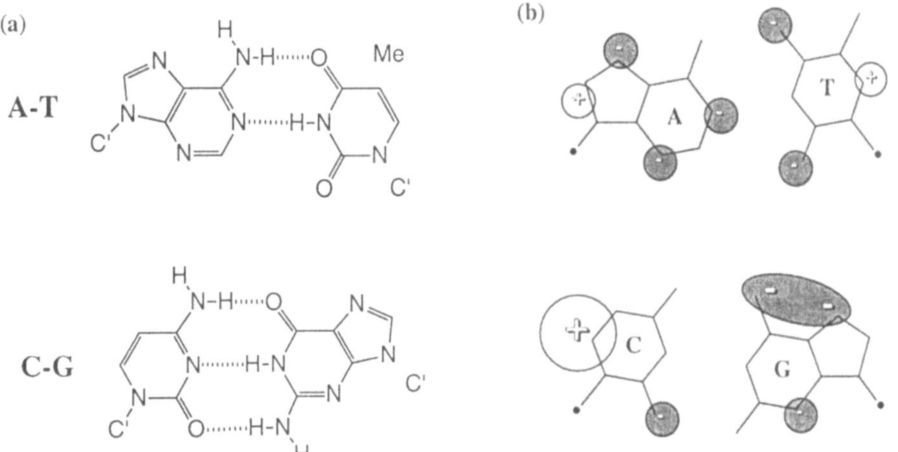

Figure 9-12   The only difference between different DNA sequences lies in the chemical structures of the bases: the sugar-phosphate backbone is identical for any sequence and can therefore be ignored in the first approximation. Thus it should be possible to account for all sequence-dependent effects in terms of

1. Steric interactions with the guanine amino group in the minor groove.
2. Steric interactions associated with the configuration of the step, whether the sequence of bases is purine-purine, purine-pyrimidine or pyrimidine-purine.
3. Steric interactions with the thymine methyl group in the major groove.
4. Electrostatic interactions associated with the different molecular charge distributions across the A-T and C-G base-pairs (Figure 9-12(b)).

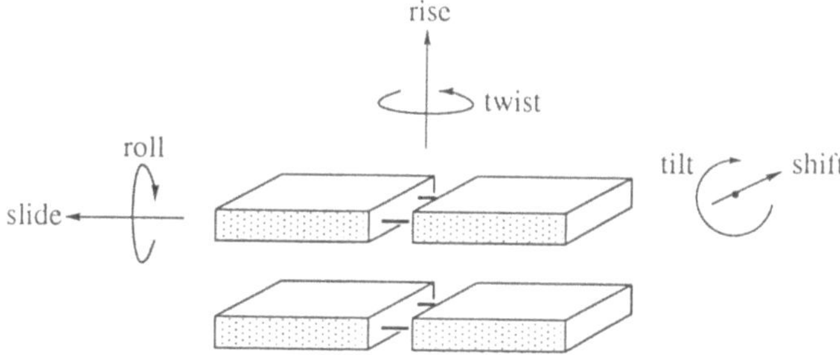

Figure 9-13   The approach to tackling this problem was set up by Calladine and Drew who showed that the conformation of the individual base-pair steps can be related to the overall three-dimensional structure of the double helix [18]. The conformation of a step is defined in terms of the six degrees of freedom illustrated above [19]. There are another six internal degrees of freedom within each base-pair, but only one of these, propeller twist, is of any significance and so these terms will be ignored for our purposes.

An analysis of DNA oligomer X-ray crystal structures shows that slide, roll and twist are the most important parameters for defining sequence-dependent changes in DNA structure [18, 20]. We have therefore examined the π-π interactions for all 10 possible base-pair steps as a function of these parameters [4]. It turns out that small changes in twist do not change the π-π interaction energies much, so we will concentrate on just slide and roll. Our approach is illustrated using two steps which have been thoroughly characterized by experiment and which have very different properties. AA/TT has a strong preference for a B-DNA conformation (slide $\approx$ 0 Å, roll $\approx$ 0°, twist $\approx$ 32°) [21], whereas CC/GG has a strong preference for an A-DNA conformation (slide $\approx$ −1.5 Å, roll $\approx$ 12°, twist $\approx$ 36°) [22].

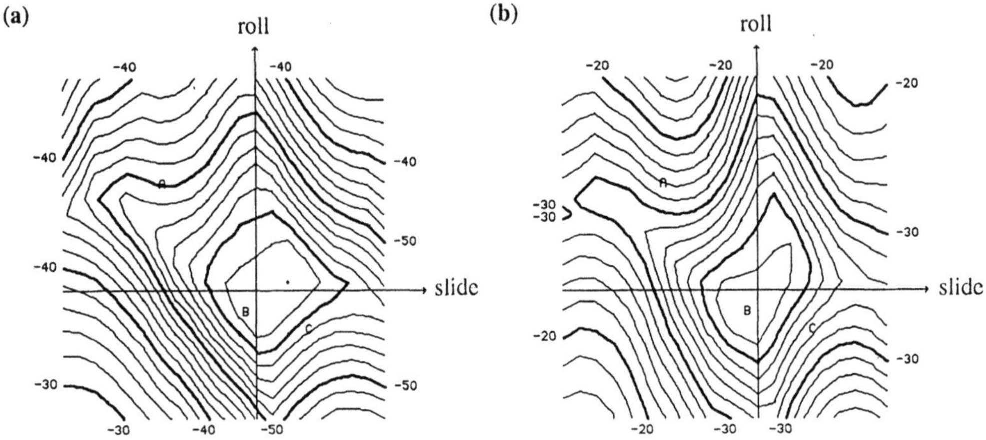

Figure 9-14   Contour plots of the π-π interaction (in kJ mol^{-1}) between two A-T base-pairs as a function of slide (–3 to 2 Å) and roll (–15° to 25°) in an AA/TT step. Helical twist = 36°, propeller twist = 15° and all other parameters are zero. The contour spacing is 2 kJ mol^{-1}. Energy minima and conformations corresponding to A, B and C-DNA are labelled. (a) van der Waals interaction. (b) total π-π interaction (van der Waals + electrostatic). The energy minimum is clearly located around the B-DNA conformation as observed experimentally. The van der Waals plot shows that a steric interaction blocks any movement into an A-DNA conformation. The origin of this effect is illustrated below.

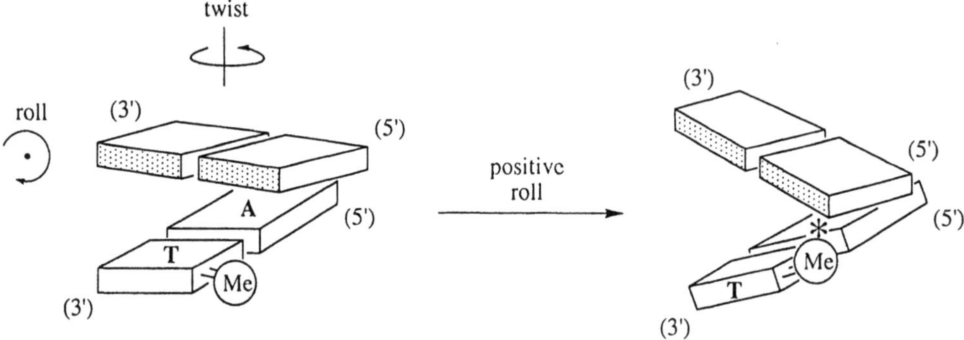

Figure 9-15   An AA/TT step viewed along the slide/roll axis. The direction of positive slide is towards the reader. The block edges which correspond to the minor groove are shaded. The thymine methyl group prevents positive roll into an A-DNA conformation. Increasing roll leads to a steric clash (asterisk) between the thymine methyl and the other base on the same strand. This interaction may explain the conformational differences between double-helical DNA which generally prefers the B-form and double-helical RNA which has only been observed in the A-form [23, 24]. In DNA, the thymine methyl groups projecting into the major groove destabilize the A-conformation, but RNA lacks these methyl groups and so is perfectly stable in the A-conformation.

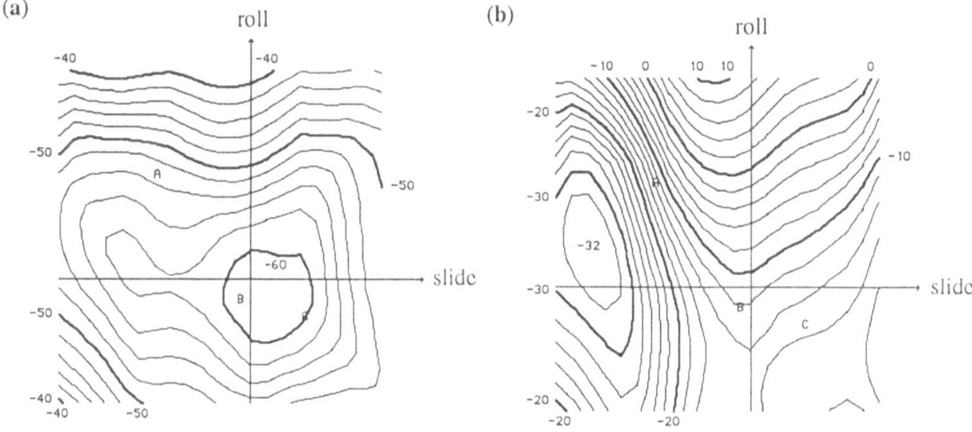

Figure 9-16   Contour plots of the π-π interaction (in kJ mol^{-1}) between two C-G base-pairs as a function of slide (–3 to 2 Å) and roll (–15° to 25°) in a CC/GG step. Helical twist = 36°, propeller twist = 0° and all other parameters are zero. The contour spacing is 2 kJ mol^{-1}. Energy minima and conformations corresponding to A, B and C-DNA are labelled. (a) van der Waals interaction. (b) total π-π interaction (van der Waals + electrostatic). The energy minimum is now located around the low slide, positive roll A-DNA conformation as observed experimentally. It is clearly electrostatic interactions which are responsible for this conformational preference. The origin of this effect is illustrated below. In general, electrostatics play a more important role in C-G containing steps than in A-T containing steps due to the much larger regions of high charge density found in C-G compared with A-T (see Figure 9-12(b)).

Figure 9-17   Atom-$\pi\sigma$ interactions in a CC/GG step. The thin lines map out the area covered by the π-electron density of one C-G base-pair. The molecular charge distribution of the other C-G base-pair is shown. ● indicates the position of the C_1' atom where the bases are attached to the sugar. In a B-DNA conformation, the π-electron density lies over the large negative charge on guanine, but in the A-DNA conformation, this repulsion is avoided and the π-electron density now lies over the cytosine positive charge giving rise to an attractive atom-$\pi\sigma$ interaction.

9.7 Summary

This work shows that it is possible to use simple chemical systems to derive general principles which may be extended to understand some properties of more complex biological systems. Of course, more detailed aspects of the behaviour of complex systems such as DNA will require consideration of cooperative interactions between the many different components of the system. For example, it is clear that the conformation of a single base-pair step influences the conformation of its neighbour and the sugar-phosphate backbone is implicated in transmitting the cooperativity [25]. Thus new approaches will have to be developed to tackle these problems. However, it is first necessary to have a good understanding of the properties of the individual components in the system and we believe that our model for π-π interactions provides this.

10 Design, Syntheses, and Supramolecular Chemistry of Smart Cascade Polymers

George R. Newkome and Charles N. Moorefield

Abstract

The construction of cascade (dendritic) polymers employing novel building blocks, or monomers, that possess tetrahedral centers of branching is discussed. Using amide-based, cascade architecture, hydrodynamic radii are observed to change as a function of environmental pH, provided that the periphery is coated with ionizable functional groups, such as, carboxylic acids or amines. These data suggest that cascade "internal void volume" can easily be expanded and contracted. The potential to perform multiple, precisely juxtaposed, chemical transformations within the interior of the cascade superstructure is demonstrated by the formation of ortho-carborane moieties equidistant from a tetrahedral core. The covalent attachment of dicobalthexacarbonyl to the cascade framework is also discussed.

10.1 Introduction

Over the past decade, numerous procedures have been developed to create dendritic (cascade) macromolecules. These molecules have the potential for molecular inclusion and recognition if the "void" inner regions possess appropriate bonding regions to recognize and discriminate specific guests. Such complementary architectures open new avenues to water soluble catalysts and unique methodology to drug and molecular delivery systems. Although this new area of polymers is still in its infancy, the preparation of these cascade materials has followed, as of yet, only a selected number of pathways [1]. We herein describe only one approach to these macromolecules and then discuss their structural variables and their potential as homogeneous catalysts.

Our preliminary synthetic route to prove the cascade construction viability was "based on the architectural design of trees" as described by Professor Tomlinson [2] in his studies of the rain forests of South America. This particular cascade approach gave rise to the family name of "Arborol", since it was derived from the Leeuweenberg model for tropical trees. This mode of growth depicts almost all of our subsequent synthetic work, and it follows a simple mathematical progression of $1 \rightarrow 3 \rightarrow 9 \rightarrow 27 \rightarrow ....$. Scheme 10-1 shows our initial work [3] in this arena; spacer groups and multiple steps were necessary to circumvent the chemical retardation caused by the presence of the neopentyl core. Subsequently, a series of new building blocks, which incorporated an appropriate distance between the core and leaving group, were designed and created to permit the facile, high yield construction of diverse polyfunctional macromolecules.

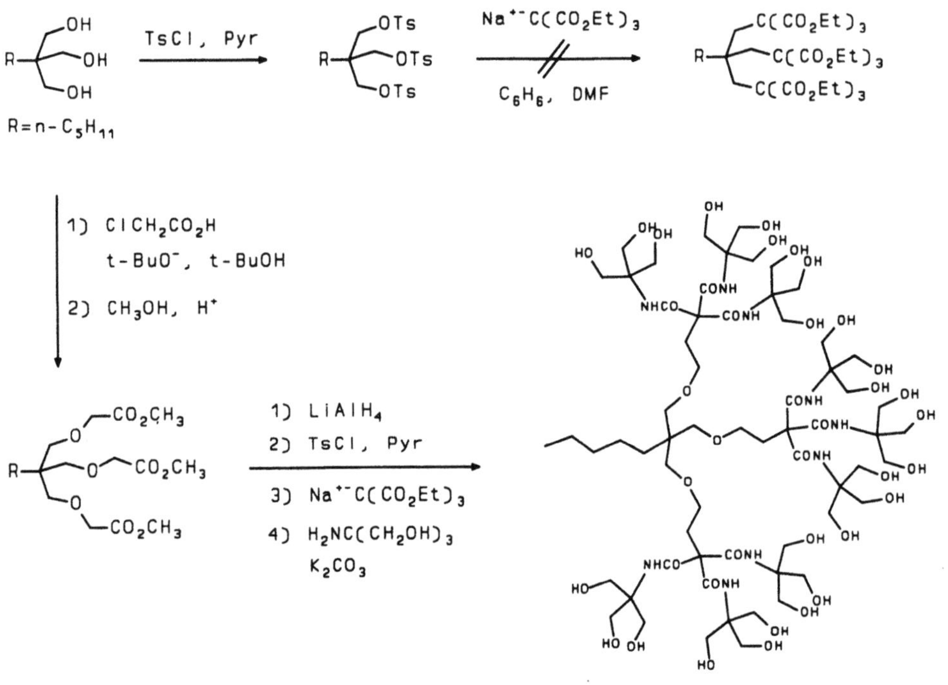

Scheme 10-1 Synthesis of a one-directional, 27-arborol.

10.2 Syntheses of Cascade Macromolecules

The generation of cascade polymers has followed two basic procedures for monomer addition: *divergent* construction beginning at the core and building outward via an increasing number of termini transformations, and *convergent* construction beginning at the eventual periphery and building inward via a constant number of transformations. Combinations of the two methods have also been utilized. Each method has advantages and disadvantages; however, we will limit our discussion to the former route.

10.2.1   Divergent, Polyamido Infrastructures

Preparation of new monomers included the four-directional core **1**, constructed in two steps by Michael-type addition of pentaerythritol to alkyl acrylates, followed by hydrolysis [4], and the monomeric building block **2**, synthesized in high yields from nitromethane and *tert*-butyl acrylate, followed by catalytic reduction [5]. The viability of these building blocks was demonstrated via the peptide-type coupling of tetraacid core **1** and amine **2**, using standard conditions [6], to afford the desired dodecaester **3**, which was hydrolyzed to the corresponding polyacid **4**, and subjected to further repetitive amidation and hydrolysis procedures to give the analogous higher generation cascade polymers (Scheme 10-2). Such divergent methodology gave rise to the family of cascade polyacids possessing an internal amido framework, which is impervious to most hydrolytic and enzymatic degradation conditions [7]. Table 10-1 depicts the family of Z-Cascade:methane[4]:(3-oxo-6-oxa-2-azaheptylidyne):(3-oxo-2-azapentylidene)$^{G-1}$:propanoic acids [8], which were synthesized by this route.

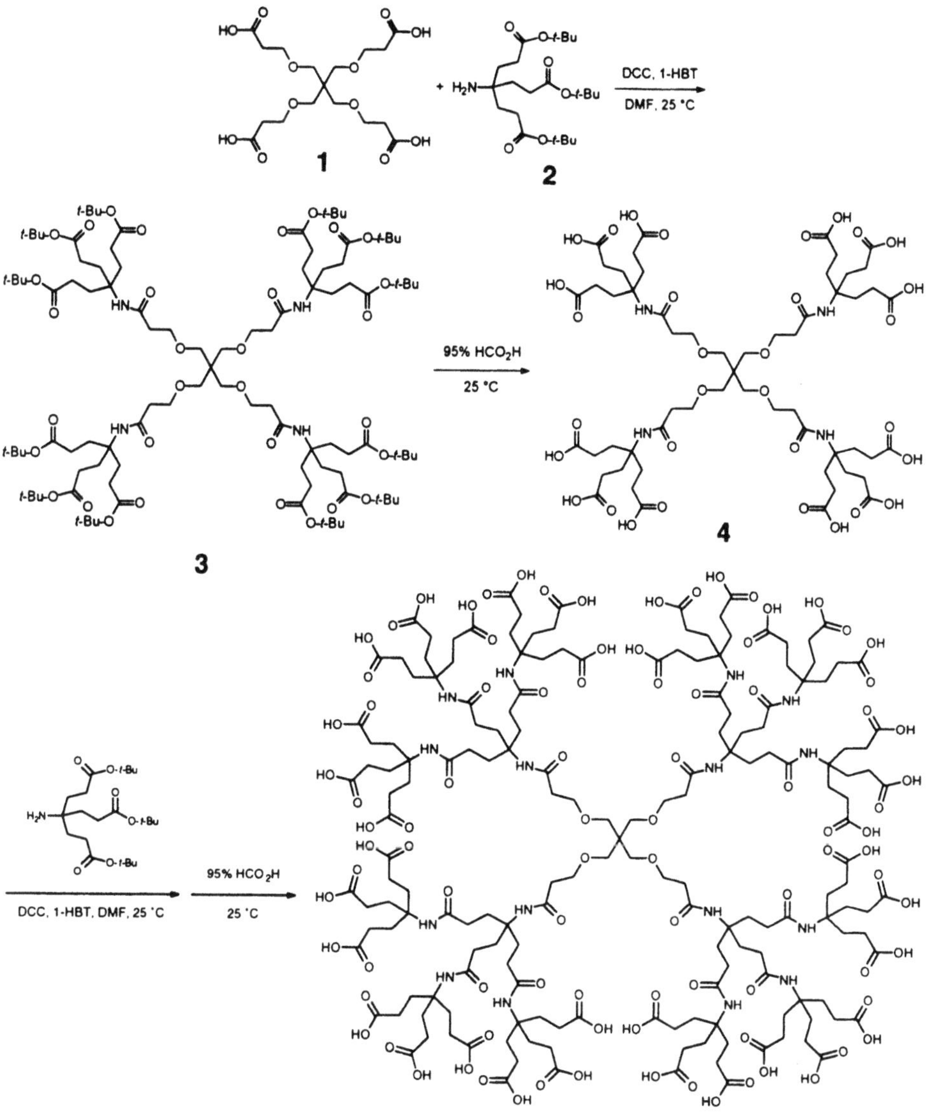

Scheme 10-2   Preparation of the amide-based cascades.

Further evaluation of the functionalization of the surface groups [9] of this polyacid family was effected by treatment with *L*-phenylalanine methylester, employing peptide coupling conditions similar to those used to build the cascade framework, thus affording good yields of the desired chiral surfaces composed of amino acid ester moieties, such as **5**, **6**, and **7** (Scheme 10-3). Facile hydrolysis of the ester groups gave the corresponding poly-amino acid surfaces [10]. It was hoped that as the number of chiral groups increased to the "dense packing limits" that an observable difference would be noted; the CD studies on the phenylalanine derivatives of the acids shown in Table 10-1 showed a straight line correlation.

Table 10-1   Generation versus molecular weight for the amide-based cascades.

Generation (G)	Number of Terminal Moieties (Z)	Molecular Weight
0	4	424
1	12	1,341
2	36	4,092
3	108	12,345
4	324	37,102
5	972	111,373
6	2,916	334,187

Using this family of monodispersed cascade macromolecules as core units, the surface functionalization was also evaluated to ascertain the "smart" characteristics of differing coatings [11]. Thus, the carboxylic acid cascade series described in Table 10-1 was treated with the masked synthons [12] amine **8** and alcohol **9** (Scheme 10-4), each prepared independently from trinitrile **10**, to generate the closely related polyamine and polyalcohol series, respectively, after termini deprotection.

These three families of amine, alcohol, and carboxylic acid terminated cascades facilitated the analysis of the dependence on environmental pH changes of hydrodynamic radii of the macromolecules as derived from diffusion-ordered two-dimensional NMR spectroscopy [13] (DOSY) data. Hence, the 36-acid terminated cascade (Table 10-1; generation 2) exhibited diffusion coefficients of 1.74×10^{-6}, 1.15×10^{-6}, and $1.26 \times 10^{-6} \mathrm{cm^2 s^{-1}}$ corresponding to hydrodynamic radii of 11.4, 17.3, and 15.8 Å at acid, neutral, and basic pH conditions, respectively. The corresponding 324-acid cascade (Table 10-1; generation 4) exhibited a molecular diameter that ranged from 45.2 Å at acidic pH to 66.2 Å at neutral pH. Such "*smart*" properties for the first three generations of each family are summarized in Figure 10-1; molecular diameter is observed to increase as the cascade periphery progresses from neutral to ionic for the ionizable macromolecules, whereas the hydroxyl terminated cascade series reveals a nearly constant diameter throughout the pH range. These data strongly suggest that the cascade void volume can be increased or decreased subject to solution pH and surface group type. Similar correlations are readily available from DOSY data and can be confirmed by size exclusion chromatography experiments.

The porosity of the same cascade series can be readily evaluated by the use of GEPOL/87 Solvent Accessible Surface [14] computation as a function of variable probes [15]. Figure 10-2 depicts the accessibility of different size probes into the inner "void" regions of four of the related cascade polyacids. The dotted line depicts an idealized spherical surface in which the probe is constrained to the surface. As the probe decreases in size from 5 Å, it has access to the variable regions within the macromolecule.

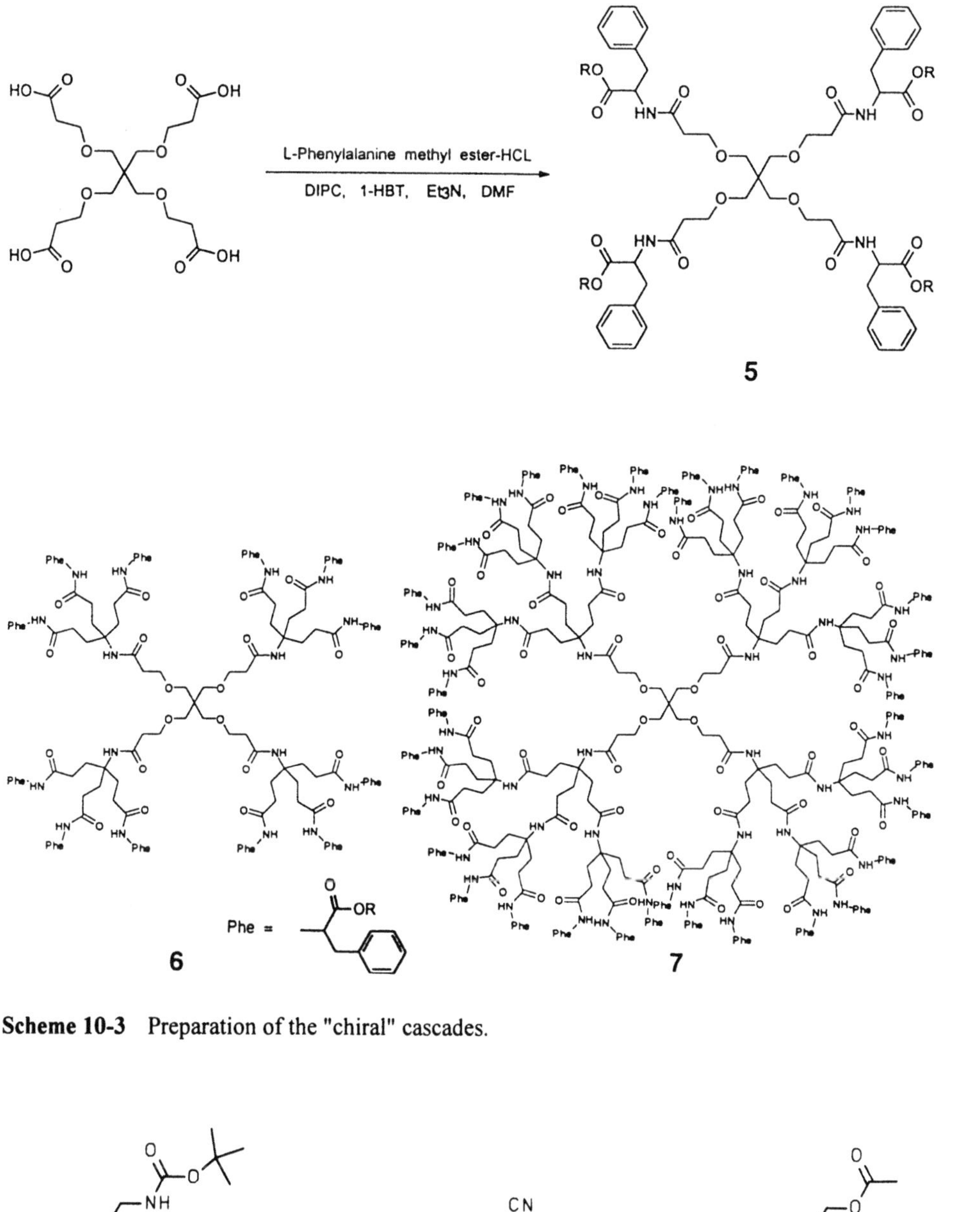

Scheme 10-3 Preparation of the "chiral" cascades.

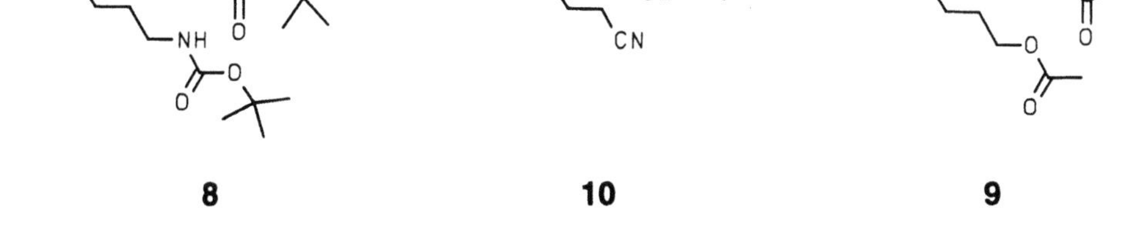

8 10 9

Scheme 10-4 Synthons for the introduction of amine and alcohol termini.

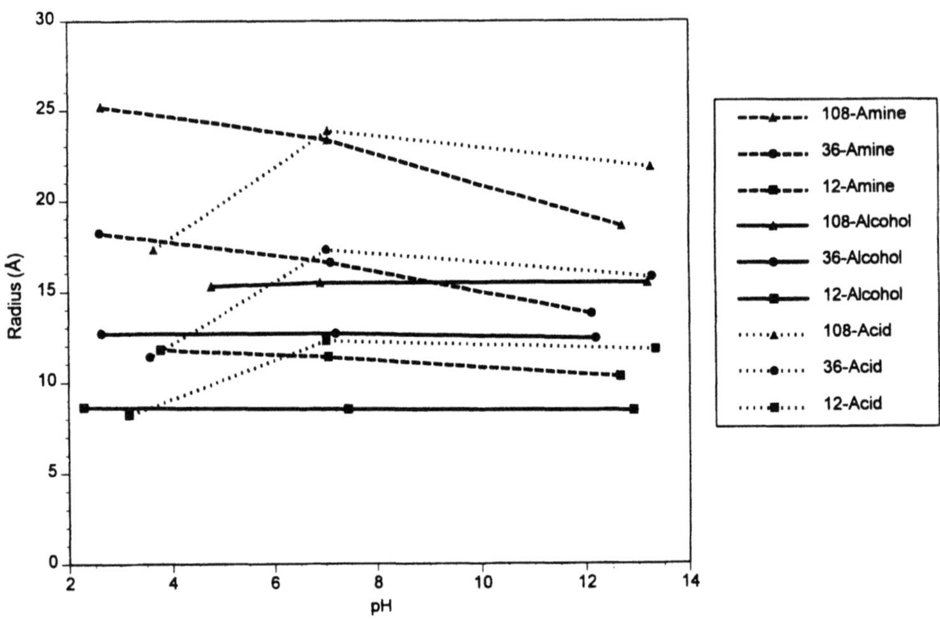

Figure 10-1   Comparison of amine, alcohol, and acid terminated cascades in acidic, neutral, and basic media.

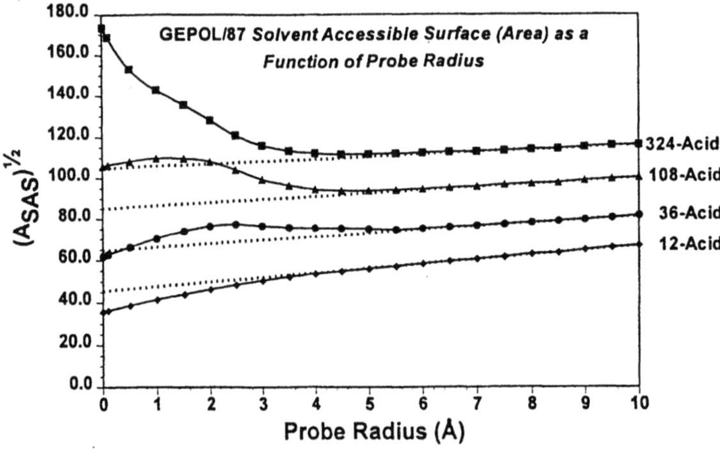

Figure 10-2   Porosity of acid terminated polyamido cascades.

10.2.2   Divergent, All-carbon Infrastructures

The penultimate all-carbon cascade macromolecule has been prepared [16] by the use of tetrabromide core **11** and the alkynide anion building block, generated from terminal acetylene **12**, both of which were prepared from a common intermediate in high overall yields. Scheme 10-5 depicts the initial C-C bond formation affording the protected tetrayne, which is subsequently reduced and deprotected, in a single step, to give dodecaol **13**. Its conversion to dodecabromide, upon treatment with the alkyne building block, generated the next tier. The simple growth via repetition of these chemical transformations gave rise to a new series of unimolecular micelles **14** [17] (Table 10-2), which we have termed Micellanes™. Although these polyols are sparingly water soluble, their oxidation with $RuO_2/NaIO_4$ afforded the corresponding $[8^2 \cdot 3]$Micellanoic™ acid {36-Cascade: methane[4]:(nonylidyne)^2:propanoic acid} (**14**).

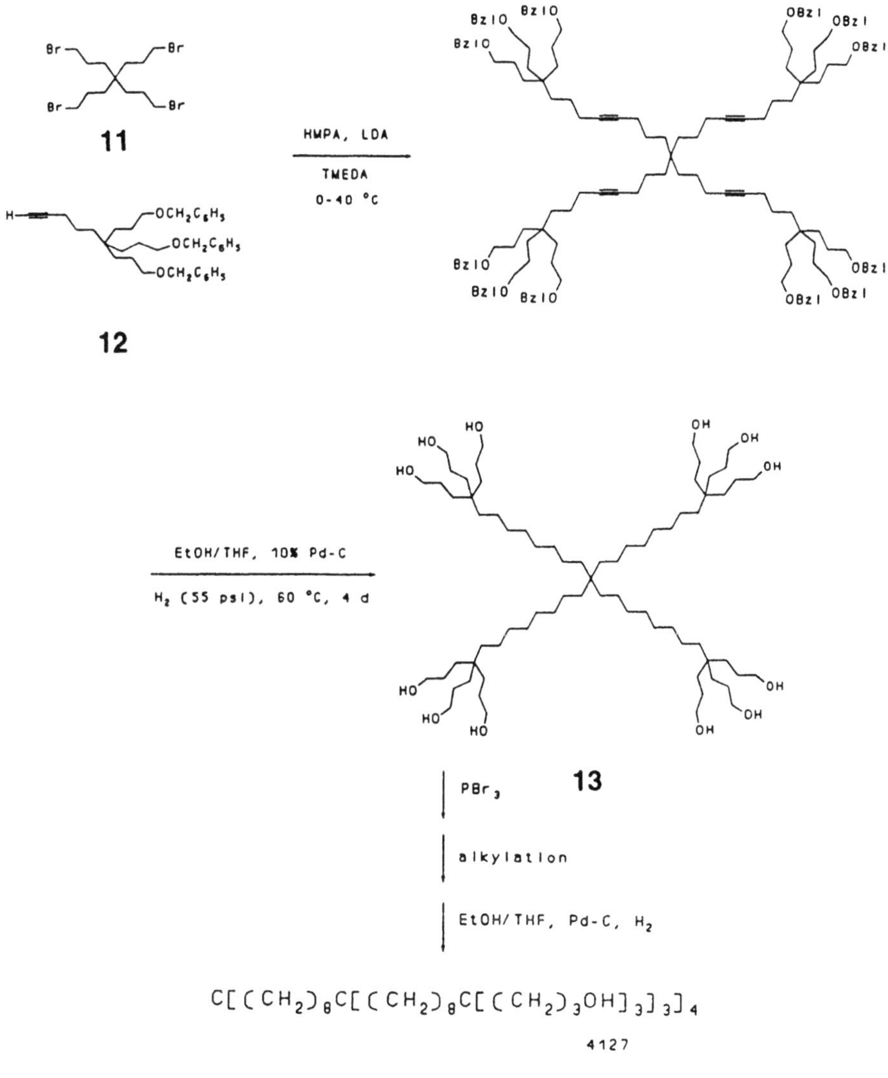

$$C[(CH_2)_8C[(CH_2)_8C[(CH_2)_3OH]_3]_3]_4$$

4127

Scheme 10-5 Construction of saturated hydrocarbon cascades.

The micellar properties of the $(NMe_4)^+$ salt of **14** were confirmed [17] via UV analysis of guest molecules traditionally used to ascertain micelle properties (i.e., phenol blue, pinacyanol chloride, naphthalene, and chlorotetracycline), as well as fluorescence lifetime decay experiments utilizing diphenylhexatriene as the molecular probe. The monodispersity, size, and lack of intermolecular aggregation were determined by electron microscopy.

Table 10-2   Generation versus molecular weight for the alkane-based cascades.

Generation (G)	Number of Terminal Moieties (Z)	Molecular Weight
0	4	304
1	12	1,386
2	36	4,630
3	108	14,363
4	324	43,563
5	972	131,163
6	2,916	393,962

14

10.3 Supramolecular Chemistry of the Unimolecular Micelles

In view of the interior "void regions" within the lipophilic interior of the Micellanes™ coupled with their smart properties, we probed the use of the molecular scaffolding to append potentially chemically active centers. Initial studies were directed toward the functionalization of the internal alkyne centers incorporated in the alkyl framework of the polybenzyl ether precursors to the Micellanoic™ acid (see Scheme 10-6; cascade **15**).

10.3.1   Metalloidomicellanes™

The chemical viability of these internal alkyne centers was easily demonstrated by their facile, quantitative reduction with H_2 using Pd/C in THF/EtOH. Thus, treatment of **15**, as well as other alkynes in this series, with acetonitrile activated decaborane ($B_{10}H_{14}$) afforded a near quantitative yield of the dodeca-*o*-carborane **16** [18], as shown in Scheme 10-6. The terminal benzyl ether protecting groups were readily removed under catalytic reductive conditions to afford the corresponding polyol. Since the polyol exhibited only slight water solubility, it was subsequently treated with chlorosulfonic acid to give the polysulfate **17** and hence instill much greater water solubility. This Metalloidomicellane™ cascade possesses 120 internal boron atoms, all equidistant from the core, as evidenced by the symmetry depicted in the ^{11}B- and ^{13}C-NMR spectra.

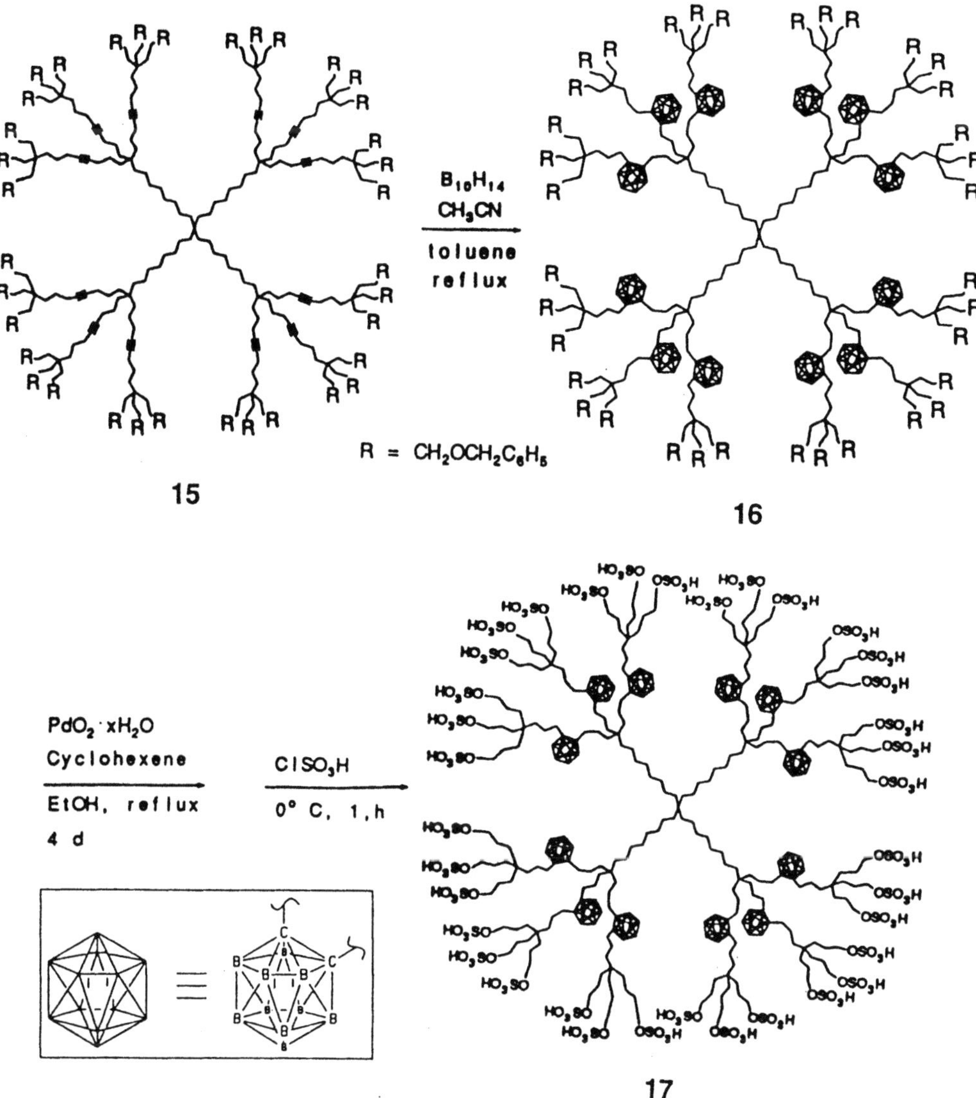

Scheme 10-6 Preparation of water soluble, boron superclusters.

10.3.2 Metallomicellanes™

On the pathway to unimolecular water soluble catalysts, the preparation of Metallomicellane™ cascades was examined. Thus, treatment of the 1st and 2nd tetra- and dodeca-alkyne cascade intermediates (see Scheme 10-6; **15**) with $Co_2(CO)_8$ afforded, in excellent yields, the desired Cobaltomicellane™ polyethers **18** and **19** [19] (Figure 10-3). These polymetal clusters represent the first examples of this new class of metal centers constrained to specific loci within an all carbon, cascade framework. The investigations of chemistry of these metallopolymers are currently in progress.

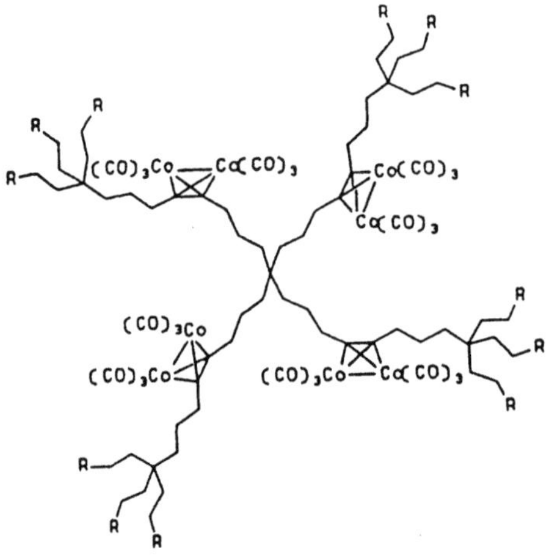

18

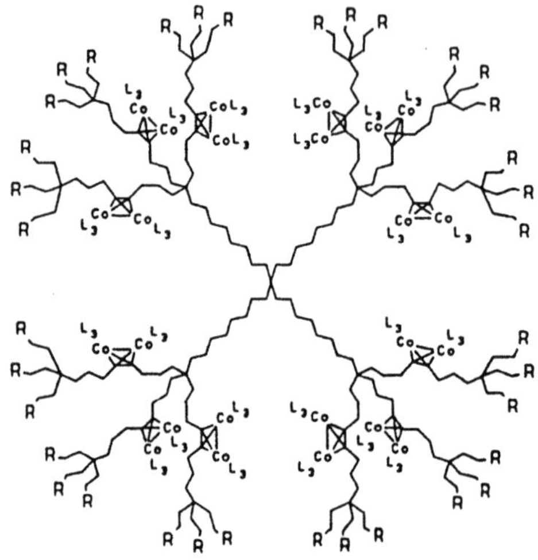

19 $R = CH_2OCH_2C_6H_5$ $L = CO$

Figure 10-3 Cobalt superclusters.

Acknowledgements

This work was supported in part by the National Science Foundation (DMR-92-17331; 92-08925), the Army Research Office (DAAHO4-93-G-0448), and the Petroleum Research Fund, administered by the American Chemical Society.

11 Polyoxometallate Cluster Anions

Michael T. Pope

Abstract

Polyoxoanions of the early transition elements vanadium, molybdenum, and tungsten comprise a very large and varied class of discrete and well-characterized species which incorporate as many as sixty metal atoms per anion [1]. Although the first examples of these substances were described in the mid 19th century, it is only with the availability of appropriate tools such as multinuclear NMR spectroscopy, coupled with single-crystal X-ray diffraction, that a clearer understanding of the extent and complexity of the field is becoming apparent.

11.1 Principles of Formation and Structure

Polyoxometallate anions are generated, at least in a formal sense, as a result of proton-driven acid/base condensation reactions. Acidification of alkaline solutions (pH 10–13) which contain the tetrahedral monoanions, VO_4^{3-}, MoO_4^{2-}, and WO_4^{2-}, leads ultimately (pH 2–3) to precipitation of the polymeric oxides, V_2O_5, MoO_3, and WO_3, the structures of which reveal five- or six-coordinate metal atoms. Solutions at intermediate stages of acidification contain polyoxoanions of varying degrees of complexity depending upon solvent, pH (or its analogue in nonaqueous solution), counterions, and the presence or absence of other solute species which can act as "heteroatoms". Examples of formation equations are

$$10\ VO_4^{3-} + 24\ H^+ \longrightarrow [V_{10}O_{28}]^{6-} + 12\ H_2O$$

$$[Cr(H_2O)_6]^{3+} + 6\ MoO_4^{2-} + 6\ H^+ \longrightarrow [Cr(OH)_6Mo_6O_{18}]^{3-} + 6\ H_2O$$

$$30\ WO_4^{2-} + 5\ PO_4^{3-} + Na^+ + 60\ H^+ \longrightarrow [Na(PO_4)_5W_{30}O_{90}]^{14-} + 30\ H_2O$$

These equations convey the essential acid/base stoichiometry of polyoxoanion formation, but do not necessarily represent the best methods of preparation. Mechanisms of formation are not, in general, known, but there is little evidence to support the existence of sequences of intermediates built up from addition of monomer units.

The expansion of coordination number from four to five or six is the key to the formation of most large polyoxoanions. Oligomers based on condensation of tetrahedral units generate linear or cyclic structures of limited variety. Thus, a few small vanadates like $V_2O_7^{4-}$ and $cyclo$-$[V_nO_{3n}]^{n-}$ (n = 4, 5, 6) are formed in weakly basic or neutral solutions [2]. The corresponding cyclic structures, M_nO_{3n}, of tetrahedrally-coordinated Mo^{VI} and W^{VI} would be uncharged. Although $cyclo$-$[M_nO_{3n}]^0$ (M = Mo^{VI},W^{VI}; n = 3, 4, 5) have been detected in the vapor phase over heated MoO_3 and WO_3 [3], these species are not observed in aqueous solution. Nevertheless, many polyanion structures can formally be "dissected" into neutral cyclic $\{M_nO_{3n}\}$ and charged $\{XO_m\}^{p-}$ components by scission of

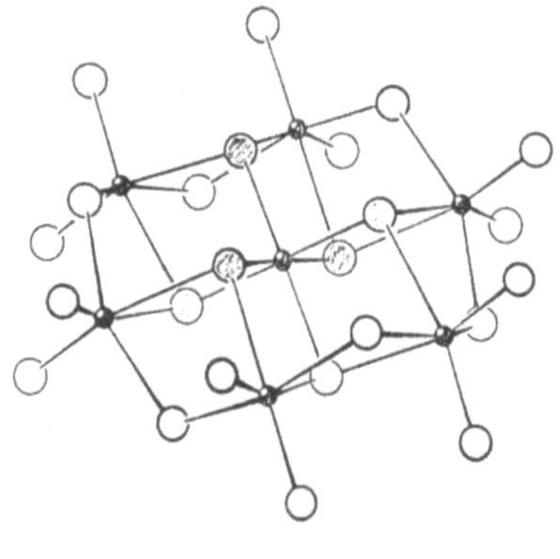

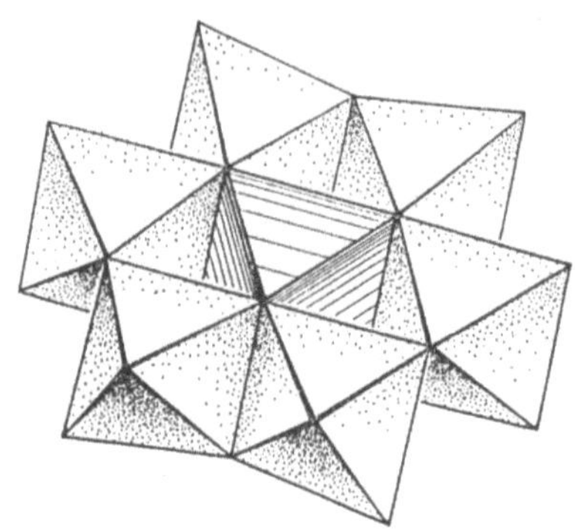

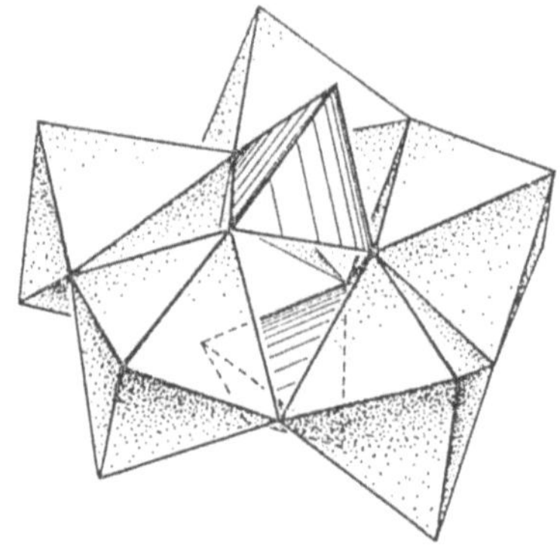

Figure 11-1 The structure of the hetero-polytungstate $[IW_6O_{24}]^{5-}$ (**1**) shown *above* as an assembly of $[IO_6]^{5-}$ (oxygens shaded) and *cyclo*-$\{W_6O_{18}\}$, and *below* in polyhedral representation, with $I^{VII}O_6$ and $W^{VI}O_6$ octahedra.

Figure 11-2 *Upper*: $[P_2Mo_5O_{23}]^{6-}$ (**2**) as an assembly of two $[PO_4]^{3-}$ (oxygens shaded) and *cyclo*-$\{Mo_5O_{15}\}$. *Lower*: Polyhedral representation of **2**.

appropriate M–O bonds [4]. In most cases, the bonds which would have to be cut in order to achieve the dissection are the longest (and weakest) in the structure. Such dissections suggest ways in which polyanions might be assembled, i.e. by *addition* of $\{XO_m\}^{p-}$ to neutral M_nO_{3n}, thereby converting tetrahedrally coordinated M atoms to "octahedral" coordination. This is illustrated for several relatively simple structures in Figures 11-1 – 11-4. It must be stressed that this is not intended to represent a *mechanism* of polyoxoanion formation, but to rationalize the rapid self-assembly processes.

For larger and more complicated polyoxometallate complexes, it is more convenient to represent the structures in polyhedral form as introduced in Figures 11-1 and 11-2. In most cases, MO_6 octahedra can be considered as the fundamental building blocks, and these are linked together by shared vertices, edges, or (less commonly) faces. Each octahedron is distorted as a result of the differences in M–O bond-lengths to terminal and bridging oxygens, i.e. the metal atoms are displaced from the centers of their coordination polyhedra toward the vertices on the surface of the polyanion structure.

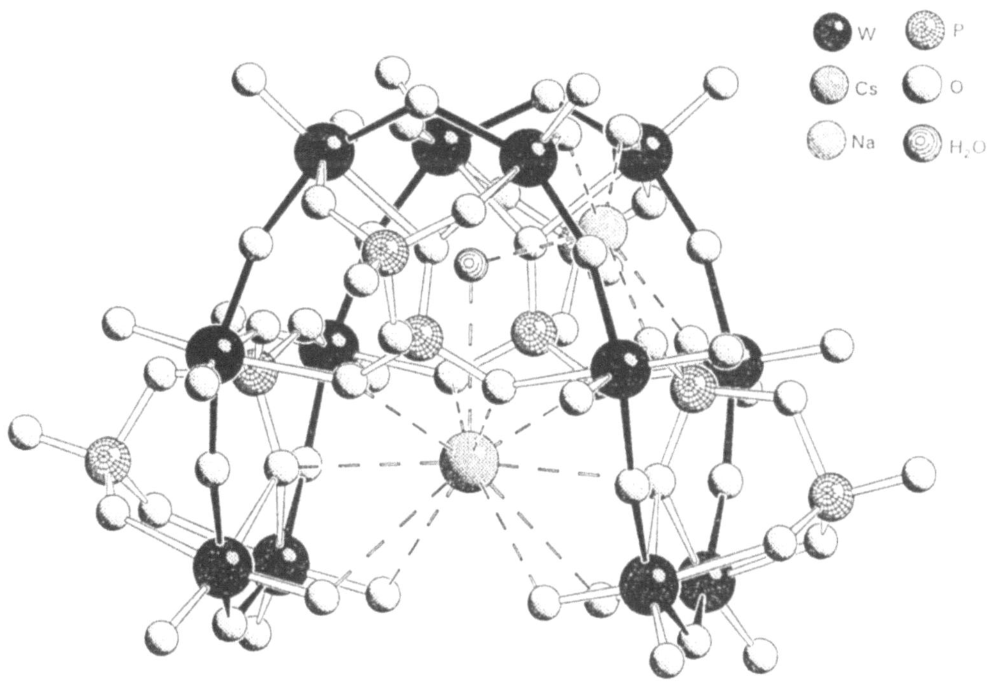

Figure 11-3 $[P_8W_{12}O_{64}]^{16-}$ (3) as an assembly of four $[P_2O_7]^{4-}$ anions and *cyclo*-$\{W_{12}O_{36}\}$ which has been folded into the conformation of the seam of a tennis ball. The positions of a cesium and sodium counterion are also indicated.

11.2 The Keggin Anion and Lacunary Structures

The structure illustrated for **4** in Figure 11-4, is commonly known as the Keggin structure [5], and plays a major role in polyoxometallate architecture. Most of the large polyoxoanions, especially the tungstates, can be viewed as modular structures based on Keggin anions or so-called "lacunary" structures derived from Keggin anions by the "removal" of one or more metal atoms and their associated oxygens. A polyhedral representation of the Keggin structure is shown in Figure 11-5 together with three derivative lacunary structures (monovacant [6] **7**, trivacant **8**, **9**) which have been isolated and characterized.

The conversion of Keggin (and other "plenary" structures) to lacunary anions corresponds to a partial degradation of the polyanion structure, and occurs as the solution pH is increased.

$$[(PO_4)W_{12}O_{36}]^{3-} + 5\ OH^- \longrightarrow [(PO_4)W_{11}O_{35}]^{7-} + \{HWO_4^-\} + 2\ H_2O$$

Keggin ("plenary") anion   Lacunary anion

$$[(AsO_4)Mo_{12}O_{36}]^{3-}$$

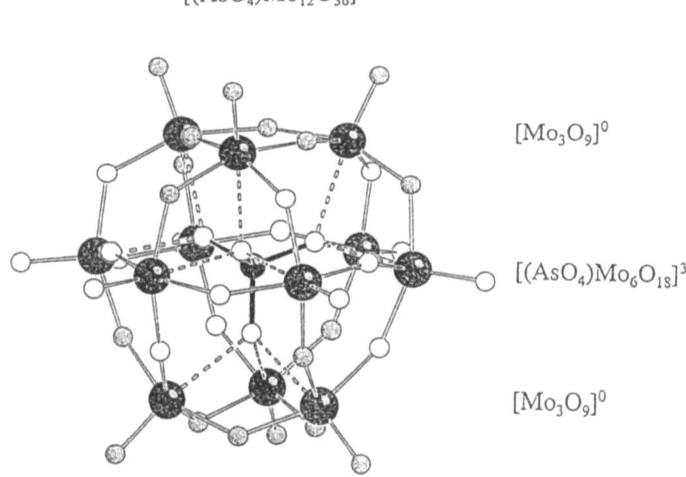

Figure 11-4   The structure of $[(AsO_4)Mo_{12}O_{36}]^{3-}$ (**4**), dissected into a sandwich of three cyclic $\{Mo_nO_{3n}\}$ layers. The oxygens "belonging" to each layer are distinguished by shading. Pieces of the whole structure can exist as separate entities, i.e. $(AsO_4)Mo_9O_{27}(H_2O)_3]^{3-}$ (**5**) (upper layer of **4** removed) and $[(H_3CAsO_3)Mo_6O_{18}(H_2O)_6]^{2-}$ (**6**) (hydrated version of central layer).

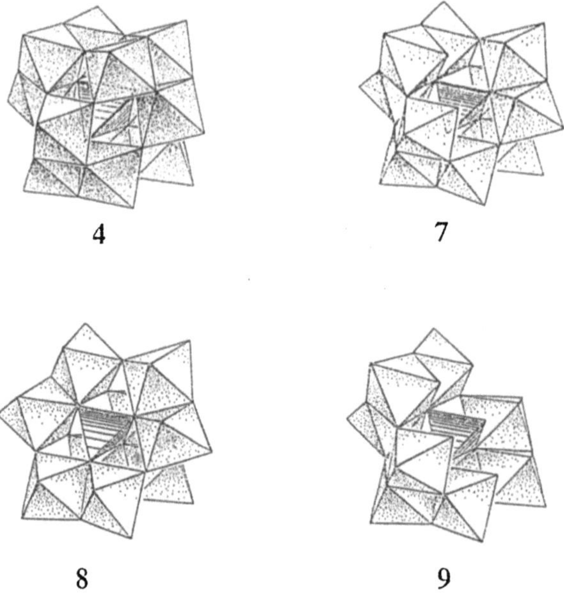

Figure 11-5   Polyhedral representations of the Keggin structure (**4**) and three lacunary or defect derivatives; (a) Keggin anion $[(XO_4)W_{12}O_{36}]^{n-}$; (b) monovacant $\alpha\text{-}[(XO_4)W_{11}O_{35}]^{(4+n)-}$ (**7**); (c) trivacant A-type $\alpha\text{-}[(XO_4)W_9O_{30}]^{(6+n)-}$ (**8**), (deprotonated version of **5**); (d) trivacant B-type $\alpha\text{-}[(XO_4)W_9O_{30}]^{(6+n)-}$ (**9**). Note that these are *represented* as though WO_6 octahedra have been removed from the Keggin anion, but stoichiometrically the loss corresponds to $\{WO\}^{4+}$ or $\{W_3O_6\}^{6+}$ moieties.

Such conversions may not always proceed smoothly, and not all possible lacunary structures are sufficiently stable or non-labile to be detected or isolated. Some 30 years ago, Lipscomb made the observation that all known polyoxometallate structures were based on MO_6 octahedra with either one or two unshared vertices [7]. This corresponds to "C_{4v}" $\{L_5MO\}$ and "C_{2v}" cis-$\{L_4MO_2\}$ coordination geometries for the metal ions [8]. The terminal oxo ligands in these units are very weakly basic and provide a non-reactive "surface" for the polyoxoanion structures. In recent years, a few examples of "Anti-Lipscomb" structures have been observed [9]. These contain one or more MO_6 octahedra with three (fac) terminal oxygens, which are sites of protonation and further polymerization. Thus, under normal hydrolytic conditions, the lacunary structures commonly observed conform to the Lipscomb restriction.

11.3 Assembly of Larger Polyoxometallates from Lacunary Structures

The lacunary structure **7** functions as a tetra- or pentadentate ligand to form both 1:1- and 2:1-complexes with 6- and 8-coordinate central atoms respectively, e.g. $[(SiO_4)Mo_{11}O_{35}(Co^{II}OH_2)]^{6-}$ and $[\{(PO_4)W_{11}O_{35}\}_2Ce^{IV}]^{10-}$. The trivacant structures **8** and **9** react analogously to yield species such as $[\{(PO_4)W_9O_{30}\}_2Co_4(H_2O)_2]^{10-}$ (**10**) and $[\{(PO_4)W_9O_{30}\}_2(C_6H_5SnOH_2)_3]^{12-}$ (**11**) (see Figure 11-6). Polyoxoanions **12** and **13** which may be regarded as fused dimers of **8** and **9** are illustrated in Figure 11-7. These structures in turn give rise to a series of lacunary anions, especially those derived from the dimer of **9** known as the Dawson structure (**12**). The latter structure (α) [10] contains two kinds of symmetry-distinct metal atoms, "cap" (of which there are six) and "belt" (of which there are twelve). Thus there are two monovacant lacunary derivatives, α_1- and α_2-$[(PO_4)_2M_{17}O_{53}]^{10-}$ (corresponding to the loss of a "belt" and a "cap" metal atom, respectively). Derivatives of both isomers are known, e.g. α_1-$[\{(PO_4)_2W_{17}O_{53}\}Co(py)]^{8-}$, α_2-$[\{(PO_4)_2W_{17}O_{53}\}_2U^{IV}]^{16-}$. The trivacant lacunary species formed by loss of three "cap" metal atoms yields derivatives analogous to **10** (Figure 11-6), e.g. $[\{(PO_4)_2W_{15}O_{48}\}_2Zn_4(H_2O)_2]^{12-}$.

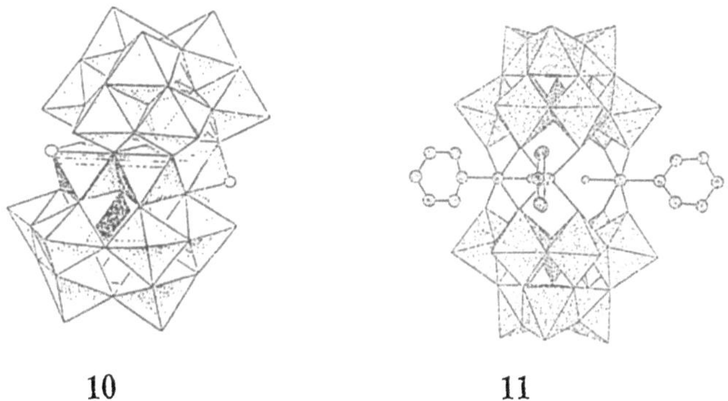

10   11

Figure 11-6   Polyhedral representations of the structures of $[\{(PO_4)W_9O_{30}\}_2Co_4(H_2O)_2]^{10-}$ (**10**) and $[\{(PO_4)W_9O_{30}\}_2(C_6H_5SnOH_2)_3]^{12-}$ (**11**), containing B- and A-type PW_9 fragments respectively.

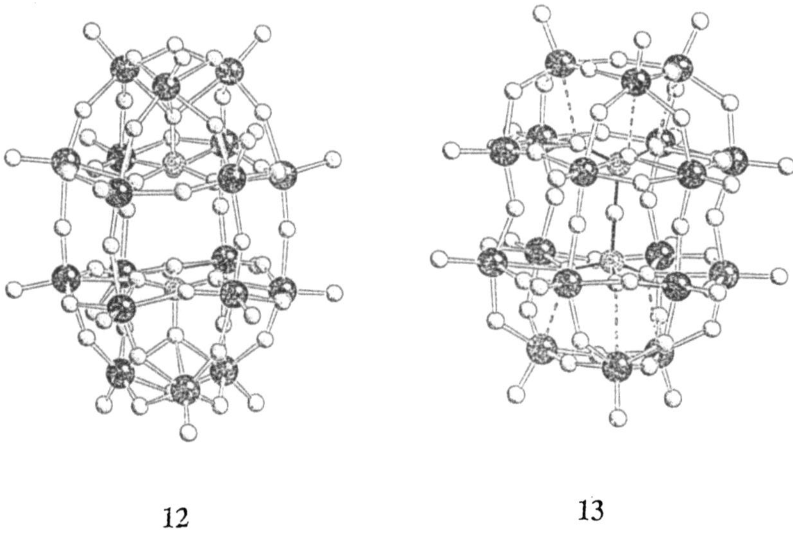

12   13

Figure 11-7   (a) The structure of the Dawson anion $[(PO_4)_2Mo_{18}O_{54}]^{6-}$ (**12**) which can be viewed as a dimer of the structure represented for **5** (or, in polyhedral form, **8**). (b) The structure of the pyrophosphate complex $[(P_2O_7)Mo_{18}O_{54}]^{4-}$ (**13**) which is a fused "dimer" of **9**.

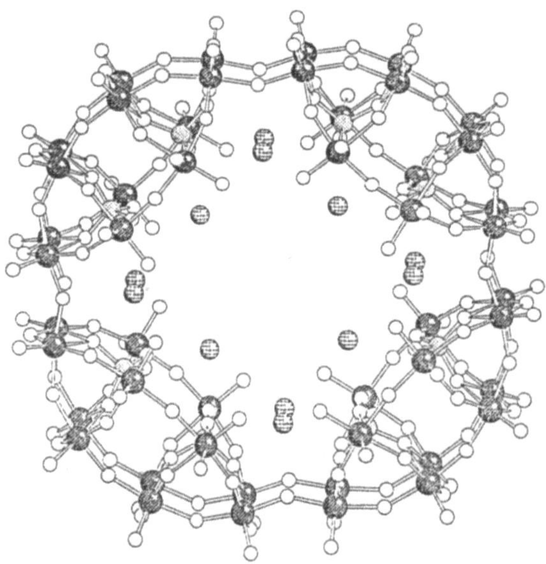

Figure 11-8   The structure of the cyclic anion $[(PO_4)_8W_{48}O_{152}]^{40-}$ (**15**) is formed by condensation of four protonated anions $[(PO_4)_2W_{12}O_{40}H_4]^{10-}$ (**14**). Of the eight "internal" potassium cations shown, four are crystallographically disordered over two positions and are represented by double spheres.

A hexavacant lacunary derivative of **12**, $[(PO_4)_2W_{12}O_{40}]^{14-}$ (**14**), may be induced to oligomerize in the presence of lithium ions to form the unusual cyclic structure of $[(PO_4)_8W_{48}O_{152}]^{40-}$ (**15**) shown in Figure 11-8. The high charge of **15** is compensated by partial protonation and by counterion-binding. The crystal structure of the $K_{28}Li_5H_7$ salt of

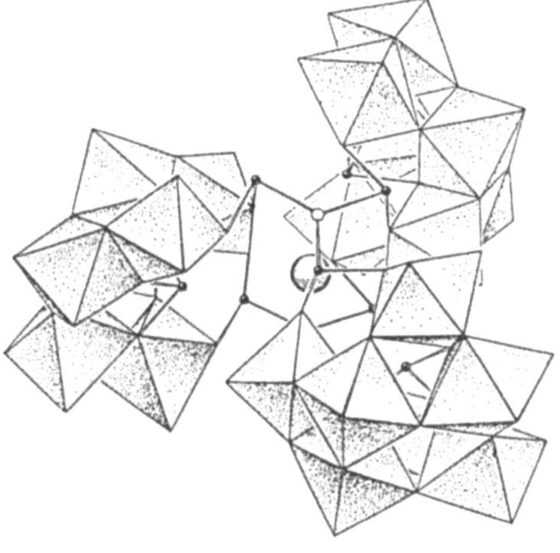

Figure 11-9 Polyhedral representation of the structure of $[NaSb_9W_{21}O_{86}]^{18-}$ ($[Na(Sb_3O_7)_2\{(SbO_3)(W_7O_{21})\}_3]^{18-}$) (**16**). The central Na^+ is shown as the large sphere enclosed in a trigonal prism formed by six Sb^{III} atoms (small dark spheres) capped by two oxygens (white sphere).

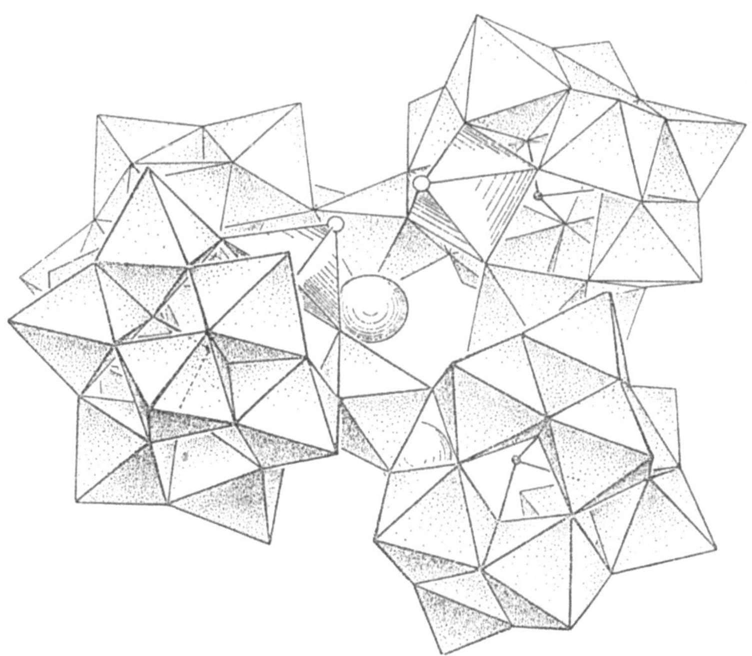

Figure 11-10 Polyhedral representation of the structure of $[(NH_4)As_4W_{40}O_{140}\{Co(OH_2)\}_2]^{23-}$ (**17**). The large sphere depicts the central NH_4^+ cation, and the square pyramids capped by small white spheres denote $Co^{II}O_4(H_2O)$ coordination moieties. Four B-type $[(As^{III}O_3)W_9O_{30}]^{9-}$ are linked by cis-WO_2^{2+} groups.

15 [11] reveals eight K^+ cations in the central cavity, and in this sense the polyanion could be viewed as a kind of cryptand, although its solution chemistry has not been explored.

Other large polytungstates shown in Figures 11-9 – 11-11 have been shown to behave as more conventional cryptates. The Sb^{III} polyanion, $[Na(Sb_3O_7)_2\{(SbO_3)(W_7O_{21})\}_3]^{18-}$ (**16**), Figure 11-9, has overall C_{3h} symmetry with the three-fold axis defined by the sodium ion and the central oxygens of the Sb_3O_7-groups. An analogous complex with $\{(As^{III}O_3)(W_7O_{21})\}$ groups can also be prepared. The relative binding affinities of **16** for alkali metal cations fall in the order $Li^+ < Na^+ > K^+ > Rb^+, Cs^+$ [12]. The tungstate groups in **16** can be viewed as lacunary derivatives of a hypothetical Sb^{III}-centered Keggin anion [13]. Another large complex containing a related lacunary fragment is $[Eu_3(H_2O)_3(SbW_9O_{33})(W_5O_{18})_3]^{18-}$ [14].

The complex illustrated in Figure 11-10, $[(NH_4)As_4W_{40}O_{140}(CoOH_2)_{32}]^{23-}$ (**17**), is constructed of four trivacant lacunary derivatives (cf. **9**) of the hypothetical As^{III}-centered Keggin anion, linked by four additional WO_2 groups into a puckered cyclic structure encapsulating NH_4^+. The complete structure has the ability to bind up to four additional metal cations in surface cavities. The anion illustrated in Figure 11-10 shows two Co^{II} cations bound in this way. The formation constants for alkali and alkaline earth complexes of **17** fall in the sequences $Li^+ < Na^+ < K^+ > Rb^+ > Cs^+$ and $Mg^{2+} < Ca^{2+} < Sr^{2+} < Ba^{2+}$ [15]. A complex similar to **17**, $[Na_2Sb_8W_{36}O_{132}(H_2O)_4]^{22-}$ (**18**), has recently been reported [9(c)]. The linking atoms are four Sb^{III}, and the $\{(SbO_3)(W_9O_{30})\}$ groups are based on a different (β) isomer of the Keggin structure than was found in **17**. *Two* sodium cations are tightly bound to **18**, but there is to date no evidence that they may be exchanged for other cations.

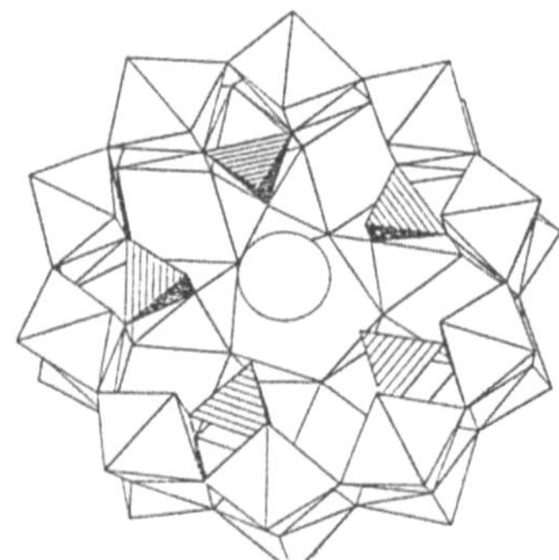

Figure 11-11 Polyhedral representations of the structure of $[Na(PO_4)_5W_{30}O_{90}]^{14-}$ (**19**), viewed almost along the C_5 axis. *Left*, complete anion; *Right*, anion with upper level of five WO_6 octahedra removed to show the enclosed PO_4 tetrahedra.

The compact tungstophosphate illustrated in Figure 11-11, $[Na(PO_4)_5W_{30}O_{90}]^{14-}$ (**19**), has five-fold symmetry. The complex can be imagined to have been constructed by fusing five hypothetical hexavacant Keggin lacunary anions, $[(PO_4)W_6O_{24}]^{15-}$. The Na^+ cation lies off the equatorial plane as shown, thereby reducing the overall symmetry from D_{5h} to C_{5v} [16]. The central cavity of **19** is *highly selective* for Na^+, and only those cations with radii very similar to that of Na^+, *viz* Ca^{2+}, some trivalent lanthanides, and U^{4+} can replace Na^+ [17].

11.4 Very Large Polyoxomolybdates

Most of the examples of large polyoxometallates given above are tungstates. In spite of virtually identical ionic radii for Mo^{6+} and W^{6+} there are surprisingly few examples of *large* isostructural polymolybdates and -tungstates. As we have seen, most tungstates are based upon fragments of the Keggin structure. Although some Keggin-type molybdates, e.g. $[(SiO_4)Mo_{12}O_{36}]^{4-}$, are well-characterized, the lacunary derivatives tend to be much less stable, and assemblies analogous to **10**, **11**, **15**, **16**, **17**, **18**, and **19** are currently unknown. Nevertheless the *largest* polyoxometallate anions currently identified are *molybdates*. All of these have structures based on the heptamolybdate anion, $[Mo_7O_{24}]^{6-}$ (**20**), shown in Figure 11-12 [18].

Lanthanide cations (La^{3+} through Dy^{3+}) and Am^{3+} form very large molybdate complexes [19]. Three of these have been structurally characterized. Two contain an anion $[(MoO_4)Ln_4(H_2O)_{16}(Mo_7O_{24})_4]^{14-}$ (**21**) (Ln = Ce [20], Eu [21]), and the third has a dimer of **21**, $[\{(MoO_4)Pr_4(H_2O)_{13}(Mo_7O_{24})_4\}_2]^{28-}$ (**22**) [19]. The structure of **21** is based on a central MoO_4^{2-} which bridges four Ln^{3+} cations. Each Ln^{3+} has 9-fold coordination (tricapped trigonal prism) supplied by four water molecules, an oxygen of MoO_4, and oxygens of three of the $[Mo_7O_{24}]^{6-}$ anions. The last species are arranged to produce a structure of overall D_{2d} symmetry. Crystals of the praseodymium complex **22** contain a dimer of this structure in which three aqua ligands of *one* of the Pr^{3+} cations in each moiety are replaced by two oxygens of an $[Mo_7O_{24}]^{6-}$ of the other. The whole structure, $[Pr_8Mo_{58}O_{200}(H_2O)_{26}]^{28-}$, has inversion symmetry.

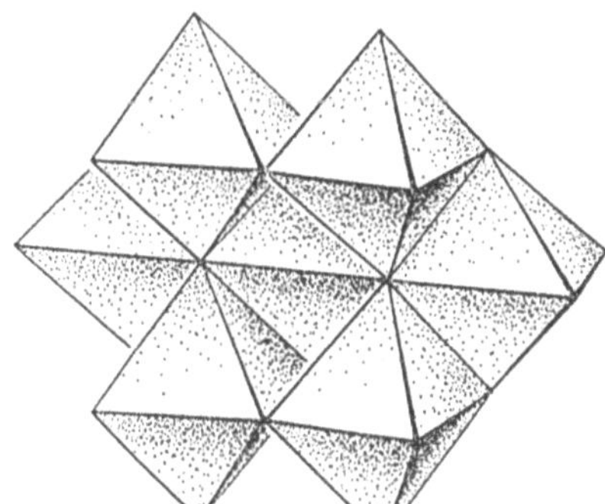

Figure 11-12 The C_{2v} structure of heptamolybdate $[Mo_7O_{24}]^{6-}$ (**20**), which is the predominant solute species in aqueous solutions of Mo^{VI} at pH 4–5.

Acidification of an aqueous solution of molybdate(VI) from pH ca. 4–5 (in which the predominant solute species is anion **20**) to pH ca. 1.8 leads to the anion, $[Mo_{36}O_{112}(H_2O)_{16}]^{8-}$ (**23**), Figure 11-13 [22], isolable as sodium, ammonium, potassium, and barium salts. The structure of **23** can be viewed as two subunits each comprising a heptamolybdate encircled by a ring of ten corner-shared MoO_4 "tetrahedra". Water molecules and oxygens of the heptamolybdate complete six-fold coordination about each of the encircling Mo atoms. The Mo_{17} moieties (in which two Mo atoms are now 7-coordinate, with approximately pentagonal-bipyramidal geometry) are linked by an additional pair of MoO_6 octahedra (labelled A in Figure 11-13) to form a complete assembly with overall inversion symmetry.

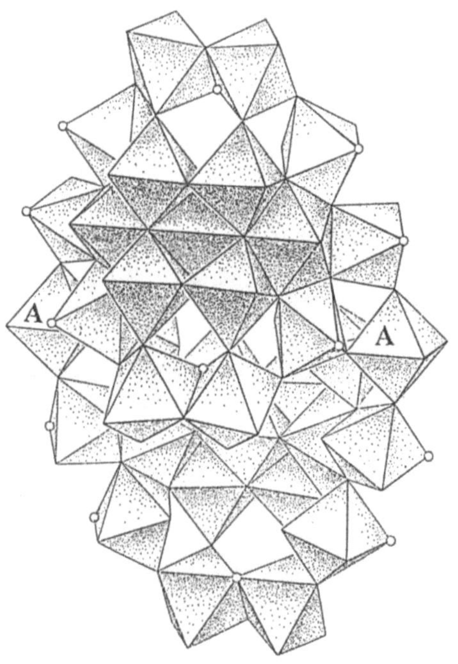

Figure 11-13   The structure of $[Mo_{36}O_{112}(H_2O)_{16}]^{8-}$ **(21)** in poly-hedral form. The anion consists of two Mo_{17} units, related by inversion symmetry and linked by the two MoO_6 octa hedra A. The heptamolybdate moiety in one such unit is emphasized by darker shading and the water molecules are indicated by open circles.

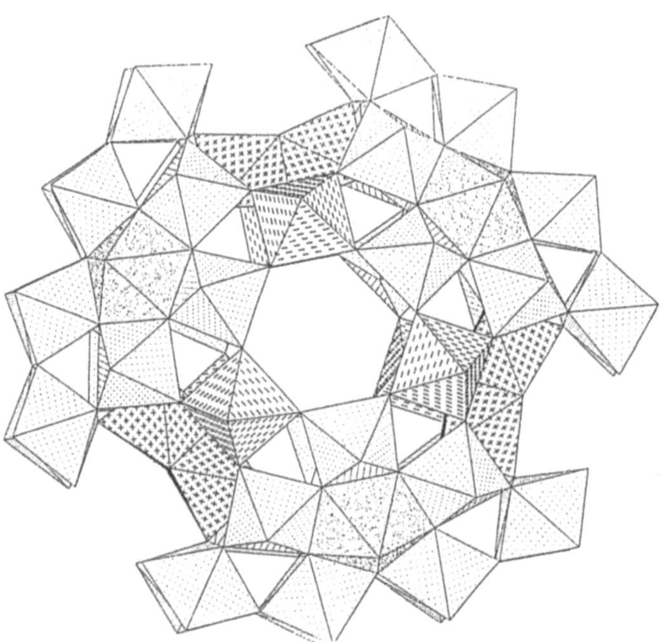

Figure 11-14 The structure of $[Mo_{45}(MoNO)_6Mo^V_6V^{IV}_6O_{180}(OH)_3(H_2O)_{18}]^{21-}$ **(25)** in polyhedral form. Three Mo_{17} units as described in the text are linked by Mo^VO_6 and $V^{IV}O_6$ octahedra.

Figure 11-13 may also be used to represent the structure of $[Mo_{32}(MoNO)_4O_{108}(H_2O)_{16}]^{12-}$ (**24**) which contains $\{MoNO\}^{3+}$ in place of $\{MoO\}^{4+}$ in the pentagonal-bipyramidal sites. The nitrosyl ligand is introduced by addition of hydroxylamine during the acidification of the molybdate. An even larger cyclic assembly of the nitrosylated Mo_{17} fragments is seen in $[Mo_{45}(MoNO)_6Mo^V{}_6V^{IV}{}_6O_{180}(OH)_3(H_2O)_{18}]^{21-}$ (**25**), shown in Figure 11-14 [23]. The three Mo_{17} groups in **25** are linked by Mo^V dimers and VO^{2+} cations to produce what is currently the largest discrete polyoxoanion that is readily generated by self-assembly in aqueous solution.

Acknowledgements

Current polyoxometallate research at Georgetown University is supported by the National Science Foundation. I thank members of my research group, especially Uli Kortz, Feibo Xin and Dr. Michael Dickman, for their recent structural studies reported here and other assistance. Several of the Figures have been generated by SCHAKAL 92 (Copyright Egbert Keller, University of Freiburg).

12   At the Boundary of the Metallic State

Günter Schmid

Abstract

The formation of a metallic particle from single atoms passes through various phases of development. An assembly of only a few atoms will electronically behave like a multinuclear metal complex with clearly separated states of energy of the valence electrons. With the increase of metal atoms, the orbital separation turns slowly into the quasi-continuous density of states of a typical metal. The synthesis of ligand stabilized transition metal clusters in the range between about 1 and 30 nm opens novel opportunities to study just this transition state between the molecular and the bulk state. The separation of single metal atoms in solution succeeds by reduction of metal complexes, mainly those of noble metals. To stop uncontrolled formation of metal particles, appropriate ligands are added. They envelop the chemically highly reactive metal clusters and colloids to give isolatable, stable chemical compounds. Following this principle, ligand stabilized gold, palladium and platinum colloids of 15–30 nm diameter as well as bimetallic colloids can be obtained. Smaller dimensioned clusters of palladium (3.6, 3.15, 2.2 nm) and of platinum (1.8 nm) can also be synthesized and characterized. They incorporate a preferred number of metal atoms, the so-called magic numbers. The boundary of the metallic state, as far as it can be recognized by the available physical methods, is reached by the two-shell clusters of about 1.5 nm in diameter. Their physical properties, which are only briefly touched in this article, clearly show the existence of only a few "last" mobile electrons. There are some first indications that such clusters can be used as quantum dots enabling single-electron transitions. Future applications of such clusters instead of the presently used semiconductors seem possible.

12.1   Introduction

Let us carry out a gedanken experiment: we reduce a piece of metal step by step to a size of only a few atoms or, in the opposite direction, we build up a metal particle with bulk properties by assembling single metal atoms. Somewhere on this way, we will pass clusters incorporating that few (or many) atoms that typical bulk properties begin to disappear or, going the other way, that "metallic" behaviour can be observed for the first time. Here, the most important question dominating all others arises: how many metal atoms do we need to observe typical bulk properties such as electric conductivity, magnetism etc. for the first or the last time, respectively? These and all other "metallic" properties are associated with the existence of free spin density or, in other words, with the transition from discrete electronic energy levels, as realized in simple complexes and low-nuclearity clusters, to a delocalization of valence electrons ending up with the quasi-continuous density of states in the bulk.

Molecular clusters, consisting of up to a dozen of atoms, are described numerously in the literature. Due to a stabilizing shell of organic ligands, they can be isolated and handled like normal molecular species in organometallic chemistry. However, with increasing numbers of metal atoms, the number of examples becomes quite small. On the other hand, only assemblies consisting of at least some dozens of metal atoms are of interest with respect to the fundamental questions mentioned above.

With the beginning (or ending) of the metallic state, metal clusters can be regarded as quantum dots. The study of quantum size effects is at present a challenge for scientists all over the world, because it can be expected that such nano-sized particles will show novel properties compared with those of the bulk and the molecular state, respectively. New generations of computers with extremely high capacities, minilasers as well as optoelectronic switches are future goals in science and technology. Clusters in the range of a few nanometers could open the door to such novel technologies which probably will never be reached by established methods such as nanolithography.

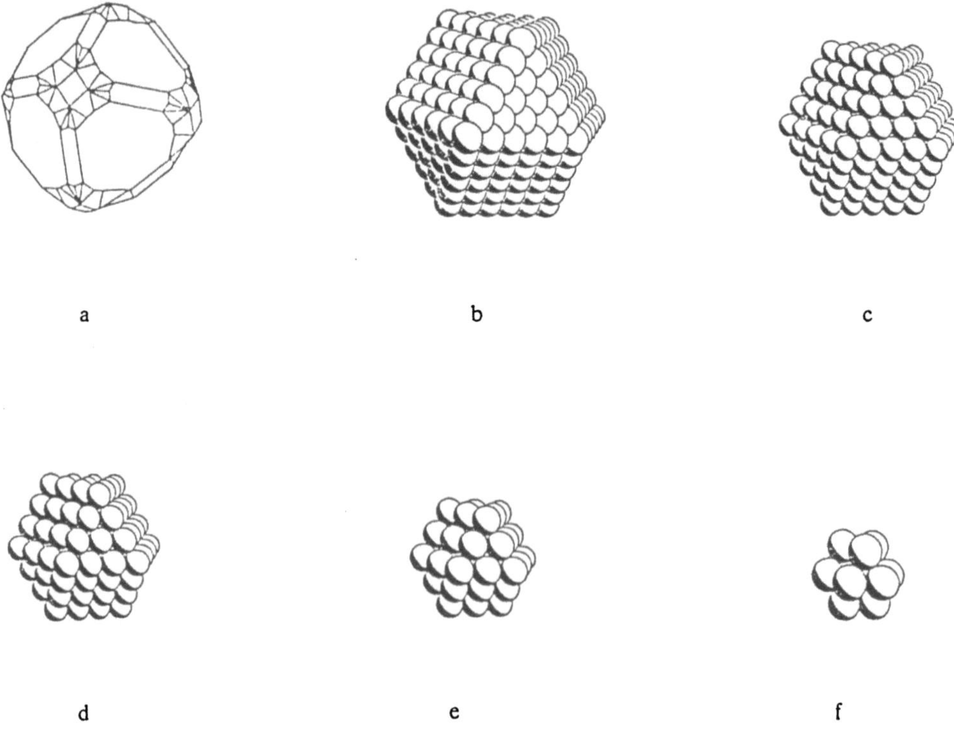

a   b   c

d   e   f

Figure 12-1   Illustration of the transition from the bulk (a) via some high-nuclearity clusters consisting of 561 (b), 309 (c), 147 (d), 55 (e) to 13 (f) cubic close-packed atoms.

The application of metal clusters as quantum dots requires quantum inclusions which can be realized by materials with band gaps larger than those of the metal clusters. Like for the smaller molecular clusters, ligands as they are used in complex chemistry should be best suited. In the following, methods to prepare ligand protected metal particles in the range of

1–30 nm will be described. However, only a few characteristic physical methods shall be discussed in order to elucidate the electronic properties of the clusters in dependence on the number of metal atoms. Finally, we will see whether some first answers with respect to quantum properties can be given. Figure 12-1 illustrates the transition from the bulk via some larger clusters to a 13 atomic small cluster of mainly molecular character.

12.2 Synthetic Aspects

In the following, some routes to prepare ligand protected metal particles with decreasing size will be described. Beginning with the largest species to be considered in this contribution, we have to discuss synthetic pathways to generate colloids. The term "colloid" is normally used in chemistry if the particle diameter is beyond about 10 nm.

Metal colloids, especially those of the element gold, were already known in the middle ages and have been used to colour glasses and china (purple of Cassius). In the last century, M. Faraday synthesized aqueous solutions of gold colloids using a method which is still in use nowadays: the reduction of tetrachloroauric acid, $HAuCl_4$, by citric acid or better by its trisodium salt in boiling water [1–8]. This procedure leads to magnificent red solutions of gold colloids in the range of about 20 ± 2 nm in diameter. Faraday's method has meanwhile been modified in various ways and so became available for other metals such as palladium and platinum. The original colloids in aqueous solution are stabilized by different species which are present in the reaction mixture: water, sodium chloride, formic acid, formic aldehyde etc. However, these accidental stabilizers are more or less quantitatively removed if the water is evaporated. As a result, the formation of gold mirrors and metallic precipitates is observed. Similar procedures happen if other metals are used.

To enable investigations of colloidal metal particles in a solid state, a much more stable protecting environment must be created. Numerous papers describe the stabilization of metal colloids using polymeric materials such as poly(vinyl alcohol), [9–11], poly(vinyl pyrrolidone) [9], copolymers of vinyl alcohols and N-vinyl pyrrolidone [12], cyclodextrines [13], colloidal silicid acid [14], and others. Such systems are well suited for various purposes, however, for the study of electronic properties, quantum size effects etc. it might be advantageous to reduce the protecting materials as far as possible. For that reason, we introduced complex ligands into colloid chemistry which are well established in complex chemistry: phosphines and nitrogen bases with hydrophilic substituents for use in aqueous solutions. Figure 12-2 shows three examples of such ligands which are usually applied by us: mono and threefold sulfonated triphenylphosphine, (a), (b), best suited for the coordination of gold colloids, and the sodium salt of the sulfanilic acid (c), which is mainly used for palladium and platinum colloids.

The procedure to envelop colloidal particles by such ligands is simple: the original aqueous phases, containing the less good stabilized particles, are treated with an excess of the corresponding ligand molecules. Without any visible change, the stronger phosphine or amine ligands substitute the weaker bonded ions or molecules. Concentration of the dilute solutions, followed by centrifugation, yields the ligand protected colloids as metallic looking powders which now can be isolated and applied to various physical investigations. In case of need, these solid materials can be redispersed in water in any desired concentration [15, 16].

Figure 12-2 Three types of ligands to stabilize metal colloids: (a) monosulfonated triphenylphosphine, (b) trisulfonated triphenylphosphine, and (c) sodium salt of sulfanilic acid.

Figure 12-3 A layer of 18 nm gold colloids, stabilized by the trisulfonated triphenylphosphine P(m-C$_6$H$_4$SO$_3$Na)$_3$ (reproduced from [17]).

To analyze size distributions and the particle structure, high resolution transmission electron microscopy (HRTEM) is normally used and is considered to be the most suitable method [*].

Figure 12-3 shows a monolayer of 18 nm gold colloids, stabilized by P(m-$C_6H_4SO_3Na)_3$ ligands. The strongly fixed phosphine molecules on the surface of the colloidal particles prevent any coalescence processes. In Figure 12-4, a single gold colloid of 13 × 11 nm is shown. The clearly recognizable ligand shell consists of a double layer of P(m-$C_6H_4SO_3Na)_3$ molecules [17].

Bimetallic particles are of increasing interest for catalytical reasons and, in addition, also for the study of alloy formation processes. Alloy-like particles are defined to consist of a homogeneous mixture of two metals. Colloids with a core of the one and a shell of another metal can also be prepared by the method described above. Both types of bimetallic species have already been described [18, 19]. However, they could be isolated in the solid state just as hardly as the monometallic species. Temperature dependent HRTEM and EXAFS studies on the shell type bimetallic colloids give valuable information about alloy formation processes [20].

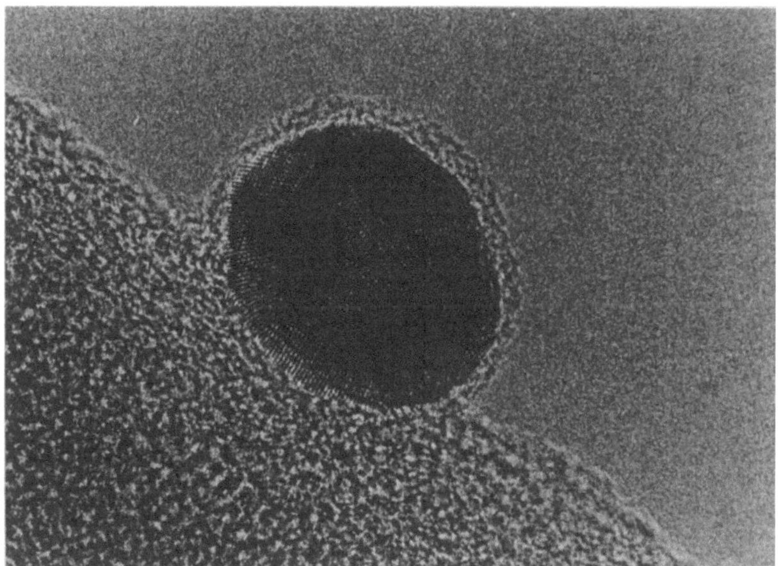

Figure 12-4   A single gold colloid of the size 13 × 11 nm, protected by a double-layer of P(m-$C_6H_4SO_3Na)_3$ molecules (reproduced from [17]).

To prepare bimetallic ligand protected colloids structured in layers, we start with the generation of the inner core, e.g. consisting of gold. In a following reaction step, the second metal, e.g. palladium, is added as a salt and is then reduced by an appropriate reducing agent. The palladium atoms now form layers on the gold germs, the thickness of which just depends on the amount of palladium salt which has been added before. After finishing the second reduction step, the stabilizing ligands – here we use p-H_2N-$C_6H_4SO_3Na$ – are added

[*] We would like to express our gratitude to Dr. Jan-Olov Bovin and Dr. Jan-Olle Malm from the University of Lund, Sweden, for their excellent work on HRTEM.

[21]. The result of such a procedure is shown in Figure 12-5. The particle consists of a gold core of 18 nm in diameter, covered by a 9 nm palladium shell which bears the stabilizing ligand molecules. The particle shows a monocrystalline arrangement of the gold and the palladium atoms as well. Energy dispersive X-ray analyses (EDX) prove the existence of separated gold and palladium spheres.

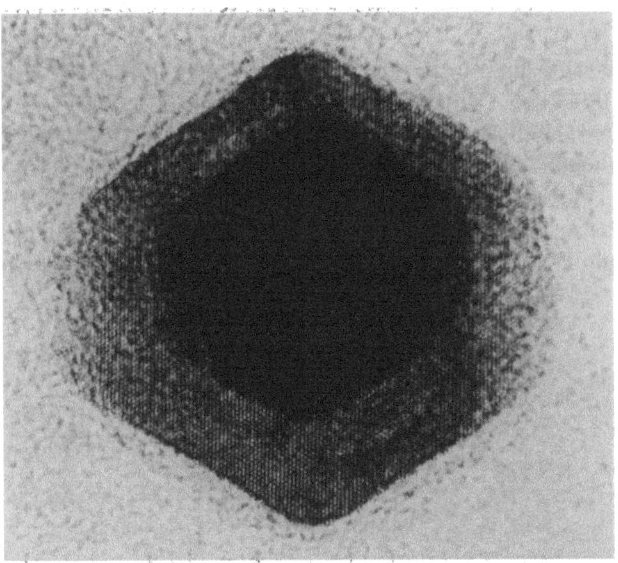

Figure 12-5   A monocrystalline bimetallic colloid, consisting of an 18 nm gold core and a 9 nm shell of palladium atoms. The ligand shell, consisting of p-H$_2$N-C$_6$H$_4$SO$_3$Na, cannot be observed (reproduced from [17]).

Instead of gold, palladium can be used as a nucleus, covered by gold. Other combinations which have been realized by us are Au/Pt and Pt/Pd.

On the way from these 15–30 nm colloids to smaller particles, the synthesis of about 2 and 3 nm palladium clusters shall be described now.

If an acetic acid solution of palladium(II)acetate is reduced by gaseous hydrogen in the presence of small amounts of phenanthroline (phen), followed by the controlled addition of oxygen, three different cluster species are formed with 3.60, 3.15 and 2.2 nm in diameter (without ligand shell) [22]. The electron microscopic study of these particles proves that the predominant part of the clusters has a monocrystalline structure and that the size of the different particles is characterized by a distinct number of atomic planes. So, the 3.60 nm clusters consist of 17, the 3.15 nm clusters of 15 and the 2.2 nm particles of 11 atomic planes. These numbers of planes are perfectly compatible with the existence of a so-called magic number of atoms in each of these cluster types. Magic numbers result from the geometric conditions in cubic or hexagonal close-packed structures as they are observed in most of the bulk metals. Each atom is coordinated in a first shell by 12 neighbouring atoms. A second shell is generated by 42 atoms, the next by 92 etc. In principle, the nth layer consists of $10n^2 + 2$ atoms. Following this model, the 17 atomic planes in the 3.60 nm clusters are formed by 2057, the 15 planes by 1415 atoms, and the 11 atomic planes by 561 atoms. In other words, these numbers correspond with eight, seven, and five-shell clusters.

Of course, HRTEM does not allow to determine the exact number of metal atoms. However, considering the fact that more than 90% of the particles observed in the microscope concur with one of the discussed numbers of planes, it becomes evident that nature tries to reach these magic numbers in so-called full-shell clusters.

The very similar chemical behaviour of the seven and eight-shell clusters, the idealized complete formulas of which can be written as $Pd_{2057}phen_{84}O_{\sim1600}$ and $Pd_{1415}phen_{60}O_{\sim1100}$, does not allow to separate them preparatively. Together, they form about 90% of the products. The missing 10% are found in the five-shell cluster $Pd_{561}phen_{36}O_{\sim200}$ [23, 24]. A very similar cluster was described before by Moiseev et al. [25]. The use of oxygen in addition to the phenanthroline molecules aims at the coverage of non-coordinated surface atoms by O_2 molecules. This prevents uncontrolled oxidation of the clusters by air-contact and relieves handling of these materials.

Figure 12-6 shows a single eight-shell palladium cluster with a more or less perfect structure of cubic close-packed atoms, just as in the bulk. In Figure 12-7, a larger area covered with uniform five-shell palladium clusters in shown.

A very similar procedure, starting with platinum(II)acetate, gives the only four-shell cluster which is known up to now, $Pt_{309}phen^{*}_{36}O_{30}$ [26]. The phen* ligand derives from phenanthroline by a 4,7-substitution by p-sulfonated phenyl groups. Again, the atoms are arranged in a cubic close-packed manner known from platinum metal.

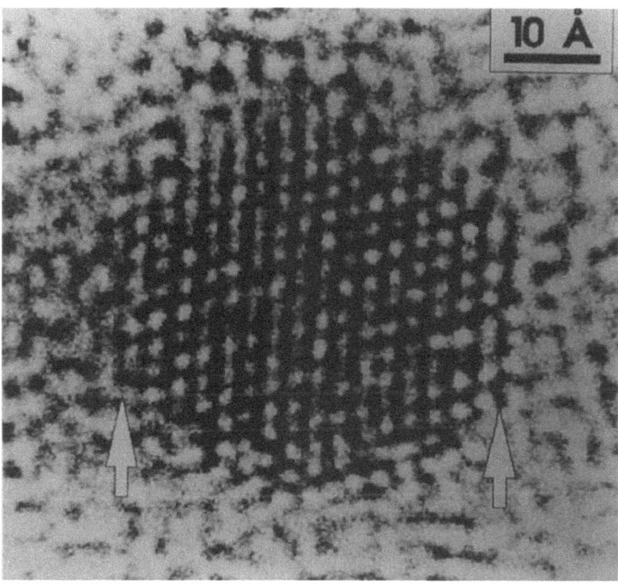

Figure 12-6   A high resolution microscopic image of an eight-shell palladium cluster showing 17 atomic fringes (reproduced from [22]).

Finally, we reach the smallest clusters to be discussed here, the 55 atomic two-shell clusters. A series of compounds of the general formula $M_{55}L_{12}Cl_x$ (M = Rh, Pt, Au) is available, if appropriate metal salts are reduced by diborane, B_2H_6, in an inert organic solvent [27–29].

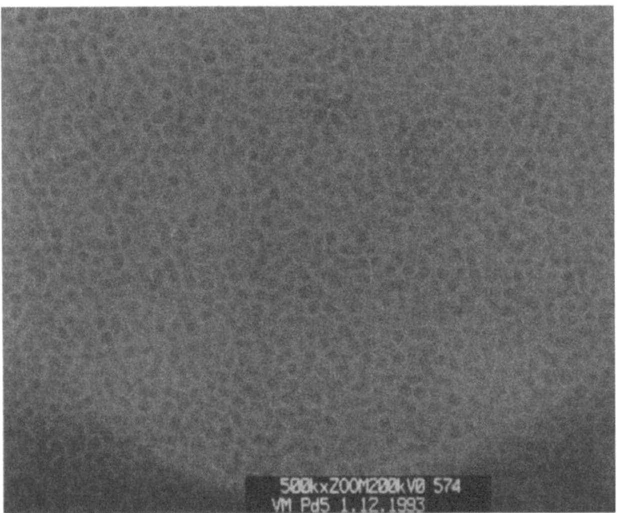

Figure 12-7   A larger area covered with very uniformly dispersed palladium five-shell clusters with the idealized number of 561 atoms.

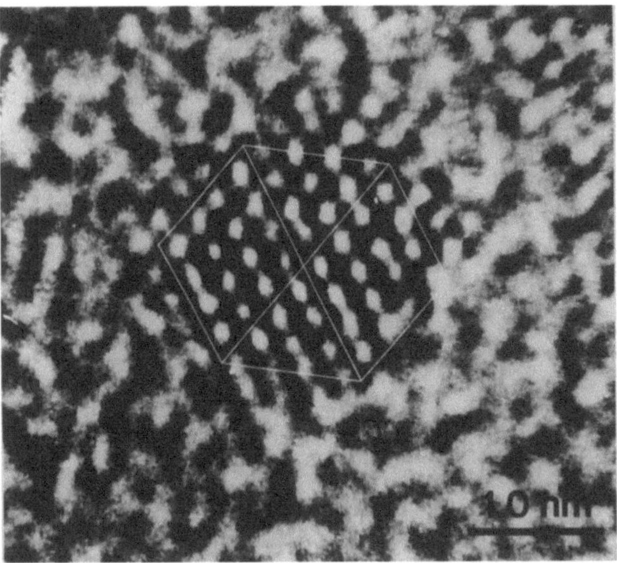

Figure 12-8   A nucleus of a single four-shell cluster of the formula $Pt_{309}phen^*_{36}O_{30}$ (reproduced from [26]).

Again, HRTEM is most suitable to elucidate the structure and to prove uniformity of these clusters. Figure 12-9 shows a single Pt_{55} particle with atomic resolution.

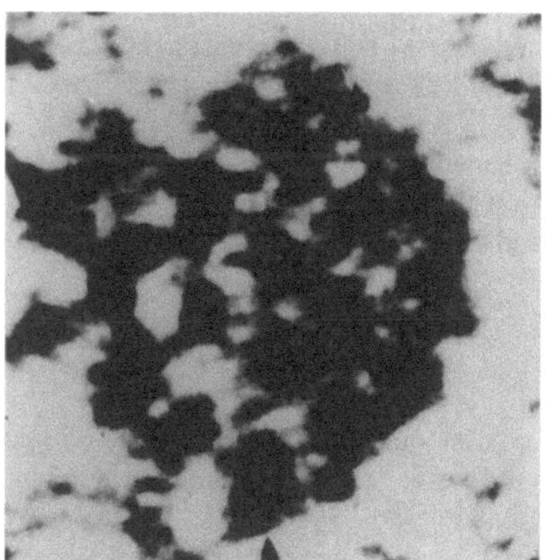

Figure 12-9   The 55 close-packed atoms of a two-shell platinum cluster (reproduced from [17]).

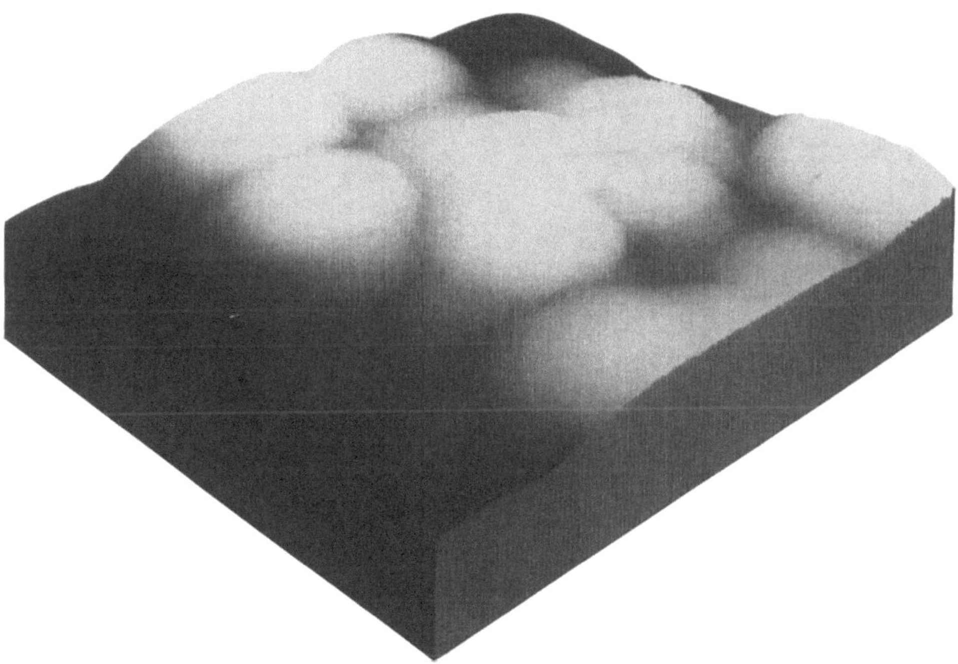

Figure 12-10   STM image of $Au_{55}(PPh_3)_{12}Cl_6$ cluster molecules on the surface of a cluster pellet.

Such 1.4 nm clusters have only short life-times in the microscope, as the size dependent melting point has already decreased to a few hundred degrees with the consequence that the energy, which is transferred to the clusters by the 400 keV electron beam, causes melting and coalescence processes in the course of seconds.

Scanning tunnelling microscopy (STM) complements HRTEM in an ideal manner. Figure 12-10 shows $Au_{55}(PPh_3)_{12}Cl_6$ molecules on the surface of a pellet. The spheric balls consist of cluster molecules including the ligand sphere [30].

12.3 Electronic Properties

Among numerous physical methods which have been applied to study the electronic properties of the clusters and colloids discussed in the previous section, namely magnetism, conductivity, Mössbauer spectroscopy, NMR, EXAFS, XPS, UV/Visible spectroscopy, specific heat, and electrochemical studies, only a few shall be mentioned to elucidate the present state of knowledge.

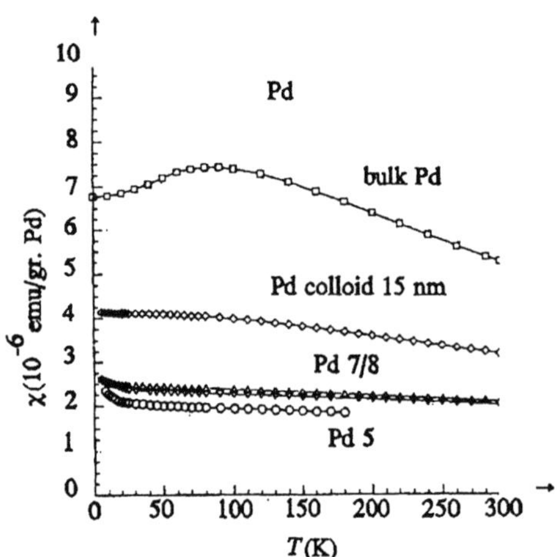

Figure 12-11 Temperature dependent susceptibility of some palladium particles of different size compared to bulk palladium. The values are normalized to the weight of the palladium cores.

Magnetic investigations have been performed with many cluster compounds. We will restrict the discussion to some palladium species. Figure 12-11 illustrates the dependence of the susceptibility of ligand stabilized palladium particles on the particle size. Palladium is well suited for the study of magnetic behaviour as, due to the band structure near the Fermi energy E_F and the position of E_F, the Pauli susceptibility at low temperature is already dependent upon temperature [31–33]. As can be seen, the curves for the five, seven, and eight-shell clusters show a significant difference compared with the curves for bulk palladium. The most remarkable effect, however, is that even the susceptibility curve of a 15 nm palladium colloid, consisting of thousands of atoms, is still far from that of the bulk. A theoretical model which has been developed by de Jongh et al. [34] results in the surprising statement that even for a particle of 1 micron there must be a reduction in χ of about 1% ! This means that on the way from the bulk to clusters the magnetic behaviour already begins to change at a size which would not be expected to have reduced bulk qualities. However, it should be stated that this result does not allow a generalization saying that 15 nm colloids have completely lost their bulk properties. On the contrary! As we will show by the next method, even a few hundred or dozen of metal atoms still possess considerable free spin density.

^{195}Pt nuclear magnetic resonance measurements on solid $Pt_{309}phen^*_{36}O_{30}$ clusters clearly prove that even the inner core of the cluster, consisting of only 147 platinum atoms, causes a so-called Knight shift [35]. The Knight shift, a high-field shift, is a function of the local density of states at the Fermi level of the 5d electrons which for its part is an exponentially decreasing function of the distance from the center to the surface. ^{195}Pt NMR experiments on small platinum particles showed that this metal exhibits one of the largest Knight shifts. Therefore, it can be considered to be one of the hallmarks of the metallic state indicating whether a cluster has metallic behaviour or not. Figure 12-12 shows the solid state ^{195}Pt NMR spectrum of the Pt_{309} cluster at 77 K.

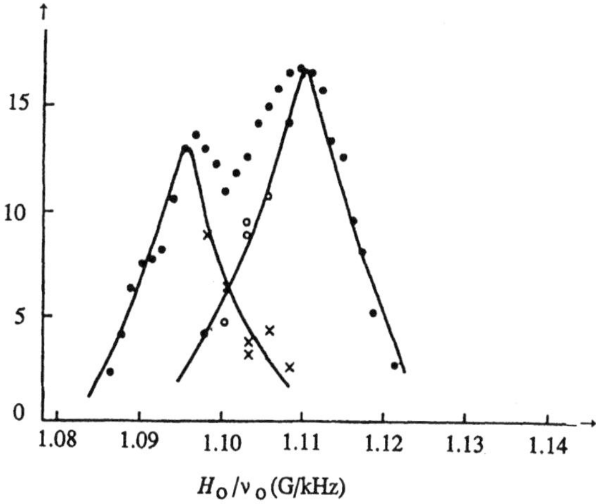

Figure 12-12 ^{195}Pt NMR spectrum of $Pt_{309}phen^*_{36}O_{30}$ at 77 K.

The low-field peak at 1.096 GkHz^{-1} is characteristic for ^{195}Pt signals in complexes and can be assigned to the surface atoms with coordinated ligands. The high-field peak at 1.110 GkHz^{-1} indicates a considerable Knight shift and must be caused by the inner cluster core of 147 platinum atoms. However, the extent of this Knight shift differs already remarkably from that of bulk platinum which is caused by a signal at 1.138 GkHz^{-1}. So, one can conclude that even an assembly of 147 platinum atoms owns enough free spin density to cause a considerable Knight shift, which is, however, characteristically smaller than that in the bulk.

This finding has recently been confirmed impressively by de Jongh et al. [36]. In a unique experiment they irradiated the Pt_{309} cluster by thermal neutrons to generate a low percentage of ^{197}Au atoms. As these atoms occupy statistically all positions in the cluster, Mössbauer spectroscopy can now be carried out. A ^{197}Au Mössbauer spectrum of such an irradiated sample shows indeed that the Mössbauer contribution of the "Au_{147}" cluster core is very close to that of bulk gold!

$Au_{55}(PPh_3)_{12}Cl_6$ has been used to perform *photoelectron spectroscopy (XPS)*. The results indicate, like the *^{197}Au Mössbauer spectra*, that even these 1.4 nm cluster particles possess considerable free spin density. In the Mössbauer spectra, the main contribution comes from the inner core atoms which cause a chemical shift very close to that of the bulk [37–40], but also clearly shifted from it. The XPS data complete this finding. The 5d spin-orbit splitting is close to the value for bulk gold, but greater than that in smaller gold clusters, such like $Au_{11}L_7Cl_3$ [38]. As can be seen from Figure 13-13, there is a strong indication of a finite density of states at the Fermi level, suggesting a certain "metallic character". In addition, the $5d_{5/2}/5d_{3/2}$ band gap of 2.4 eV is also close to the bulk. For the Au_{11} cluster it is only 1.9 eV. These XPS results obviously prove that all atoms of the Au_{55} cluster nucleus seem to be involved in the metallic binding.

Finally, some recent results concerning the intra- and intercluster *conductivity* of pressed ligand stabilized clusters shall be considered [41]. *Low frequency impedance spectroscopy* (*IS*) allows the observation of electronic relaxation processes. A powdery $Au_{55}(PPh_3)_{12}Cl_6$ sample, which is pressed to a pellet applying 2.5×10^8 Pa of pressure, reaches a density of about 90% of a close-packed structure. In such a pellet of 0.2–0.4 mm thickness and 5 mm in diameter, the neighbouring Au_{55} quantum dots are isolated from each other by the ligand shells and, at the same time, they are in contact via their ligands. Temperature dependent IS measurements (253–333 K) result in two semicircles in the Argand diagram. They can be interpreted as an ideal Debye resistance/capacitance parallel link (R_2/C_2) (lower frequency process 2) in series with an additional Cole-Cole resistance/constant phase element (R_1/CPE) parallel link (higher frequency process 1). The Ohmic part is the same for both processes: $R_1 \approx R_2$ or $R_{total} = 2 R_1$. The specific conductivities are $\sigma_1 = \sigma_2 = 3.23 \times 10^{-4}$ $Ohm^{-1}m^{-1}$ at 293 K. As both processes are thermally activated, the activation enthalpies E_{A1} and E_{A2} can be calculated as 0.15 ± 0.03 eV and 0.16 ± 0.03 eV, respectively. Within the experimental error, both values are equal. The higher frequency Cole-Cole process can be interpreted as the relaxation of an *intercluster* process, as has been done already to explain former conductivity measurements [42]. This interpretation is supported by the existence of a distributed CPE capacitance instead of a single capacitance. It is caused by a distribution in the jump times and in the distances between the clusters. The electron transfer from one cluster to a neighbouring one works by tunnelling through the fluctuating ligand shells.

The Debye process 2 must be interpreted as an *intracluster* step, coupled with process 1. In contrast to the broad Cole-Cole-resonance in the immitance spectra, the Debye process causes a sharp resonance. From these experimental results, some important conclusions can be drawn: the Au_{55} cluster acts as a quantum box where two electrons occupy a kind of a cluster valence orbital. Its energy can be calculated from the experimental capacitance $C_2 = $ 48 pF and the experimental density of electronic states $D_{ex} = 10.75$ eV^{-1}. Using the formula for a three-dimensional electron gas in a quantum box $x = \hbar\pi(2mE_F)^{-1/2}$ ($m = $ electron mass, $E_F = $ Fermi energy, $x = $ lateral dimension of the quantum box), the energy for a doubly occupied electronic state results as $E_{ex} = 0.186$ eV.

At resonance, the two electrons occupying the "cluster orbital" in the ground state participate in a single-electron tunnelling process between clusters and cause doublication of the conductivity. Figure 13-13 illustrates these results. Figure 13-13a elucidates the two activation steps. They can be compared with a chemical disproportionation process, Figure 13-13b shows how the total conductivity σ_{total} is doubled under resonance conditions.

A cluster-pellet as used for these measurements is already a handy device with quantum properties at room-temperature! It can act as a tunnel resonance resistance (TRR), the resistance of which is divided into two under resonance conditions. Two(!) cluster molecules can in principle act as an absolutely minimized electronic switch. Rows of clusters could be regarded as 1D quantum channels of correlated electrons.

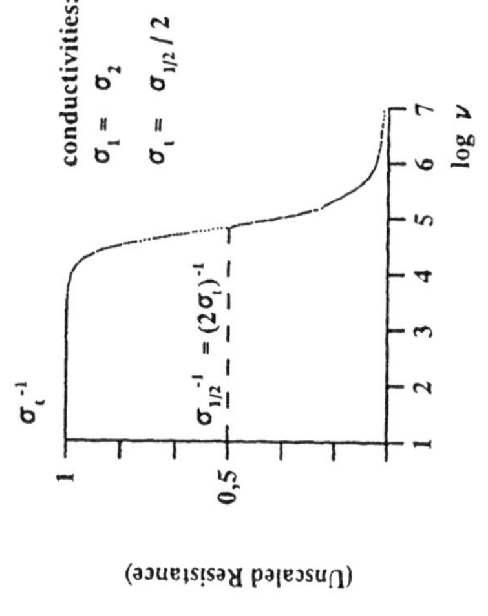

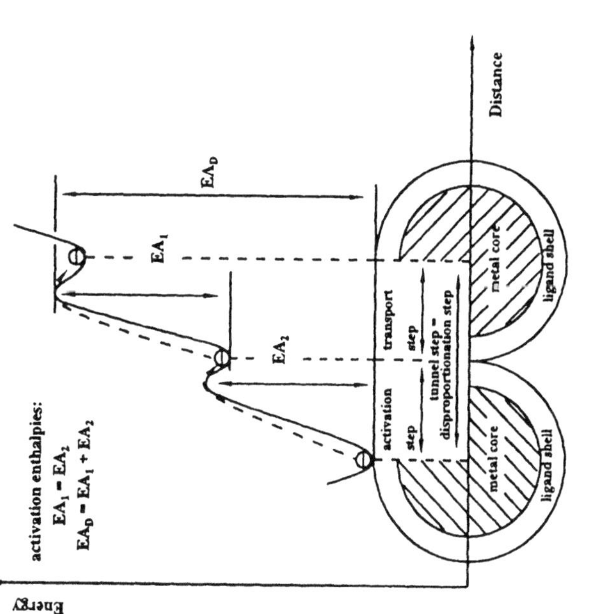

Figure 12-13   a) Illustration of the activation steps between two cluster molecules.
b) The conductivity σ_t is doubled under resonance conditions $\sigma_1 = \sigma_2$.

The question, how the frequency of about 60 kHz, applied to the macroscopic sample, couples with the frequency of the metallic electrons in single clusters (they must be different by orders of magnitudes) seems to find a simple answer: it is found that $\nu_{Dmacro} = 2 \ |x| \ \nu_{Dmicro}$ (D = Debye). This means that just the dimension of the quantum box couples the macroscopic and the microscopic frequency. Following this relation, the microscopic frequency results as 10^{13}–10^{14} Hz, a realistic value. So, if ν_{Dmacro} is increased by a factor of 2, the Cole-Cole frequency ν_C is reached and intercluster tunnelling processes can start. Figure 12-14 helps to explain these fundamental findings. If the energy of an excited single-electron exceeds the Coulomb barrier, it cannot be localized in the quantum box further and it will pass through the ligand shells into neighbouring clusters.

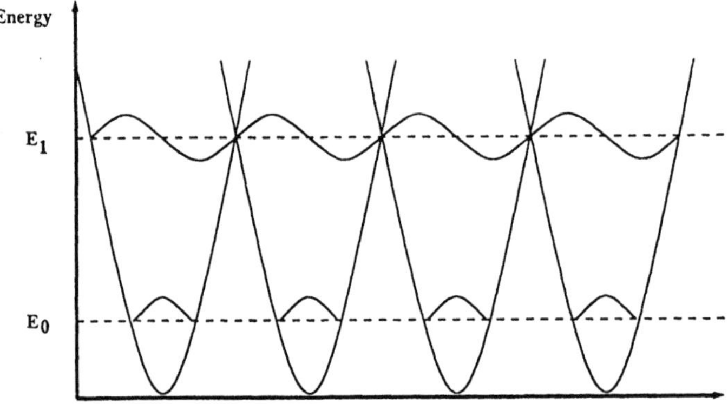

Figure 12-14 Elucidation of the electron transition from the ground state to an excited state in a one-dimensional array of ligand stabilized clusters.

12.4 Conclusions

The synthetic potentials in cluster chemistry open the door to novel materials which could play a decisive role in the future of quantum devices. Ligand stabilized transition metal particles in the range between 1 and 30 nm represent the transition between the bulk and the molecular state. The physical properties of these clusters indicate that already colloids in the upper part of this range have left the perfect bulk state. On the other hand, even 1.4 nm M_{55} clusters still show enough free spin density to belong to the family of metals, although they range at the very end of the transition from bulk metal to covalent cluster molecules. The world of microelectronics would be revolutionized if clusters could be handled as quantum boxes, making single-electron transitions possible in practice. Compared with established techniques, the application of 1 nm clusters in microelectronics would lead to an increase of the capability by many orders of magnitude. First results seem to support these ideas and look very promising. For the first time, chemically synthesized cluster molecules could be used for single-electron tunnelling processes. It will need still great efforts to realize these goals; however, they are within reach.

13 Physical Properties of High-nuclearity Metal Cluster Compounds: Model Systems for Uniform-sized Metal Particles

L. Jos de Jongh

13.1 Introduction

Polynuclear metal cluster compounds form an interesting class of systems that, until recently, had been scarcely studied by the physics community. They consist of macromolecules, each macromolecule being composed of a "core" of a certain number (n) of metal atoms, surrounded by a "shell" of ligands. For an introduction to the chemistry of these materials, we refer to the pertinent article by G. Schmid (see p. 149-162 in this volume). An example of a large nickel carbonyl cluster is given in Figure 13-1. The macromolecule can be an ion or a neutral molecule. The important point is that, since we are dealing with chemical compounds, the macromolecules in a given compound are identical. Consequently, *the solid formed can be seen as a macroscopically large assembly of identical metal particles, embedded in a dielectric matrix.* A major problem with other physical or chemical methods employed so far to obtain metal clusters, e.g. by condensation in atomic beams, by deposition in rare gas matrices, or in colloids and catalysts, is that one is plagued by a large distribution in particle size. This is a clear drawback if one wants to make a systematic investigation of physical properties as a function of cluster size, since then a variable but homogeneous particle size is an absolute necessity. In the metal cluster compounds, this goal can be realized.

At present, there exist already a few hundred metal cluster compounds, with metal atoms of almost every transition element in the periodic table (Fe, Co, Ni, Cu, Mo, Ru, Rh, Os, Ir, Pt, Au, Ag, Pd, etc.). In particular, the metal carbonyl clusters form a very rich class of cluster compounds [1]. For the same metal, the number n of metal atoms can be varied by considering different cluster compounds. Our group of physicists at Leiden is collaborating with a number of chemist colleagues who synthesize the compounds, notably with G. Schmid (Essen), D. Fenske (Karlsruhe), A. Ceriotti (Milano) and G. Longoni (Bologna). In the last few years, the maximum number n_{max} of atoms in the metal core has been greatly increased. About ten years ago, $n = 55$ was realized in the "Schmid-clusters" [2] of general formula $M_{55}L_{12}Cl_x$, where the metal atom M is Au, Pt, Rh, Ru or Co, the ligand L is an organic molecule (e.g. PPh_3, PMe_3), and some of the metal atoms are ligated by chlorine. The structure of the "magic number" clusters, of which the Au_{55} cluster core is an example, is shown in Figure 13-2. Subsequently, G. Schmid and coworkers [3] succeeded in synthesizing the five-shell palladium cluster $Pd_{561}Phen_{36}O_{100}$, a macromolecule which contains a metal core of 561 palladium atoms, and afterwards [4] the four-shell cluster $Pt_{309}Phen_{36}O_{30}$. These cluster molecules are in fact so large that with X-rays the structure of the metal core itself can already be seen separately! Recently, a palladium cluster compound consisting of a 50:50-mixture of 7- and 8-shell palladium clusters was synthesized by G. Schmid. The size-evolution of these multi-shell, cuboctahedral cluster cores is illustrated in Figure 13-2. The conclusion appears to be justified that we are seeing an evolution towards larger and larger metal cluster compounds in the near future. In fact, there is an obvious extension to metal colloids (for which the number of metal atoms per particle is typically in the range $n \geq 10^3$), and the chemists mentioned use similar techniques of ligand-stabilization to obtain a homogeneity of particle size in metal colloids that approaches that of the metal cluster compounds. For more information on the chemistry, we refer to the article by G. Schmid, and to a book that appeared recently [5].

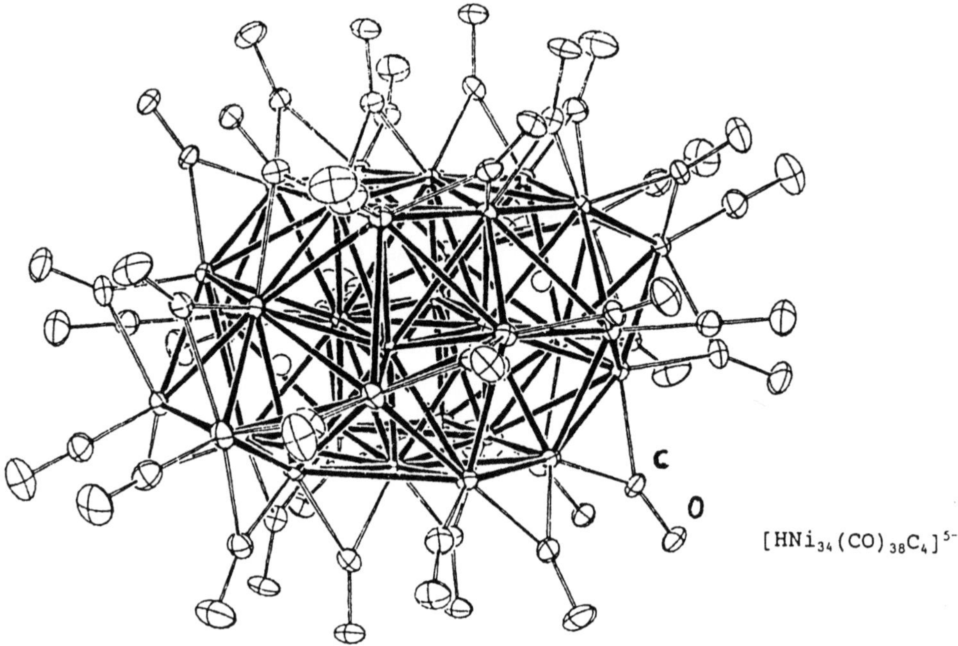

Figure 13-1   An example of a high-nuclearity nickel carbonyl metal cluster macromolecule. The nickel framework is indicated by the heavy lines [1].

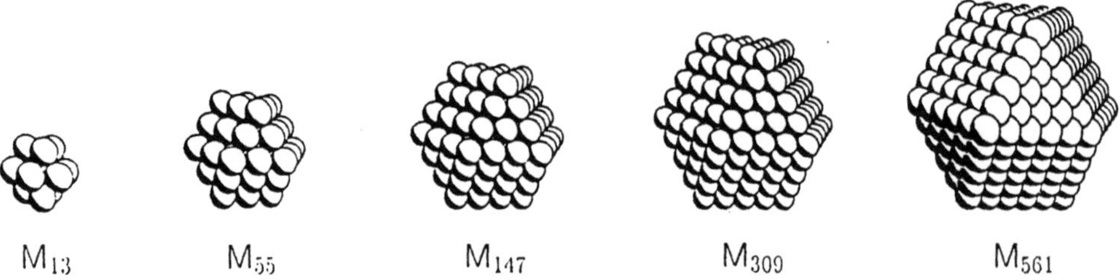

Figure 13-2   Magic number clusters manganese obtained by surrounding a given atom by successive shells of atoms (the illustration shows cuboctahedral packing).

The typical size of these clusters (diameters of 1–10 nm) puts them clearly into the class of *mesoscopic systems*. The metal cluster molecules may in principle also be exploited to obtain *nanocrystalline metals* by taking away the ligand shells surrounding the metal cores, e.g. by electrochemical methods. The catalytic properties of the bare and ligated cluster molecules are obviously of great interest. They are investigated by chemists we collaborate with.

13.2 Fundamental Physical Questions

For us as physicists, these materials form a rich playground, by means of which a substantial number of interesting and fundamental physical problems may be studied. We mention the following examples:

- **Quantum size effects**
  For electrons with energies comparable to Fermi energies (E_F) in metals, a nanometer-sized metal cluster is a quantum well, since the De Broglie wavelength of the electron is comparable to the cluster size. Thus, the electronic energy-level structure for the metal cluster will be discrete (quantum size effect), implying that the physical properties will differ from those of the bulk metal, as long as the thermal energy k_BT is smaller than, or comparable to, the distances δE between energy levels. Furthermore, the statistics of the energy level distribution in conglomerates of mesoscopic particles is an interesting problem in itself. The relationship to statistical theories for nuclei (the random matrices theories advanced by Wigner [6] and Dyson [7]) was pointed out long ago by Gorkov and Eliashberg [8], and continues to be a subject of study [9]. Recently, the connection to quantum chaos (nonintegrable systems; quantum billiards) has become apparent [10], as well as the relation to the t-J Hubbard model and the Sigma-model [11].

- **Size-induced transitions from molecular to bulk-metal behaviour**
  A metal cluster of a few atoms is basically a molecule with widely spaced electronic energy levels. For the bulk metal, the spectrum is a (quasi) continuum. Clearly, with increasing cluster size, the properties should change at some point from molecular to bulk-like. The question at what size such a transition will occur has intrigued physicists involved in cluster science since the very beginning of the field. As we shall explain below, it depends on the criterion accepted for "metallic" behaviour, and thus on the thermodynamic function considered, as well as on the temperature region involved.

- **Cluster molecules as building blocks for nanostructures**
  By packing cluster molecules into solids (as occurs naturally by chemical synthesis of the metal cluster compounds), one creates a three-dimensional (3d) array of quantum wells. Likewise, it should be possible to construct 1d quantum wires and 2d quantum sheets by packing the cluster molecules in the form of a chain or a monolayer on suitable substrates. By changing the size of the metal clusters, the energy-level structure in the quantum wells may be modified. By varying the size of the ligands which separate the metal cores, the degree of electrical insulation between metal cores may be changed. Such 1d-, 2d- or 3d arrays should give rise to novel (and in principle tuneable!) electronic, optical, and electrical properties.

- **Single-electron tunnelling and related effects**

  In sufficiently small (nanometer-sized) particles, the selfcapacity of each particle, as well as the capacity involved with interparticle charge transfer, are so large that charging/single-electron tunnelling effects become predominant [12]. Preliminary studies on metal cluster compounds have already evidenced such phenomena [13].

- **Doping metal cluster compounds to obtain new molecular conductors and superconductors**

  The electronic structure of a crystalline metal cluster compound will be qualitatively similar to that of solid C_{60}. As a consequence of intercluster charge transfer, the discrete, single molecular energy levels appropriate for the individual cluster molecule, will each broaden into a miniband. In the undoped systems, the minibands up to and including the HOMO-derived miniband will all be completely filled (full-shell structure). The widths of the minibands are determined by the intercluster charge transfer, and if the latter is not too large, the HOMO-and LUMO-derived minibands will not overlap, leaving a semiconductor energy gap (HOMO = Highest Occupied Molecular Orbital; LUMO = Lowest Occupied Molecular Orbital). We think this is a basic argument as to why (almost) all cluster solids are semiconductors. Using the same concept, however, one may argue that doping may produce either holes in the HOMO-derived miniband, or electrons in the LUMO-derived miniband (as in K_3C_{60}), yielding metallic behaviour (and in K_3C_{60} even superconductivity!). We expect that similar games as with K_xC_{60} may be played with the metal cluster compounds.

13.3 Some Results of the Physical Studies

In our group, the physical properties of the metal cluster compounds are studied by means of extensive series of (solid state) pulse-NMR experiments, Mössbauer Effect Spectroscopy (MES), specific heat experiments, magnetic measurements, electrical conductivity, and dielectric measurements. Parallel to this work, photoelectron spectroscopy and EXAFS studies were performed by partners in a European collaboration. Since several review papers have already appeared [14–17], we shall mention here only some of the most important results obtained. The majority of the experiments were carried out on the Au_{55}, Pt_{309}, and Pd_{561} "Schmid"-clusters (we use this short-hand notation for the full chemical formulae). A major reason for studying these materials is that they are among the largest metal clusters presently available, and, in addition, are not air sensitive.

One of the most important questions which had to be addressed concerns the *influence of the ligand shell* on the properties of the metal atoms of the cluster core. Indeed, the same ligands which are so crucial for keeping the metal cores apart, will at the same time alter the properties of the metal core, as compared to those of a bare metal cluster of the same size and symmetry. Fortunately, we have made much progress in answering this question experimentally. Parallel to our efforts, Local Density Functional (LDF) calculations were performed by quantum theorists (N. Rösch, München; G. Pacchioni, Milano) who participate in our EC network. They have been able to calculate and compare the level structures and magnetic moments of bare *and* ligated nickel clusters, even as large as up to 44 nickel atoms and 48 carbonyl ligands [18]!

Both, the LDF calculations and our experiments, agree with the following picture. We may divide the metal atoms of a metal cluster core into surface metal atoms (to some of which the ligands are bonded) and inner-core metal atoms (in the interior). The metal-ligand interaction is found to have a strong effect indeed, but to be mainly limited to the surface metal atoms. The surface atoms of a ligated cluster are definitely nonmetallic, whereas the inner-core atoms constitute a minute piece of bulk metal, with strong quantum size effects in particular for small clusters, evolving towards bulk properties with increasing size. Thus, the Au_{55}, Pt_{309}, and Pd_{561} ligated cluster compounds provide us with arrays of Au_{13}, Pt_{147}, and Pd_{309} metal particles, respectively (the inner-cores, compare Figure 13-2). Obviously, the average level separations, δE, in the electronic spectra of these particles should still be relatively large. By rule of thumb, one may estimate δE by dividing the bulk metal Fermi energy E_F by half the total number of valence electrons in the cluster. For the 3d metals in question, there are about 10 valence electrons per atom and E_F is of the order of 5 eV, yielding $\delta E/k_B \approx 100$ K for a particle of 100 metal atoms.

The experimental evidence for the above conclusions was based on a combination of Mössbauer Effect Spectroscopy (on Au_{55} and Pt_{309}), NMR (on Pt_{309}), specific heat (on all three), and magnetic susceptibility (on Pd_{561}) measurements. The Mössbauer spectra clearly distinguish between surface and inner-core metal-atom sites. The Isomer Shift (IS) and Quadrupole Splitting (QS) parameters obtained from these spectra for the surface sites are very different from the bulk-metal values, being close to literature values for non-conducting metal salts [19–21]. For Au_{55}, the parameters for the 13 inner-core atoms are already close but not yet equal to the bulk values. Quite recently, we have been able to extend the MES experiments to Pt_{309}. A special trick was needed here, since there is no platinum isotope known that is suitable for MES. By irradiating the Pt_{309} cluster sample with neutrons, a small fraction of the Pt_{309} clusters (1 in 10^6) was transformed into $Pt_{308}Au$, where the gold nucleus could be used as a Mössbauer source (for the absorber, gold foil was taken). The most important result of this study [22] was the finding that the MES parameters for the inner-core sites were undistinguishable from the corresponding values known for the bulk metal. Since the IS parameter is a measure of the 6s electronic charge density seen by the Mössbauer nucleus, this leads to the conclusion that, at the inner-core metal-sites, the 6s charge density is already extremely close to the bulk value. In other words, as regards this particular property, a Pt_{147} cluster core shows bulk-metal behaviour.

As mentioned, however, different criteria for "metallic" behaviour can be considered, depending on the physical property of interest. As mentioned, the Mössbauer IS depends on the total 6s charge density seen by the nucleus (the integrated 6s density of states), and will therefore be relatively insensitive to quantum gaps in the electronic energy-level spectrum, induced by the quantum size effects. On the other hand, physical properties like the electronic contributions to the specific heat and susceptibility, the NMR Knight shift (K_s) and nuclear spin-lattice relaxation time (T_1), do depend on the Density Of States (DOS) close to E_F. For such quantities, therefore, the criterion for "metallic" behaviour (as juxtaposed to molecular) is obviously the presence of a quasi-continuous DOS close to E_F. Such a quasi-continuum will be reached as soon as the thermal energy $k_B T$ is larger than the level-spacings δE close to E_F, in other words, for large enough clusters and/or for high enough temperatures. (It should be noted that level statistics may also produce density fluctuations.)

The experimental NMR spectra [23, 24] for the Pt_{309} cluster could be analyzed in terms of separate contributions from surface atoms and inner-core sites. The surface contribution was definitely nonmetallic, whereas the inner-core contribution showed metallic properties from room temperature down to about 80 K. That is to say, the NMR signal for this contribution showed a considerable Knight shift, and the T_1 followed the temperature dependence given by the Korringa law ($T_1 \propto T^{-1}$), known to be well-obeyed in bulk metals. Interestingly, below about 50 K, these metallic properties were also lost for these inner-core sites. Analysis of the observed behaviour in terms of quantum size effects gave satisfactory results, with a value $\delta E/k_B = 50$ K, in reasonable agreement with the rough estimate for δE mentioned above.

We think that the combination of Mössbauer (MES) and NMR studies, which we have briefly summarized, constitute the first unequivocal experimental evidence that for transition-metal clusters the transition from molecular to bulk metal behaviour occurs in the size region of about 100 atoms. The NMR study on Pt_{309} shows in addition such a metal-non-metal transition as a function of temperature, since above 50–80 K metallic-like properties are seen, whereas below these temperatures the appearance of quantum gaps is found.

Several problems remain to be answered, however. One of these stems from the fact that the low-temperature specific heat data [25, 26] on both, the Pt_{309} and the Pd_{561} clusters, gave evidence for a linear term in the specific heat in the range below 0.5 K, while for Au_{55} such a behaviour was not observed. It seemed a logic conclusion to interpret this linear term as an electronic contribution. For bulk metals, it is well-known that the (quasi-) continuous DOS around E_F yields an electronic specific heat, $C_e = \gamma T$, for $T \ll E_F/k_B$, with the coefficient γ proportional to the DOS at E_F. The NMR evidence for quantum gaps of the order of 50 K in the energy spectrum, however, prohibits such a simple explanation of the specific heat in terms of bulk behaviour with a reduced DOS (the experimental γ values are about 1/3 of the bulk values). In our publications so far [24, 26] an explanation in terms of a random distribution of the size of the HOMO-LUMO gap throughout the metal cluster compounds has been proposed. The point is that such a distribution of level-separations is mathematically described by the well-known Poisson distribution function, which is the only distribution that yields a finite probability for the occurrence of (near) zero-level separations (the *average* separation may still be substantial). All other distribution functions known in level statistics (orthogonal, unitary, symplectic) reduce to zero in this limit, see e.g. [27, 28]. Although the assumption of a Poisson distribution could explain satisfactorily both the specific heat and the NMR data (even quantitatively), it shifts the problem to the question as to why such a random distribution in the HOMO-LUMO gap would be present in these cluster solids. A tentative explanation that we have suggested is based on the random packing (like in a glass) of the cluster molecules in these Schmid-compounds. This randomness would lead to randomness in the broadening of the molecular energy levels by the intermolecular charge transfer, and thus to a random value of the HOMO-LUMO derived gap around E_F.

In contrast to NMR and MES, physical quantities like the experimental electronic specific heat and susceptibility are not site-specific, being averages over the contributions from all different sites. Nevertheless, these quantities are proportional to the average electronic DOS of the cluster, and thus should differ markedly from bulk values when, for instance, the surface metal atoms have a strongly reduced DOS (the fraction of surface atoms is of course quite large in a small particle). Such a strong reduction is in fact predicted to result from the metal-ligand charge transfer in ligated metal clusters by the LDF calculations mentioned above. Our experimental specific heat [25, 26] and magnetic susceptibility data [29] for the Pt_{309} and Pd_{561} metal cluster compounds indicate a DOS at E_F of about one third of the known bulk values. The ratio of surface to inner-core metal atoms is roughly 50/50 in these clusters. Thus if we assume a zero DOS at E_F for the surface atoms, and a slightly reduced DOS for the inner-core atoms, the experimental results can be accounted for.

For the nickel carbonyl clusters, the reduction of the DOS at E_F induced by the ligand-metal interaction, is predicted by the LDF calculations to be accompanied by a complete disappearance of the magnetic moment of the nickel atoms at the surface of the cluster. Since bare nickel clusters are calculated to have magnetic moments per atom comparable to the bulk value (0.6 μ_B/atom), the ligation of a bare cluster should lead to a very strong reduction of the magnetic moment of the cluster as a whole. Indeed, our previous magnetic measurements have shown only quite weak magnetic moments even for very large nickel carbonyl clusters. An extensive magnetic study of a single-crystal sample of a large nickel carbonyl cluster ($Ni_{38}Pt_6$) has now been performed. It confirms the complete quenching of the magnetic moments of the surface atoms due to the ligands, as predicted by LDF theory. This study has recently been submitted for publication.

Other magnetic studies which we performed were concerned with a series of palladium clusters and colloids. Also in this case, evidence for strong reduction of the average DOS at E_F could be obtained. Although bulk palladium is nonmagnetic, it is a strongly Stoner-enhanced paramagnetic metal, and its magnetic susceptibility at low temperatures is quite high compared to the Pauli-spin susceptibility of usual metals. In our experiments, we found large, size-dependent reductions of the susceptibility with respect to the bulk, which could be explained in terms of a combination of surface effects and a (size-dependent) reduction of the DOS (and thus of the Stoner-enhancement factor).

We end this brief report by noting that a book, devoted to the physical and chemical studies on metal cluster compounds performed in the course of our collaborative efforts together with our chemist colleagues, has been prepared and is scheduled to appear in June 1994 [30].

Acknowledgements

The research at Leiden summarized here has been performed mainly by D. van der Putten, D. A. van Leeuwen, J. Baak and F. M. Mulder, under the supervision of H. B. Brom, R. C. Thiel and J. M. van Ruitenbeek. The work is supported in part by the "Stichting voor Fundamenteel Onderzoek der Materie" (FOM), which is sponsored by the "Nederlandse Organisatie voor Wetenschappelijk Onderzoek" (NWO). The financial support of the Commission of the European Communities under contracts ST2J-0084 and SCI-CT92-0788 is also gratefully acknowledged.

14 Towards Nanoscale Molecular Magnetic Materials

Dante Gatteschi, Andrea Caneschi, Luca Pardi, and Roberta Sessoli

Abstract

The magnetic properties of large metal ion clusters are reviewed. These compounds provide a chemist's way to nanoscale magnetic materials. The synthetic strategies used to produce large spin clusters are discussed, together with theoretical approaches to the interpretation of their magnetic properties. The principal classes of large clusters reported so far, polyoxo-vanadates(IV), poly-iron and poly-manganese oxo complexes, are reviewed.

14.1 Introduction

In a conference dedicated to complexity, large magnetic clusters must necessarily find some space. In fact, these materials are characterized by a structural complexity, analogous

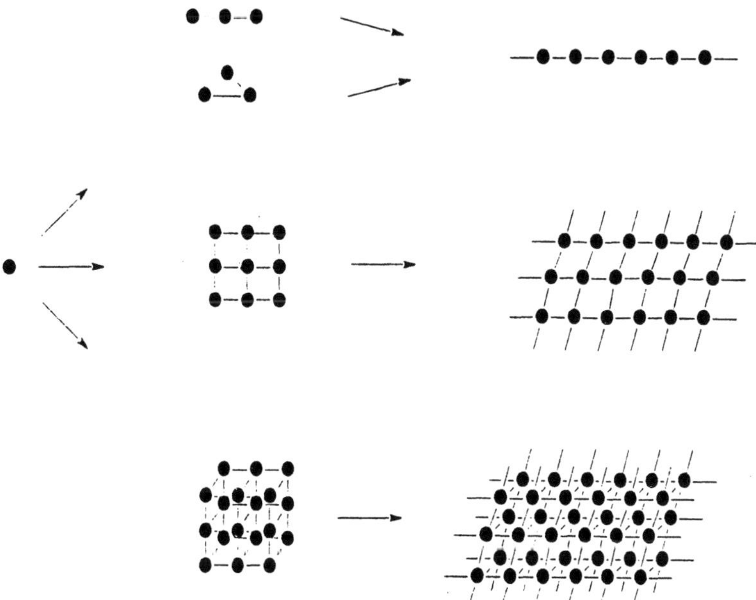

Figure 14-1 The ideal process of building up an infinite array of coupled spins. Clusters which are pseudo-one, -two-, and -three-dimensional are shown.

for instance to that met in large metal clusters, also reported in these proceedings. Further, it is possible to associate to each magnetic centre a spin, and the assembly of spins also forms a complex structure. For the same geometrical structure, the spins introduce additional complications associated with the different ways in which they orient themselves relative to each other. Finally, it will be important to understand not only the static organization of the spins, but also their dynamics. Spins by themselves are ideal objects which have long been investigated without necessarily attaching them to real objects. Nowadays, it is possible to have them available and to investigate the magnetic properties of the clusters in order to have first hand information on the preferred spin orientation and on their dynamics.

The first difficulty to be met in the investigation of large magnetic clusters is that of synthesizing the systems themselves. There are essentially two ways of producing them. The first ideally breaks an infinite lattice into smaller pieces ending with particles of the dimensions of a few nanometers (top-down approach). The second uses an opposite approach, assembling together small pieces to end up at the same result (bottom-up approach). An ideal scheme for the latter approach is outlined in Figure 14-1. Complex structures similar to fragments of one-, two-, and three-dimensional arrays can thus be realized.

The second difficulty is associated with the interpretation of the magnetic properties of large spin clusters. In fact, the properties of simple paramagnets, in which the individual centres are essentially uncoupled, are well understood using a treatment which is the extension to magnetic particles of that appropriate to ideal gases [1]. On the other hand, the cooperative behaviour of infinite assemblies of spins, leading to ferromagnetism, antiferromagnetism, ferrimagnetism, weak ferromagnetism, etc., can be reasonably understood taking advantage of the translational symmetries of the lattices [2]. Finite, but large spin clusters within which the individual spins are coupled represent a challenging problem, because the number of energy levels rapidly increases with the number of spins, and it is not possible to use translational symmetry in order to reduce the size of the

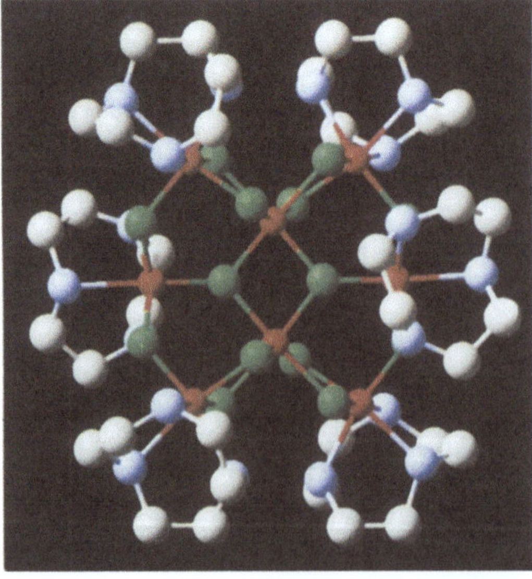

Figure 14-2   Structure of the octanuclear iron(III) cluster Fe$_8$.
Iron atoms are drawn in red, oxygens in green, carbons in white, and nitrogen atoms in pale blue.

matrices for calculating their diagonalization. As an example, let us consider a cluster as the one [3] depicted in Figure 14-2 which comprises eight iron(III) ions, $S = 5/2$. The total number of states is $(2S+1)^8 = 1679616$. Therefore in principle, one should calculate and diagonalize a matrix with these dimensions. In Section 3, we will indicate how it is possible to use symmetry arguments in order to reduce the dimensions of the matrices. Nevertheless, it is still necessary to compute 81 matrices, with dimensions ranging from 1 to 4155.

If things are that complicated, why then do we become interested in large spin clusters? For the same reason which gave impetus to focusing on nanoscale electronics: Matter at the mesoscopic scale is a brand new world, where novel properties can be expected. Some of the properties which are worth investigating at the nanometric scale are giant magneto-resistance, magneto-caloric effects, and optical properties. When a conductor is exposed to an external magnetic field, its resistance changes due to the interaction with the conduction electrons. This effect is exploited by magnetic heads to read information stored on computer hard disks. Typically, the variation of resistance is by 1 or 2% in a weak magnetic field. Now it has been discovered that nanoscale magnetic particles dispersed in non-magnetic metallic matrices show much larger variations in resistance (Giant Magneto Resistance, GMR) ranging from 8 to 25–30%, in fields of ca. 1 T [4]. If it will prove possible to reduce the field to observe GMR, a new generation of information storage systems might develop.

The magneto-caloric effect is the cooling effect which is observed when a magnetically disordered system is saturated in an external field and then the field is removed. When the field is applied, the system orders magnetically, losing entropy. When the field is removed, the system cools. This effect has been long used for cooling samples at very low temperatures. It has now been found that large spin clusters may have optimal magneto-caloric effects at higher temperatures, requiring smaller fields. In particular, a nanostructured $Gd_3Ga_{3.25}Fe_{1.75}O_{12}$ which is prepared by solution chemistry has been found to yield a magneto-caloric effect which is three to four times better at 15 K and 1 T than the best material available up until now [5].

The optical properties of materials also show dramatic changes as a function of particle size. For instance $\gamma\text{-}Fe_2O_3$ particles are usually opaque, and this is unfortunate, because if light could pass through the magnetized particles new opportunities would open up for magneto-optical applications. Now it has been found that $\gamma\text{-}Fe_2O_3$ particles of nanometric size embedded in a matrix of an ion-exchange resin are superparamagnetic and optically transparent [6].

Beyond the excitement for possible applications, the area of nanometric scale magnetic particles is of large theoretical interest because it will allow to answer fundamental questions such as: where does bulk behaviour begin in the ideal process of assembling individual particles, be they atoms or molecules? When a bulk magnet is progressively reduced in size, it becomes a superparamagnet, where the barrier to reorientation of magnetization is comparable to thermal energy. Now in the reverse process of growing larger and larger particles, when does superparamagnetic behaviour begin to be observed? Which shape effects of the particles are to be expected? All these questions are not only philosophical, but they may have important consequences for our understanding of the properties of metalloproteins, like ferritin, which contains superparamagnetic particles of iron oxides and which assembles them adding one iron ion after the other [7–9]. Other biological examples of large metal ion particles are present in magnetotactic bacteria, which use superparamagnetic magnetite particles in order to optimize their feeding opportunities [10]. Finally, similar particles are present in the brain of homing pigeons. They are expected to be associated with their unique ability to orient themselves.

From a more fundamental point of view, small magnetic particles may be associated with the problem of the Copenhagen interpretation of quantum mechanics [11]. The key proposition for this interpretation is that the language of classical physics is adequate for macroscopic objects independent of the quantum origins of the macroscopic properties. Therefore, macroscopic objects are either in one place or in the other and they do not tunnel between two different states.

Now, the problem is that of defining a macroscopic object. Mathematically we might say that an object is macroscopic if it contains a number N of particles, with N tending to infinity. However, we know that all real objects have a large but finite number of particles. Of particular interest are nanoscale materials, which contain 10^3–10^4 particles. Compared to genuine quantum objects they are macroscopic, but they are not yet large enough to grant classical bulk behaviour. Therefore, in these objects one can expect to observe quantum interference effects. The lack of observation of these effects might be an indication that quantum mechanics is not a complete and universal theory of the physical world and that it is not possible to extrapolate its assertions to macroscopic objects.

Quantum interference effects are observed in macroscopic objects like superconducting devices based on the Josephson effect for instance. However, in the last few years, it has been recognized that also the magnetization of large spin clusters can show quantum effects like tunnelling between two equivalent wells [12]. Therefore, the controlled synthesis of large magnetic molecules may provide new important insights in this field, too.

Having illustrated the reasons for interest in nanoscale magnetic particles, we now wish to describe the classes of compounds which have been used in the last few years in order to obtain nanoscale magnetic molecular materials using relative synthetic strategies. We will concentrate on some of those for which the magnetic properties have been intensively investigated.

We will completely neglect the metal particles which have been the focus of keen interest and detailed reports [13, 14]. Instead, we will concentrate on those systems which provide molecular objects, i.e. systems containing discrete molecules, even if they may be very large, which can be crystallized and can be investigated with the usual techniques of molecular chemistry.

The outline of the paper will be the following: in Section 2, we will review the synthetic strategies to obtain nanoscale magnetic molecular materials, both organic and inorganic; in Section 3, we will provide a cursory introduction to the theory of magnetic properties of large spin clusters, in particular focusing on the transition from simple paramagnetic to bulk magnetic behaviour; in Sections 4–6, we will describe the magnetic properties of various classes of compounds; and finally, in Section 7, we will draw the conclusions concerning the current state of the art and future developments.

14.2 Synthetic Strategies to Nanoscale Magnetic Molecular Materials

Currently, there are essentially two approaches to obtain nanoscale molecular magnetic materials, an inorganic one, where metal ions are assembled – with or without the help of organic ligands – to form large molecules, and a purely organic one, where the magnetic bricks which can be used for building up the magnetic structure are organic radicals.

14.2.1   Organic Clusters

One important class of compounds is that of polycarbenes, synthesized by the groups of K. Itoh and H. Iwamura in Japan [15,16]. These compounds are rather unstable and must be obtained by photochemical techniques insulating them in matrices, at liquid nitrogen temperature. They have been assembled in larger and larger clusters. The maximum number reached so far is nine [17] and the corresponding nona-carbene is shown in Figure 14-3.

Each carbene has two unpaired electrons in a ground triplet state, and the coupling between the individual carbenes is ferromagnetic. As a result, an N-carbene has a ground N state. EPR and magnetic data do not provide any evidence of thermally populated excited states. The magnetic properties are those of a simple paramagnet, with a ground multiplet which is split by the dipolar interactions between the various centres. Therefore, the dimensions must be enlarged before new types of magnetic behaviours may be expected.

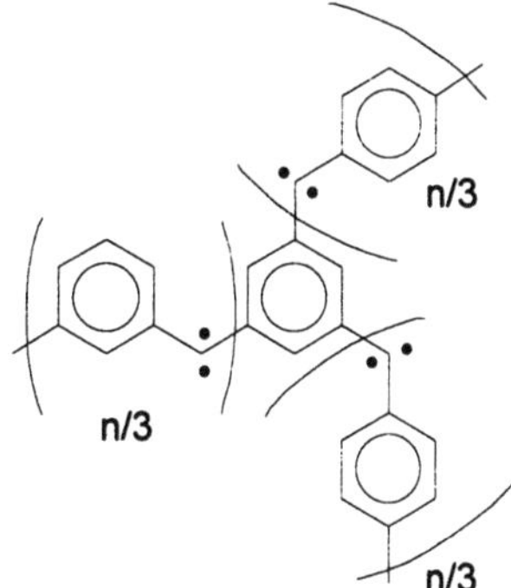

Figure 14-3   Scheme of the structure of a nona-carbene.

There are also other classes of organic polyradicals which have been investigated. Of particular interest are those which can be classified as dendrimers [18] (see also Chapter 10 by G. R. Newkome and C. N. Moorefield in this book). Dendrimers can be assembled in successive synthetic approaches according to the general scheme sketched in Figure 14-4.

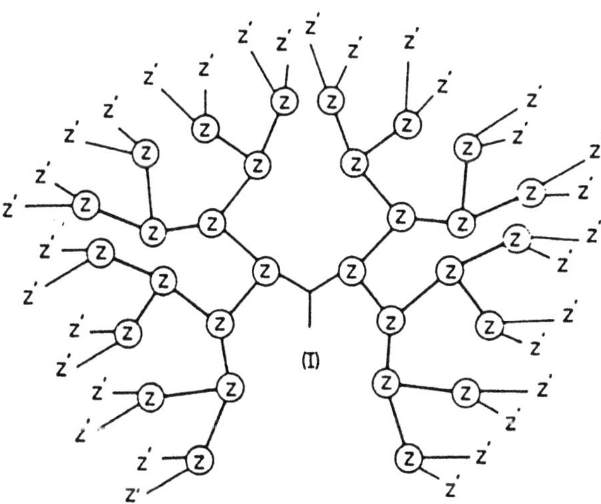

Figure 14-4   Scheme of the formation of dendrimers.

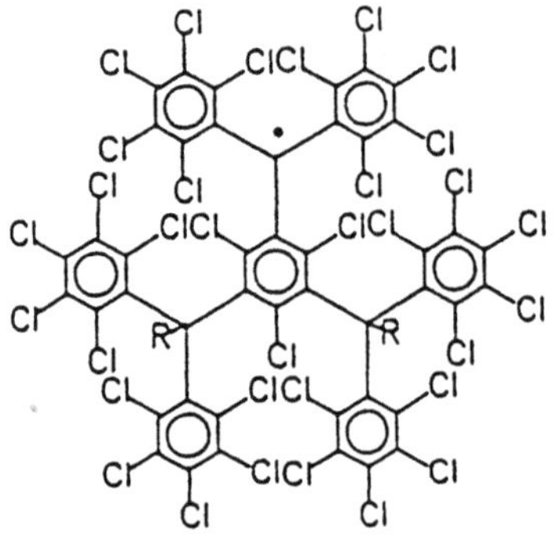

Figure 14-5   Structure of 2,4,6-trichloro-
$\alpha,\alpha,\alpha',\alpha',\alpha'',\alpha''$-hexakis(pentachlorophenyl)mesitylene.

When one wants to assemble magnetic dendrimers, it is necessary to choose individual bricks which are paramagnetic. One example is provided by 2,4,6-trichloro-$\alpha,\alpha,\alpha',\alpha',\alpha'',\alpha''$-hexakis(pentachlorophenyl)mesitylene [19] whose structure is shown in Figure 14-5.

In this case, the particles are still rather small, and the maximum spin ground state so far observed is $S = 3/2$. However, this is a rapidly expanding field, and rapid improvements can be expected, similar to those which provided the first genuine examples of organic ferro-magnets.

14.2.2   Inorganic Clusters

At present, inorganic approaches have been more fruitful, and this is by no means unexpected, because transition metal ions have long been known to be able to form bulk magnets, and therefore they must also form magnetic particles under appropriate conditions.

In order to obtain large metal-ion clusters, oxo ligands are particularly well suited, given the tendency to form polymeric oxides and hydroxides. Clusters containing many metal ions have long been known in the form of both, polycations and polyanions [20], which are related by the equilibria of the reactions shown in Scheme 14-1.

$$\text{aquoions} \rightleftarrows \text{polyoxo-hydroxo cations} \rightleftarrows$$
$$\rightleftarrows \text{oxides-hydroxides} \rightleftarrows \text{polyoxo-hydroxo anions} \rightleftarrows$$
$$\rightleftarrows \text{oxo-hydroxo anions}$$

Scheme 14-1

In both cases, the metal ions must show a tendency to aggregate but they must also stop at some finite stage rather than continue to form infinite lattices. For polyanions, the stop of the tendency to grow is provided by the metal-oxygen groups connected by double bonds which stretch out of the growing particles. These groups are rather unreactive, because the oxygen with a double bond is largely electron deprived and much less basic than O^{2-}, therefore they form a protective shell for the growing particle which is stabilized this way. Polyoxometallates form preferentially with the d^0 ions vanadium(V), niobium(V), tantalum(V), and molybdenum(VI), which easily form strong $M = O$ bonds and are large enough to give complexes with coordination numbers larger than four. With tetrahedral complexes, it is difficult to achieve geometries of the polymetallate clusters which block the growth of the particles [20].

Although polyoxometallates of the ions indicated above are non-magnetic, it is possible to form clusters comprising magnetic ions by partial reduction. Further, they can give rise also to heteropolyoxometallates incorporating magnetic ions. However, the examples of large clusters comprising only magnetic ions were rather scarce up to a few years ago, the only notable exception being $[V_{18}O_{42}]^{12-}$, which comprises magnetic vanadium(IV) ions, $S = 1/2$. Matters have changed drastically in the last few years, mainly due to the efforts of A. Müller and coworkers in Bielefeld [21–26].

Polyoxocations can be easily stabilized by additional ligands, like carboxylates, alcolates, etc., which provide a hydrophobic environment for the growing particle, and therefore blocks its growth. Several different cations can give rise to such compounds. Those which have been more intensively investigated in the last few years are iron and manganese clusters. The reason is that both ions are involved in polynuclear aggregates in biological systems, like photosystem II, ferritin, etc., and much synthetic work has been devoted to the development of suitable inorganic models. However, iron and manganese do not conclude the list of suitable choices for a large cluster, and interesting examples are provided by copper(II), nickel(II), cobalt(II), etc.

Transition-metal complexes with sulphur ligands have long been studied in an effort to model the active site in non-heme proteins [27] involved for instance in photosynthesis, mitochondrial respiration and nitrogen fixation. Metal-ion clusters with 2 to 6 layers have been synthesized. Particularly well-known is the series of tetranuclear cubane models of the prosthetic group of ferredoxins.

A series of high nuclearity metal chalcogenides have also been prepared in which the structure and the solubility are enhanced using tertiary phosphines as terminal ligands [28–30]. Recently, nanometer scale clusters have been reported in which monocyclic and bicyclic structures are built up by assembling edge sharing Fe_2X_2 (X = S, Se) diamonds in which no terminal ligands are present [31–33]. In these systems, structural stabilization is provided by the Na^+ ions around which the cyclic structures are wrapped. The reaction between $FeCl_3$ and Li_2X in ethanol and in the presence of an appropriate promotor (sodium acetanilate NaPhNCOMe) gives rise – depending on the reaction conditions – to three

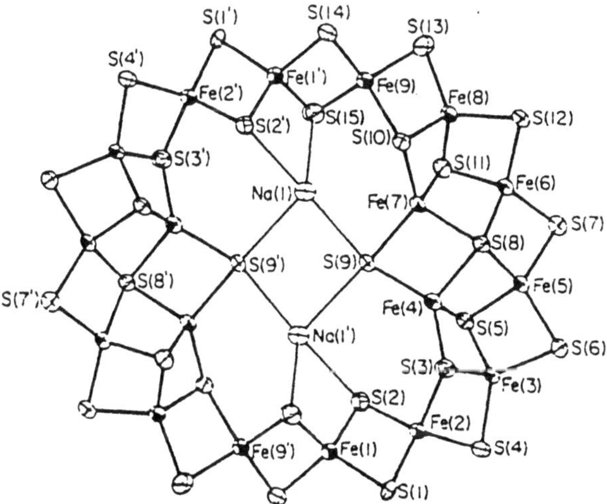

Figure 14-6   Structure of $[Na_2Fe_{18}S_{30}]^{8-}$.

different compounds, two of them containing two isomers of the $[Na_2Fe_{18}S_{30}]^{8-}$ anion, while the third contains the bicyclic $[Na_9Fe_{20}S_{38}]^{9-}$ anion. The presence of Na^+ ions in the reaction environment seems to be crucial. Substitution of Na^+ with other alkali ions gives rise to completely different final products. The presence of the inorganic polymeric species $(FeS_2)_n^{n-}$ as a precursor of the cyclic species has been spectroscopically detected. The template effect of Na^+ is in this case the driving force that leads the reaction to the final products, providing the electron poor centres to which the electron rich chalcogenide part of the $(FeS_2)_n^{n-}$ can bind.

From the structural point of view, the two $[Na_2Fe_{18}S_{30}]^{8-}$ systems exhibit the same topology, being two toroidal shaped clusters with lateral dimensions of 12×16 Å and a thickness of 3.3 Å. The structure of one of the two is shown in Figure 14-6. They are formulated as $Fe(III)_{14}Fe(II)_4$ with an average oxidation state of 2.78, confirmed by Mössbauer data. The magnetic data show that in these clusters there is a dominant antiferromagnetic coupling, leading to a low value of the effective magnetic moment at low temperature.

The bicyclic $[Na_9Fe_{20}S_{38}]^{9-}$ cluster has a prolate ellipsoidal shape with dimensions 11.2×17.4 Å. It can be formulated as $Fe(III)_{18}Fe(II)_2$. Mössbauer spectra and temperature dependence of the magnetic susceptibility are consistent with a singlet ground state as a result of an overall antiferromagnetic coupling.

14.3 Magnetic Properties of Large Spin Clusters

In a simple paramagnet, all spins in the lattice are independent of each other and they behave collectively as a gas. In a magnetically ordered solid, all spins in the lattice are strongly correlated to each other and respond collectively to external perturbations. In a ferromagnet, in which all spins tend to be parallel to each other, domains are spontaneously formed in the ordered phase in order to minimize the magnetic energy. A domain is an area

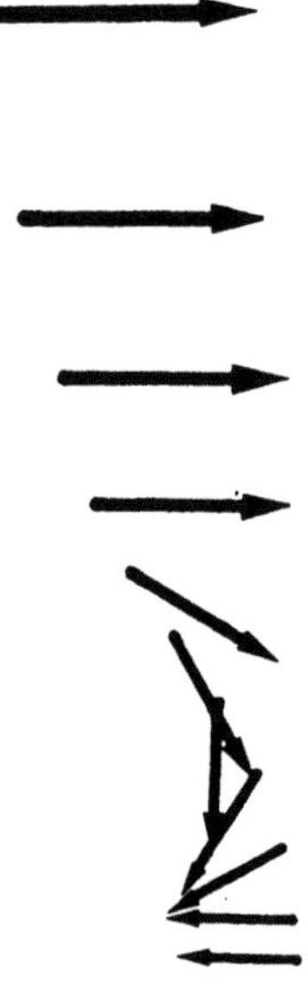

Figure 14-7   Bloch walls in a one-dimensional magnetic material.

in the lattice in which all spins have essentially the same direction. It is separated from other domains, in which the spins are still ordered, but oriented parallel to other directions, by areas in which the spin direction rapidly changes. A simple example, one-dimensional ferromagnetic domains, is shown in Figure 14-7.

The area in which the preferred direction rapidly changes is called a Bloch wall. Energy is minimized if the change from one orientation to the other does not occur abruptly but is distributed over a number of intermediate spins. The size of the Bloch walls depends on the magnetic anisotropy of the system, i.e. on the energy barrier which the magnetization must overcome in order to change its orientation. The anisotropy depends on several factors, one being the volume of the system, to which it is directly proportional [2].

When the size of the system becomes smaller than those of the domains, the formation of several domains is no longer energetically favourable and the particle becomes a single domain. On further reducing the size of the particle, the barrier to reorientation becomes comparable to thermal energy, and the magnetization is no longer fixed in space, but fluctuates freely. As a result, the time average of the magnetization of a particle is zero, like in a paramagnet. The average time, τ, during which a particle is in a given orientation can be expressed as:

$$\tau = \tau_0 \exp\left(\frac{KV}{k_B T}\right) \tag{14-1}$$

where K is the anisotropy energy per unit volume, V is the volume of the particle, and τ_0 is the time of residence in a non-activated behaviour. A particle in these conditions is said to be superparamagnetic. A similar behaviour can be observed in antiferromagnets, too.

A particle will show a superparamagnetic or a regular collective behaviour depending on the time scale, τ_{exp}, of the technique used for investigating it. For instance, for susceptibility measurements τ_{exp} is ca. 10^{-2} sec, while for Mössbauer spectroscopy τ_{exp} is ca. 10^{-8} sec. Therefore, a particle may be normal ferromagnetic (antiferromagnetic) for the latter and superparamagnetic for the former.

Equation (14-1) is important, because it provides a means to determine the dimensions of a particle if τ is measured and K is known. This approach has largely been used for instance to determine the dimensions of the inorganic core of ferritin, the iron storage protein. Mammalian ferritin is a roughly spherical, 130 Å diameter protein which has a central cavity of ca. 70 Å in which up to 4500 iron ions can be stored as mixed oxo-hydroxides. This core is an antiferromagnet, but given its dimensions it shows superparamagnetic behaviour in the Mössbauer spectra [8].

Large molecular clusters are paramagnets in which however the individual spins do not behave independently but show cooperativity effects limited by the small size of the cluster itself. It can be expected that, when the size of the clusters becomes large enough, eventually the superparamagnetic region will be attained.

Clusters are complex paramagnets, and the interpretation of their magnetic properties is by no means simple. In fact, while the spin levels for a simple paramagnet are $(2S+1)$, where S is the appropriate spin value, they become $(2S+1)N$ for a cluster of N identical spins S, and the magnetic properties depend on the relative populations of all of them. In principle therefore, it is necessary to calculate all levels, in the presence of an external magnetic field, and then use them in order to calculate the magnetic properties.

Several approaches have been used up to now, ranging from brute force, which calculates all levels, using as bases simple product functions of the individual S_z eigenfunctions, to more sophisticated approaches which exploit symmetry as far as possible. For instance, we have used a formalism based on irreducible tensor operators, ITO, which exploits the symmetry of the total spin functions characterized by $S = \sum_i S_i$, and eventually allows the introduction of point group symmetry as well [34].

Let us take again into consideration the example of Figure 14-2. The states can be classified according to the total spin states S. In this way, the unique 1679616×1679616 matrix can be split into 21 matrices, corresponding to S values ranging from 0 to 20, as shown in Table 14-1. Some matrices are very simple, as for instance that for $S = 20$, which is 1×1. This is obvious since there is only one way of arranging all individual spins parallel to each other. However the order of the matrices, i.e. the different ways in which the individual spins can be arranged to give a value of S, rapidly increases with decreasing S, and for instance for $S = 5$ the order is 16576. In order to further reduce the order of the matrices, it was necessary to use the point group symmetry appropriate to the cluster. Using D_2 symmetry the dimensions of the matrices become as indicated in Table 14-1.

Table 14-1   Symmetry classification of the total spin states for Fe_8 in the D_2 group.

S	A	B_1	B_2	B_3	Total
20	1	0	0	0	1
19	2	1	2	2	7
18	10	6	6	6	28
17	22	18	22	22	84
16	60	50	50	50	210
15	118	108	118	118	462
14	243	225	224	224	916
13	419	401	420	420	1660
12	717	690	686	686	2779
11	1088	1061	1092	1092	4333
10	1614	1578	1568	1568	6328
9	2174	2138	2184	2184	8680
8	2841	2799	2780	2780	11200
7	3401	3359	3420	3420	13600
6	3927	3885	3854	3854	15520
5	4139	4097	4170	4170	16576
4	4155	4122	4076	4076	16429
3	3704	3671	3750	3750	14875
2	3019	3001	2940	2940	11900
1	1899	1881	1960	1960	7700
0	703	703	630	630	2666

The largest matrix is now only 4155×4155, and the problem may be tackled with a suitable work station.

The matrices are calculated in a parametric way, according to the Hamiltonian:

$$H = \sum_{ij} J_{ij}\, S_i \cdot S_j \qquad\qquad (14\text{-}2)$$

where S_i and S_j are two individual spins, and J_{ij} is a parameter describing the nature and the intensity of the interaction between the two spins. A positive sign of J_{ij} indicates antiferromagnetic coupling, while a negative sign is appropriate for ferromagnetic coupling.

A general computer program, CLUMAG, has been written which uses the above approach. With a RISC work station we can calculate rigorously the energy levels, and the magnetic properties, of clusters of 18 spins $S = 1/2$ or of 8 spins $S = 5/2$.

In Figure 14-8 we show the energy levels for Fe_8, calculated for a given set of exchange parameters related to the exchange interactions between the different iron(III) ions [35].

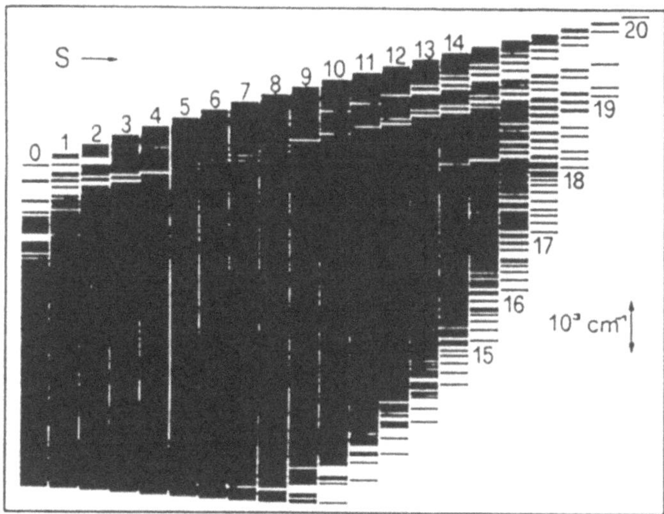

Figure 14-8   The calculated energy levels of Fe_8.

It is apparent that the levels tend to merge in a continuum, as must be expected on passing from isolated spins to increasing numbers of coupled spins. The same levels are shown in Figure 14-9 as a three-dimensional plot of the density of states. It is apparent that for intermediate energies and intermediate spins the densities of states become very high. However at the extremes of the energy range, the levels do show some gap between each other, showing that the system is not yet completely bulk in nature.

Another series of considerations is appropriate for Fe_8. In general, the ground state for a pair of coupled spins can be easily predicted on the basis of the sign of the exchange interaction connecting them: for antiferromagnetic coupling the ground spin state is $S = S_1 - S_2$, for ferromagnetic coupling it is $S = S_1 + S_2$. However, when we pass to higher numbers, the conclusions may not be as straightforward. In fact, if we put three spins on the vertices of an equilateral triangle and demand that they are all antiparallel to each other in pairs, we immediately see that this cannot be done. If we put two spins antiparallel to each other, the third will not be able to be antiparallel to both of them.

The spin is said to be frustrated, a term borrowed from psychology, because it is under the influence of two contradicting stimuli [36]. At a qualitative level, this means that it is not possible to predict by simple hand-waving arguments the nature of the ground state for such a type of configuration. In general, the ground state becomes largely degenerate, which means that there are several different ways of reaching configurations of minimum energy.

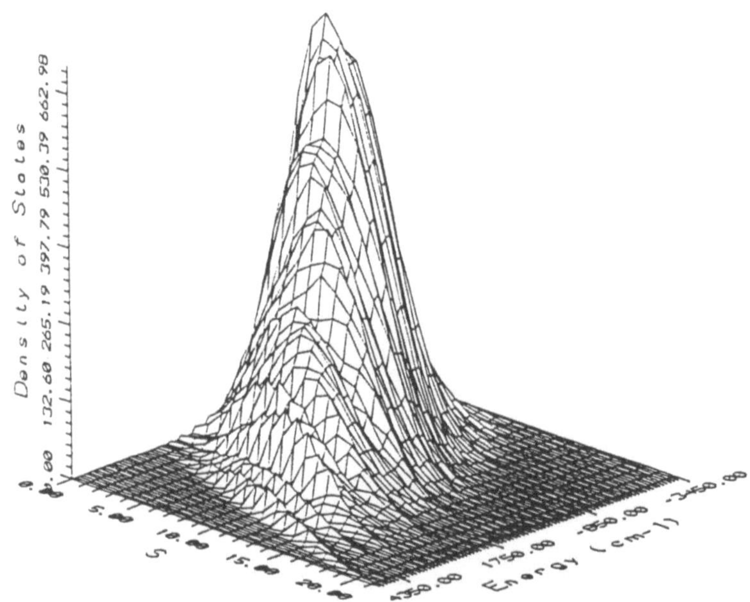

Figure 14-9   Calculated density of states of Fe_8.

For instance for a triangle of spin $S = 1/2$, the ground state corresponds to two degenerate levels with $S = 1/2$. Spin frustration has been largely studied in infinite lattices, where it has been found to be responsible for many interesting magnetic properties. It is also worthwhile to note that coupled spin systems can be viewed as an example for non-reductionism. Although it does not require any interaction in addition to those which are needed to justify the couplings of pairs of spins, the behaviour of three spins cannot be obtained as a simple sum of the properties of the pair plus the additional spin.

Fe_8 has many triangles in the topology of the magnetic interactions, as evidenced by the presence of μ_3-oxo bridges in the structure reported in Figure 14-2. Therefore, large spin frustration effects must be expected. As a consequence, assuming that the coupling between pairs of spins is antiferromagnetic, as generally observed in iron(III) dinuclear complexes, it will not be possible to predict the ground state. We will show below how in fact the ground state is not $S = 0$, as one could naively expect for an odd number of spins antiferro-magnetically coupled, but $S = 9$ or 10.

14.4 Polyoxovanadates(IV)

The magnetic properties of polyoxovanadates(IV) have already been extensively reviewed [37]. Therefore, we will not cover them here except to say a few words about using such systems as nanoscale magnetic materials. In fact in clusters like V_{18}, shown in Figure 14-10, which are the largest so far reported containing only magnetic ions, the dimensions, measured by the longest distance between two vanadium ions, are 7.84 Å, i.e. almost in the nanoscale region [38, 39].

The spins are antiferromagnetically coupled and they generally combine to produce a ground state with the smallest possible spin value ($S = 1/2$ for odd and $S = 0$ for even

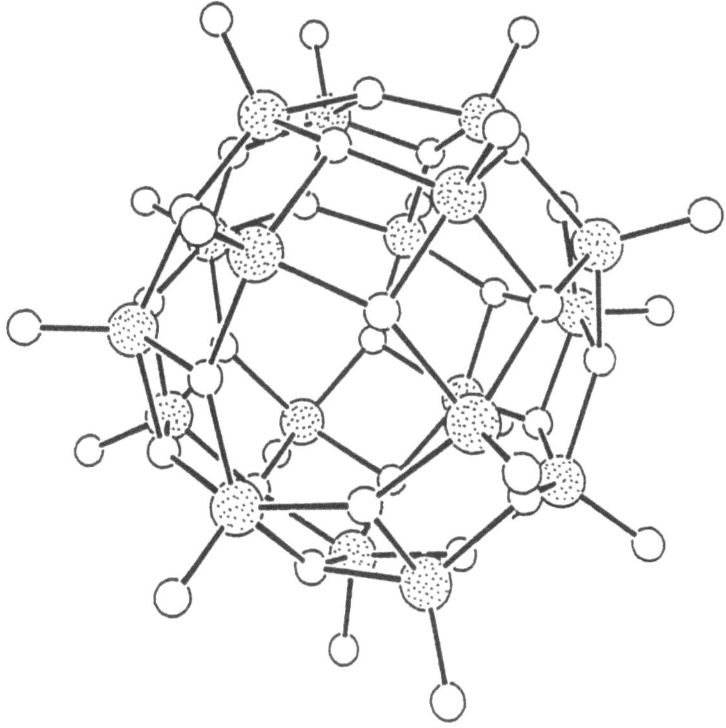

Figure 14-10   Structure of the V_{18} cluster (see 14.2.2 and [38]). The vanadium atoms are represented by dotted circles.

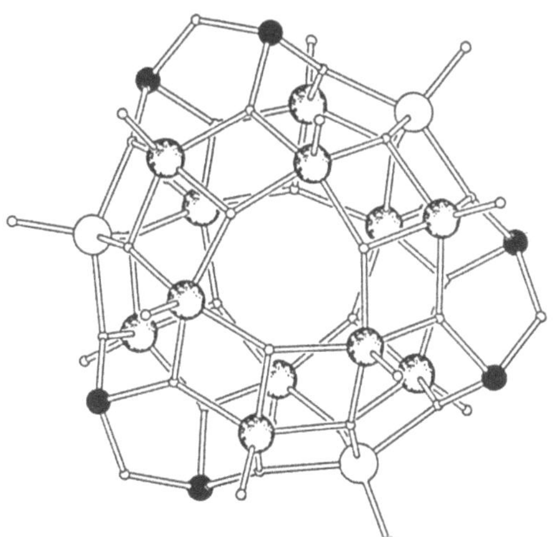

Figure 14-11   Structure of the V_{15} cluster (see text and [40]). The filled circles represent arsenic atoms, while small empty circles stand for oxygen atoms. The vanadium ions defining the six-membered rings are evidenced by the dots.

numbers of spins). Further, the ground state is in general separated by a sizeable gap from the excited levels, so that it seems that spin systems with ground state $S = 1/2$ require very large numbers before starting to show properties which are similar to those of bulk magnets. This is not surprising because $S = 1/2$ are the spins with the largest quantum nature, which decreases on increasing the value of S.

An interesting effect however was observed in $[V_{15}As_6O_{42}(H_2O)]^{6-}$, V_{15}, which has the structure shown in Figure 14-11. The vanadium ions are assembled in three different layers, of six, three, and six ions, respectively. The magnetic couplings within these layers are different from each other, therefore different magnetizations are observed in the layers at different temperatures. Therefore, this cluster can be considered as a model of magnetic multilayer structures [40].

14.5 Poly-Iron Oxo Complexes

Many new compounds have been reported recently in which iron ions, either in the valence state +2 or +3 are connected by different types of oxygen bridges (oxo, hydroxo, alkoxo, phenoxo). We will take into consideration clusters with $N \geq 8$.

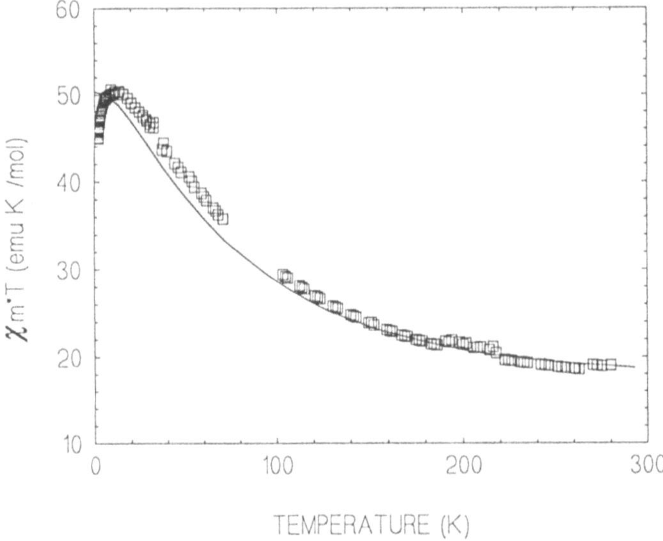

Figure 14-12   Temperature dependence of χT for Fe_8. The solid line represents the calculated values

A cluster with eight iron ions has been shown in Figure 14-2. The temperature dependence of the magnetic susceptibility, χT, of Fe_8, $\{[(tacn)_6Fe_8(\mu_3\text{-}O)_2(\mu_2\text{-}OH)_{12}]\text{-}Br_7(H_2O)\}Br\cdot H_2O$, where tacn is 1,4,7-triazacyclononane, is shown in Figure 14-12. The room temperature value is much smaller than expected for eight uncoupled spins indicating a dominant antiferromagnetic coupling as expected. On decreasing temperature χT increases to a maximum of 52 emu mol^{-1} K at 10 K, and then decreases slightly. This indicates that the ground state of the cluster is $S = 9$ or $S = 10$, and that there must be several thermally populated levels at the lowest temperatures we could reach. Magnetization at 1.8 K shows saturation in a field of 20 T. The maximum value reached is ca. 18 μ_B, indicating that a level of at least $S = 9$ must be the lowest under these circumstances.

The same measurements repeated at 4.2 K show a very weak increase of the magnetization above 18 μ_B, suggesting that the $S = 10$ state becomes populated at higher temperatures. Attempts were made to reproduce the experimental susceptibility and magnetization data, using the procedure outlined above and the CLUMAG program. The results can be considered as satisfactory, as shown in Figure 14-12, even though it was not possible to proceed to a least squares minimization due to prohibitive computer times needed. The ground state has $S = 9$ and the first excited state at 0.5 cm^{-1} has $S = 10$.

[Fe(OMe)$_2$(O$_2$CCH$_2$Cl)]$_{10}$, Fe$_{10}$, has the structure [41] shown in Figure 14-13. It contains ten iron(III) ions octahedrally coordinated. It looks like the wheel of a fun fair, a Ferris wheel in the US, and so it was called a ferric wheel. Each ion is connected to another one by two μ_2-methoxo and one μ_2-carboxylato bridges. The average nearest neighbour Fe-Fe distances are 3.028(4) Å, while the average longest distances across the ring are 9.80 Å. Therefore, this compound almost reaches nanometric dimensions. In Figure 14-14, a space filling representation of the structure shows how a small hole of 2–3 Å is present in the centre of the molecule.

Beyond the appealing symmetric aspect of the molecule, Fe$_{10}$ is very interesting from a magnetic point of view, because rings like this have long been used as models and intermediate steps to calculate the magnetic and the other thermodynamic properties of one-dimensional magnetic materials [42]. In fact, a one-dimensional magnetic material is characterized by having an assembly of spins in a row which are coupled to two nearest neighbours. The same condition applies to Fe$_{10}$ even if of course it is limited to a finite number of spins. In the calculations for infinite chains ring models were preferred to finite segments in order to take into account the conditions at the boundary. Therefore, the magnetic characterization of Fe$_{10}$ should shed some light on the magnetic properties of one dimensional magnetic materials.

The temperature dependence of the magnetic susceptibility [43] of Fe$_{10}$ is very simple, passing through a maximum at 65 K and then decreasing rapidly to zero. This behaviour is typical for antiferromagnetic coupling between nearest neighbours as must be expected, given the nature of the metal ion and the geometry of the bridges. Quantitative calculations suggest a value for J of 9.6 cm^{-1}, a value which compares well with those previously reported for dinuclear complexes with similar bridges. These results clearly show that the ground state has $S = 0$, and that there is a finite gap to the first excited levels, contrary to what is observed in chains of one dimensional magnetic materials.

From susceptibility alone it is not possible to obtain direct information on the nature and the energy separation of the excited levels, but only an indirect one, derived from a satisfactory fitting of its temperature dependence. However, magnetization measurements performed at 0.6 K, both with static and pulsed magnetic fields, provided an exceptional direct insight into the nature and the energies of the excited states.

In fact, the static magnetization, shown in Figure 14-15, is practically zero at low fields, in agreement with the fact that at that temperature only the ground non-magnetic $S = 0$ state is populated. However, on increasing the external field, the magnetization starts to increase, and at 4.2 T it reaches the value of ca. 2 μ_B, the saturation value for an $S = 1$ level. On increasing further, the magnetization starts to increase again, and in a field of ca. 8.4 T it reaches the value expected for an $S = 2$ state. A further increase of the field leads to reaching the value typical of $S = 3$ in a field of 12.6 T. Pulsed experiments, which allowed us to reach fields of 45 T, confirmed the presence of sharp transitions, at regular field

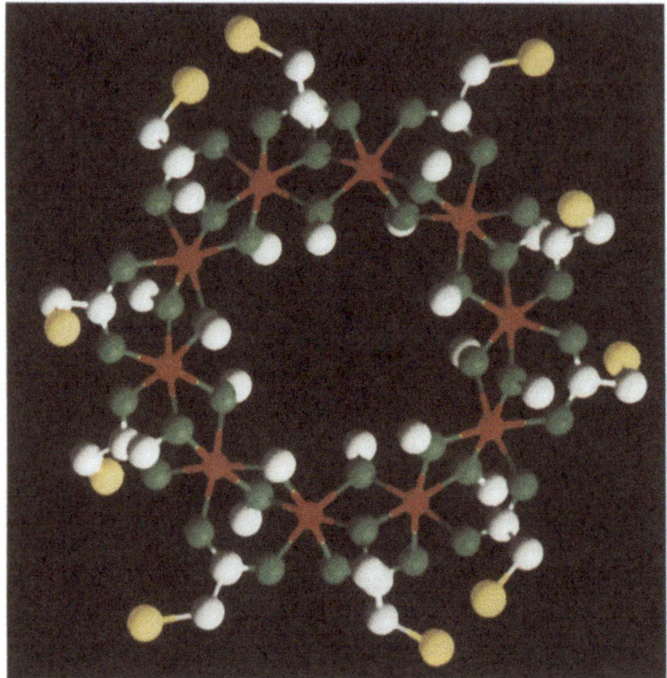

Figure 14-13 Structure of Fe_{10}.
Iron atoms are red, oxygens are green, carbons white and chlorine atoms are yellow.

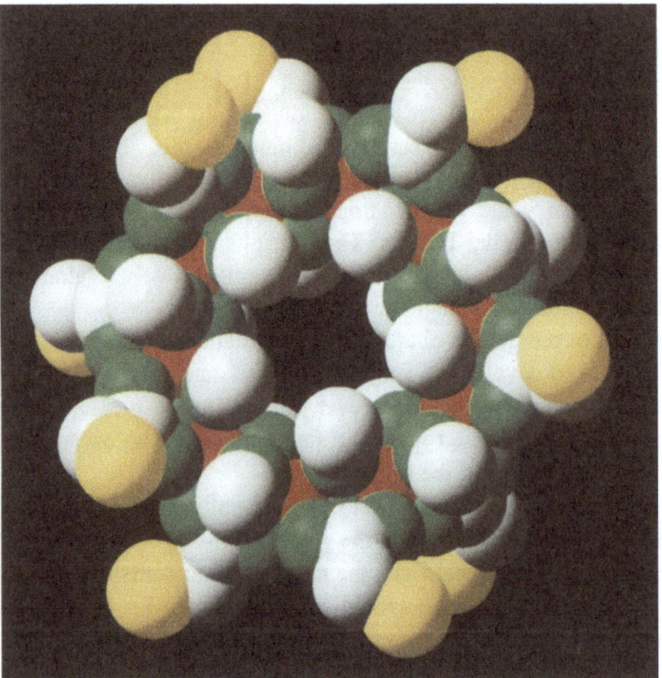

Figure 14-14 Space filling structure of Fe_{10}. The attribution of colours is the same as in Figure 14-13.

intervals, corresponding to the cross-over in the ground state from $S = 0$ to $S = 1$, up to $S = 8$ to $S = 9$. The interpretation of these data is straightforward: in a magnetic field, the components of the different S excited states decrease their energies with slopes proportional to S. In sufficiently high fields, the spin with the largest S will eventually become the ground state. What we have monitored in our experiments, has been the successive cross-over of the lowest lying levels in the order of increasing S values.

Fe_{10} is a paramagnet and not an antiferromagnet, but the observed behaviour here is similar to that observed in some antiferromagnets which – by the effect of an external field which reverses some of the spins from their preferred orientation – become metamagnets. In this case we observed 9 different "metamagnetic" transitions in the ferric wheel.

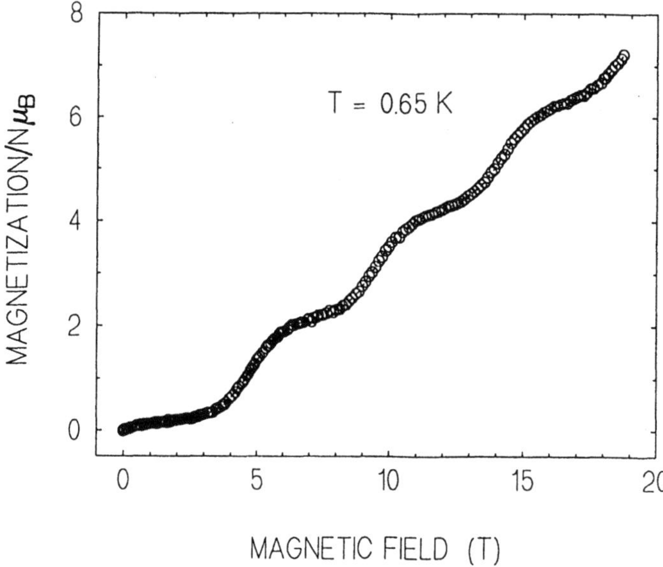

Figure 14-15 Magnetization of Fe_{10}.

The largest cluster so far reported is $[Fe_{19}(\mu_3\text{-}O)_6(\mu_3\text{-}OH)6(\mu_2\text{-}OH)_8(heidi)_{10}(H_2O)_{12}]^{3+}$, Fe_{19}, where H_3heidi is $N(CH_2COOH)_2(CH_2CH_2OH)$, whose structure [44] is shown in Figure 14-16. Particularly interesting is the central core of seven iron ions, $Fe_7(\mu_3\text{-}OH)_6(\mu_2\text{-}OH)_4(\mu_3\text{-}O)_2]^{7+}$ which is found in the molecule. The same core is found in the analogous cluster $[Fe_{17}(\mu_3\text{-}O)_4(\mu_3\text{-}OH)_6(\mu_2\text{-}OH)_{10}(heidi)_8(H_2O)_{12}]^+$, Fe_{17}, which cocrystallizes in the same unit cell as Fe_{19} . This arrangement of metal ions and oxo ligands is very similar to that observed in the layered diaspre type lattice of goethite, $FeO(OH)$, which is the usual end-product of the hydrolysis of iron salts. It is certainly tempting to speculate whether the controlled hydrolysis obtained by the addition of the heidi ligand has frozen the process of growth of goethite at an early stage.

The temperature dependence of the magnetic susceptibility [45] of $Fe_{17}+Fe_{19}$ is qualitatively very similar to that of Fe_8, with a χT value which at room temperature is much lower than expected for 36 uncoupled iron(III) ions, and increases steadily with decreasing temperature, reaching a maximum of 218 emu mol^{-1} K. Magnetization at very low temperatures saturates readily to a value of 65 μ_B. Since this value is given by the sum of the contributions of the two clusters, Fe_{17} and Fe_{19}, it is not possible to extract from one value the two individual contributions. If we split it into two identical halves, the ground state of each cluster should be $S = 33/2$. Since presumably the two clusters will not have the same magnetization, one of them should have a higher value of S in the ground state. This seems to be at the moment the largest value of S reported for a ground state. The compound might seem therefore an ideal candidate for observing superparamagnetic behaviour, but, to our dismay, no evidence of that was found either in the Mössbauer spectra or in the ac susceptibility measurements. Apparently, a large spin is not a sufficient condition for having superparamagnetic behaviour, but anisotropy must play an important role. We will comment further on this in the section on polymanganese oxo clusters.

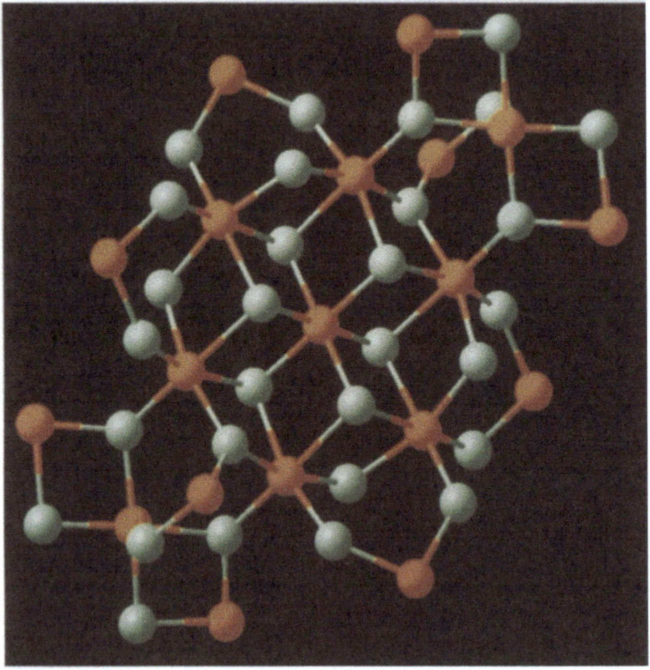

Figure 14-16 Structure of Fe_{19}. The iron atoms are drawn in red while the oxygen atoms are green. Only the atoms involved in the bridges are shown to evidence the topology of the magnetic interactions.

14.6 Poly-Manganese Oxo Complexes

Many manganese clusters have recently been investigated [46–53], after it was realized that dinuclear and tetranuclear mixed valent species are present in many biological systems, ranging from catalyzes to photosynthetic centres responsible for water oxidation.

A cluster with ten manganese ions [49] has a particularly interesting structure because it is related to the CdI_2 lattice of chalcophanite and lithiophorite. Therefore, also in this case, like for Fe_{17} and Fe_{19}, the finite clusters start to have some structural resemblance with infinite lattices, and with minerals in particular. Only room temperature magnetic data have been reported, showing net antiferromagnetic interactions within the clusters.

Another cluster of ten manganese ions was reported by Zubieta et al.[48]. It comprises eight manganese(III) and two manganese(II). Only room temperature magnetic measurements have been reported so far, and they provide a clear indication of antiferromagnetic coupling within the cluster.

By far the most investigated manganese clusters are those with the general formula $[Mn_{12}O_{12}(carboxylate)_{16}]$, where the carboxylates can be acetate or benzoate [54–60]. The structure of one of these clusters is shown in Figure 14-17. It can be described as formed by an external ring of eight manganese(III) ions, each with spin $S = 2$, and by a cubane-type structure comprising four manganese(IV), spin $S = 3/2$ on the vertices of a tetrahedron. The manganese ions are bridged by μ_3-oxo groups, which are expected to be determinant as far as exchange interactions are concerned, but also carboxylato bridges are present.

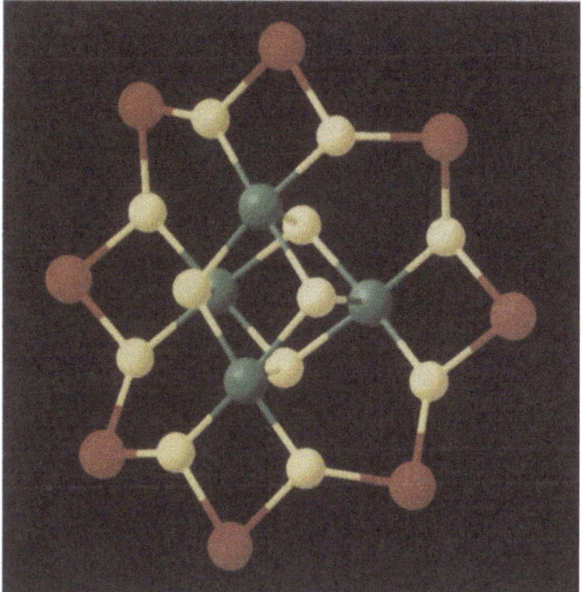

Figure 14-17 Structure of Mn_{12}. Only the metal and bridging oxygen atoms are reported for the sake of clarity. Manganese(III) atoms are reported in red, manganese(IV) in green, while oxygen atoms are yellow.

The Mn_{12} clusters are stable in solution, so that some chemistry can be done with them. For instance, cyclic voltametry provided a reduced species. Also, it was possible to substitute four manganese(III) ions with four iron(III) ions [58]. Rather surprisingly, the substitution was selective, in the sense that the iron ions entered the a sites, taking advantage of the small structural differences between them and the b sites.

The magnetic properties of all these compounds have been investigated, but those which are most interesting are those of the parent Mn_{12} clusters. We will focus in the following on the acetate for which there are more data available. The structure of Figure 14-17 shows that also in this case there are many triangles connecting manzanese ions, and therefore spin frustration effects must be anticipated. The χT value at room temperature is much smaller than expected for uncoupled spins, indicative of dominant antiferromagnetic coupling within the cluster. However, χT increases on decreasing temperature, as observed for Fe_8 and $Fe_{17}+Fe_{19}$, and reaches a plateau of 54 emu mol^{-1} K, indicative of an $S = 10$ ground state [56]. This value was confirmed by magnetization measurements. It is possible to rationalize such a state by assuming that all manganese(III) spins are up and the manganese(IV) spins are down. This preferred spin orientation would not be the result of ferromagnetic coupling between manganese ions of the same valence and antiferromagnetic coupling between those of different valence, but rather of spin frustration. In fact, we could not perform a complete calculation of the energy levels in this case, because they are far too numerous. However, a simplified calculation allowed us [57] to satisfactorily reproduce the χT values at low temperatures, assuming – on the basis of the comparison with simple compounds – that the strongest antiferromagnetic coupling must be observed between manganese(III) and manganese(IV) ions bridged by two oxo groups.

The very unusual feature of this compound is its huge magnetic anisotropy, of the Ising type, which is observed in single-crystal magnetic measurements [56], Figure 14-18. The origin of this anisotropy is immediately understood thanks to the EPR spectra obtained by exciting the transitions with a far infrared laser at frequencies ranging from 245 to 526 GHz (8.17–17.54 cm^{-1}) in strong magnetic fields. The spectra at 4.2 K show one band, which shifts to low field regularly on decreasing the exciting frequency. The transiton field goes to zero for a frequency of ca. 10 cm^{-1}. This is a clear indication that the band is due to a transition between two levels separated by that energy. Further, the evidence is that the transition is observed when the field is parallel to z, and no other transition is observed. This is a clear indication that the transition involves the $M = -10$ and $M = -9$ states of $S = 10$. Under these circumstances the system becomes anisotropic in a weak field, because the ground $M = \pm 10$ doublet splits with slopes $\pm 10\,g\mu_B B$ for a field parallel to the tetragonal axis of the cluster, and is not split at all perpendicular to that, since the transverse magnetic field has a zero matrix element within the two components.

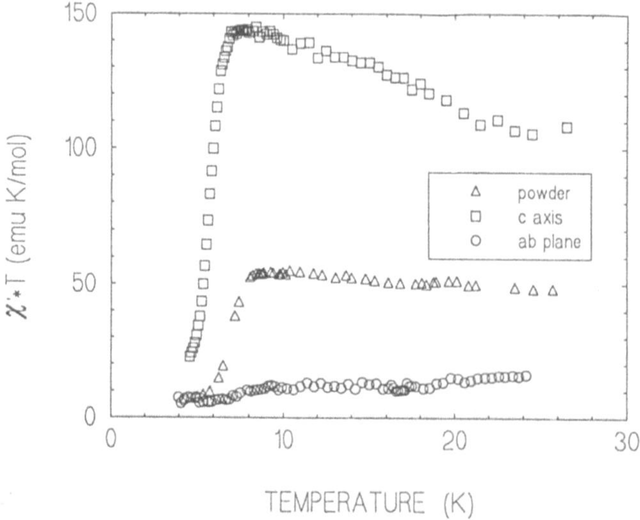

Figure 14-18   Magnetic anisotropy of Mn$_{12}$.

It is perhaps useful to notice here that the effects can be observed in Mn$_{12}$ because the sign of the zero field splitting parameter, D, which determines whether the ± 10 or the 0 spin component is lowest in energy, is negative. If instead of $M = \pm 10$ the $M = 0$ component were the ground state in the absence of an external field, the anisotropy would have been much smaller, and of the XY type.

Under these conditions Mn_{12} becomes an ideal compound to give rise to pseudo bulk behaviour, because it has a large ground spin state and a huge magnetic anisotropy. Therefore, it can be expected that the magnetization will have to undergo a large energy barrier to invert its direction, and as a consequence the relaxation times must become extremely long. This is exactly borne out by ac susceptibility measurements, which show a frequency dependent out of phase component below 10 K [57]. In an ac experiment a magnetic sample is put within two coils within which an electric current produces an oscillating magnetic field (at frequencies ranging from a few Hz to a few kHz, usually).

If the relaxation time of the magnetization is short, the sample has time to follow the external field, and a response is observed only in phase. If on the other hand the relaxation time becomes longer than the inverse of the frequency of the field, also an out-of-phase response will be observed. From the shape of the out-of-phase response it is possible to calculate the relaxation time of the magnetization at various temperatures [59].

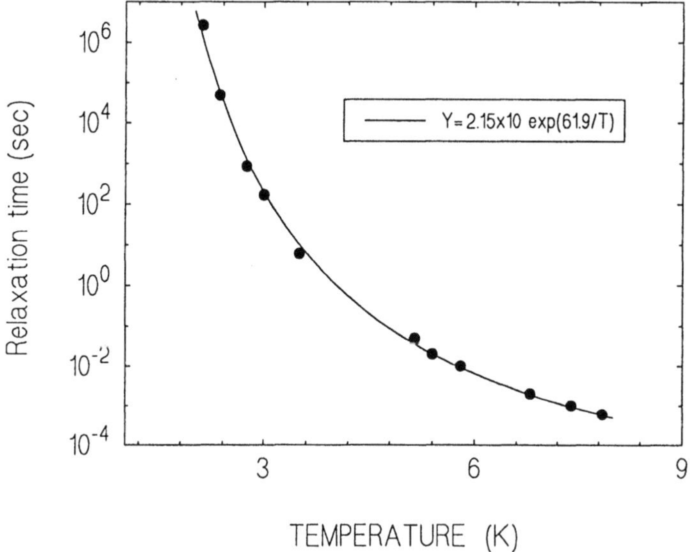

Figure 14-19   Temperature dependence of the relaxation time of Mn_{12}.

The results are plotted in Figure 14-19. At very low temperature the relaxation of the magnetization becomes of the order of one month, and the measurements can be performed by saturating the sample with an external field, removing the field and than monitoring the value of the magnetization. The long relaxation time makes the magnetization bistable at low temperature and gives rise to a large magnetic hysteresis, similar to that observed in bulk hard magnets. The value of the pre-exponential factor in Equation (14-1) is at least two orders of magnitude larger than usually accepted for superparamagnets, showing that here after all we are monitoring a different behaviour.

In Figure 14-19, the relaxation times are fitted with an exponential curve, identical to that of Equation (14-1) which is appropriate to superparamagnets. The question is now whether this really is superparamagnetic behaviour. Excluding the presence of transitions to three-dimensional magnetic order in the lattice of Mn_{12}, ruled out by the frequency dependence of the maximum in the out-of-phase component of the ac susceptibility and by specific heat measurements, the observed relaxation times can be justified with a model which is the extension of the Orbach model for simple paramagnets [60]. What makes this compound unique is its large spin state in the ground state, and the huge anisotropy. For sure, it is almost impossible to observe both in a regular paramagnet. For instance, if we assume a spin $S = 1$, in order to have relaxation times comparable to those of Mn_{12}, the parameter relative to the zero field splitting of the $M = +1, 0, -1$ spin components should be 50 cm^{-1}, instead of the 0.5 cm^{-1} observed for Mn_{12}.

There are still other experiments to be performed in Mn_{12}. For instance, it would be desirable to measure the relaxation time of the magnetization below 1 K, in order to see if the temperature dependence is still exponential, or if other mechanisms, such as tunnelling become important in that temperature range.

14.7 Conclusions

The efforts of the last years of synthesizing new large molecular clusters of metal ions have produced many new aesthetically pleasing molecules. Beyond this, it is being confirmed that in the new mesoscopic scale molecules may show novel magnetic properties which deserve to be intensively investigated. Now even more than in the past, an interdisciplinary effort is necessary in order to fully understand all the properties of these new materials. On one side, synthetic chemistry must relay on more accurate strategies to control the growth mechanism of the clusters. The model of biological systems like ferritin will for sure provide some hints on how to develop the required improvements. Most probably, it will be more and more necessary to perform the reactions not in the simple isotropic environment of a solution, but within restricted micro reaction environments such as can be provided by cavities within zeolites, micelles, etc.

In order to investigate the magnetic properties, it will be necessary to use many different and sophisticated techniques. Here, we referred to dc and ac susceptibility measurements down to very low temperatures, to magnetization measurements in static and pulsed fields, to magnetic resonance experiments at very high frequency, to Mössbauer spectroscopy, and to calorimetric measurements, but many more can be imagined.

Finally, or better, last but not least, new suitable theoretical models will have to be designed in order to quantitatively understand the thermodynamic properties of these materials, if one does not want to be limited to approximate considerations for ever. It is our firm opinion that these new materials deserve all the efforts we discussed, because they really provide a new world.

Acknowledgements

The financial support of the Human Capital and Mobility Programme of the Commission of the European Communities (Network "Magnetic Molecular Materials", Grant ERB-CHRXCT920080) is gratefully acknowledged.

15 Philosophische Aspekte der Chemie
Ihr Wesen: Universalität und Beständigkeit des Wandels *

Achim Müller und Herbert Hörz

"Chemistry provides not only a mental discipline, but an adventure and an aesthetic experience. Its followers seek to know the hidden causes which underlie the transformations of our changing world, to learn the essence of the rose's colour, the lilac's fragrance, and the oak's tenacity, and to understand the secret paths by which the sunlight and the air create these wonders.
And to this knowledge they attach an absolute value, that of truth and beauty. The vision of Nature yields the secret of power and wealth, and for this it may be sought by many. But it is revealed only to those who seek it for itself." [1]
Sir Cyril N. Hinshelwood (Centenary Address to the Chemical Society)

Abstract

At present the heuristic function of philosophy for chemistry is insufficiently perceived by nearly all scientists and can only be realized within a mutual dialogue between philosophers and chemists. The philosopher should not venture to elaborate on this subject without having previously achieved a profound insight into the realm of chemical reaction processes and theoretical coherences. We have dealt in detail with the ubiquitous character of chemistry embedded in a fascinating continuity of permanent change and have attempted to clarify the relation between the "sleeping propensities" of material systems and the change-ability of their qualities. "Chemistry" takes place somewhere between potentiality (dynamis) and actuality (energeia).

Included chapters are: *Introduction; Justified Reductions or Reductionism?; Quality Change, the Relevant History in Time and Sleeping Propensities in Material Systems; Linearity and Nonlinearity of Anthropogenic and Evolutionary Events and Relevant Philosophical Thinking; Conclusions; Supplement (The Distribution of Material Systems in the Different Spheres and Their Interactions as well as the Autobiography of an Iron Atom*; also included: Relevant list of book titles).

---

* mit Anhang (15.6):
  Die Verteilung materieller Systeme auf die Sphären und deren Wechselwirkungen sowie die Geschichte eines Eisenatoms

15.1 Einführung

*"So ist die Materie für uns im allgemeinen alles das, was unsere Sinne auf irgendeine Weise
affiziert; und die Eigenschaften, die wir den verschiedenen Stoffen zuschreiben, gründen
sich auf die verschiedenen Eindrücke oder auf die unterschiedlichen Veränderungen, die sie
in uns hervorrufen."* [2]
P. T. d'Holbach

*"If, in some cataclysm, all of scientific knowledge were to be destroyed, and only one
sentence passed on to the next generations of creatures, what statement would contain the
most information in the fewest words? [.............] all things are made of atoms [.............]
attracting each other when they are a little distance apart, but repelling upon being
squeezed into one another. In that one sentence, you will see, there is an enormous amount
of information about the world [.............]."* [3]
R. P. Feynman

Es gibt eine umfangreiche Literatur zu philosophischen Problemen der Physik und
Biologie, nicht aber der Chemie. Das könnte den Eindruck erwecken, als ob Philosophieren
über chemische Probleme nicht möglich oder nicht notwendig sei, da die elementaren
Fragen bei der Physik und die komplexeren bei der Biologie angesprochen seien.
Philosophie bemüht sich um Begriffsanalyse und Denkweisen, aus deren kritischer und
konstruktiver Sicht sie neue Ideen generieren will. Der Sachverstand des Chemikers, verbunden mit dem Interesse an der Einordnung seines Wissens in umfassendere Zusammenhänge, ist für die Diskussion philosophischer Aspekte der Chemie ebenso erforderlich wie
die weitere Entwicklung der theoretischen Positionen der Philosophie unter
Berücksichtigung von Erkenntnissen der Chemie. In der ersten Hälfte des 19. Jahrhunderts
dominierte die spekulative, idealistische romantische Naturphilosophie, die Fakten nicht
achtete, große Teile der Naturwissenschaften. Allmählich also brachte spekulatives
Philosophieren, verbunden mit unkritischer Betrachtungsweise oder gar Mißachtung
experimenteller Arbeiten, die Philosophie bei Naturwissenschaftlern in Mißkredit. Viele,
vor allem experimentell arbeitende Naturwissenschaftler, versuchten deshalb in der zweiten
Hälfte des 19. Jahrhunderts, auch kognitiv, unabhängig von der Philosophie zu sein. Dazu
Liebig [4]:

*" Ich selbst brachte einen Teil meiner Studienzeit auf einer Universität zu, wo
der größte Philosoph und Metaphysiker des Jahrhunderts zur Bewunderung
und Nachahmung hinriß; wer konnte sich damals vor Anstekkung sichern? Auch
ich habe diese an Worten und Ideen so reiche, an wahrem Wissen und gediegenen Studien
so arme Periode durchlebt, sie hat mich um zwei kostbare Jahre meines Lebens gebracht;
ich kann den Schrecken und das Entsetzen nicht schildern, als
ich aus diesem Taumel zum Bewußtsein erwachte."*

Nach unserer Auffassung führen jedoch experimentelle Arbeiten mit der relevanten
theoretischen Interpretation zu Fragen nach der Existenzweise der Welt und damit zu
philosophischen Aussagen über die Mechanismen des Weltgeschehens.

Die Allgegenwärtigkeit (Universalität) chemischer Prozesse hat Bedeutung für die gesamte Wissenschaft. Diesen Sachverhalt sollte nicht nur der seinem Spezialistentum verfallene Naturwissenschaftler, sondern auch der über Natur und Wissenschaft reflektierende Geisteswissenschaftler zur Kenntnis nehmen. Das Nichtzurkenntnisnehmen der angesprochenen Universalität der Chemie führt dazu, daß einerseits Chemiker philosophische Probleme, die der Begriffsgeschichte und der Analyse von Materieformen entspringen, nicht registrieren (möchten) und andererseits Philosophen glauben, die Materiestruktur ohne jedwede Kenntnis der Chemie ergründen zu können. Auf solche immanenten Zusammenhänge verwies Kekulé, einer der kreativen Altmeister der Chemie, der *"Chemie als die Lehre von den stofflichen Metamorphosen der Materie"* bezeichnete (vgl. auch Kapitel 15.4).

Eine *Philosophie der Chemie* könnte als *Weggefährtin* einer (experimentellen) Chemie, die sich stets um wie auch immer geartete Wechselwirkungen kümmert, die oben erwähnten Mängel beheben helfen.

Unsere Bemerkungen zu den philosophischen Aspekten der Chemie beziehen sich u. a. auf die notwendige Unterscheidung zwischen wissenschaftlich berechtigten Reduktionen komplexer Systeme und einem möglichen philosophischen Reduktionismus, der es nicht nur unterläßt, Teile wieder zum Ganzen zusammenzufügen, sondern die Detailsicht zur Gesamtsicht erklärt. Reduktionen komplexer Systeme sind allerdings erforderlich, um die Verhaltensweise der Systemelemente zu untersuchen und verschiedene Seiten der agierenden Systeme in ihren Eigenschaften und Reaktionen zu erkennen. Der Naturwissenschaftler, speziell der Chemiker, versucht ständig Systemelemente und deren spezifisches Verhalten zu erfassen, sei es, daß es sich um ein *spezielles* "Reaktionsverhalten" an einer funktionellen Gruppe eines Moleküls oder um eine *spezielle* – unter idealisierten Bedingungen gemessene und interpretierte – spektroskopische Eigenschaft handelt. Die Sicht auf die physikalisch erfaßbaren Momente chemischer Prozesse wird jedoch zum philosophischen Reduktionismus, wenn dabei die allgemeinen, philosophisch relevanten Strukturcinsichten der Chemie nicht genutzt werden, um die philosophische Strukturtheorie zu präzisieren.

Strukturwandel in der Chemie ist als Qualitätswandel und als die Realisierung von Möglichkeiten aus einem durch das existierende Bedingungsgefüge, d.h. ein durch materielle Ingredienzien aufgebautes Möglichkeitsfeld, zu analysieren. Es lassen sich nämlich allen (chemischen) materiellen Systemen Propensitäten * – Verwirklichungstendenzen, die in An-

---

* Propensität läßt sich etwa synonym mit Potentialität verwenden. Die Begriffe *Potential, Affinität* und *Potentialität* sind Bestandteile der Terminologie der Naturwissenschaften bzw. der Philosophie der Naturwissenschaften. Das *thermodynamische Potential,* das wegen seiner Extremaleigenschaften eine ähnliche Rolle spielt wie das "mechanische Potential" (die potentielle Energie in der allgemeinen Mechanik), ist für isotherm-isobare Vorgänge *die freie Enthalpie G* bzw. das *Gibbssche Potential* (für isotherm-isochore *die freie Energie F* – bzw. das *Helmholtzsche Potential*). *Das chemische Potential* μ eines Systems, das aus N gleichartigen Teilchen besteht, wird auch als seine auf ein Mol bezogene *freie Enthalpie* definiert, allgemein jedoch in *Mehrkomponentensystemen* entsprechend *partiellen Ableitungen der thermodynamischen Funktionen* bzw. *Potentiale G, F, H, U* nach der Teilchenzahl unter relevanten Bedingungen. Für den oben angegebenen Fall eines *Mehrkomponentensystems* gilt $\mu_i = (\delta G/\delta n_i)_{P, T, n_j}$. Das *chemische Potential* der Substanz i ist hierbei eine intensive Größe, die angibt, um wieviel sich G pro Mol der Komponente i ändert, wobei sich die Molzahl so wenig ändert, daß die Zusammensetzung des Systems praktisch konstant bleibt. Damit ist sie nicht die Eigenschaft einer Substanz, sondern eine Beschreibung der Eigenschaft dieser Substanz im gegebenen System. Der Name resultiert daraus,

lehnung an die Poppersche Definition [5] auf den *chemischen Affinitäten* (man denke hier auch an deren metaphorische Bedeutung in Goethes *Wahlverwandtschaften*) beruhen – zuordnen, die der Chemiker im Versuch entdecken kann und die in der chemischen Evolution unter Einbeziehung von Zufallsmomenten zum Tragen gekommen sind. Propensitäten entsprechen – bezogen auf den letzten Fall – objektiven, sich realisierenden Möglichkeiten. Das entspricht der evolutionstheoretischen Position von der Existenz realer Möglichkeitsfelder (vgl. [6]), aufgespannt durch Bedingungskomplexe, innerhalb derer sich das Geschehen durch die Verwirklichung einer bestimmten Möglichkeit realisiert, und damit das gegebene Feld reduziert wird und neue Felder aufgebaut werden. (Schlafende) *Propensitäten* sind noch nicht realisierte Möglichkeiten.

Zugleich sind die geschilderten probabilistischen Übergänge von einem Zustand in den anderen, wie z.B. im Evolutionsprozeß, durchweg nicht linear. Der Erkenntnisprozeß des Übergangs vom linearen zu einem dem wirklichen Geschehen angemesseneren nichtlinearen Weltbild ist bei weitem noch nicht abgeschlossen. Nicht-Linearität umfaßt die Auf-

---

daß bei spontan ablaufenden chemischen Prozessen – entsprechend elektrischen und mechanischen Vorgängen – bestehende *Potentialunterschiede* ausgeglichen werden.

Eine chemische Reaktion unter den angegebenen Bedingungen kann nur eintreten, wenn $G = \Sigma v_i G_i$ (v_i : jeweilige Molzahl bzw. stöchiometrischer Koeffizient) für die Reaktionsprodukte einen niedrigeren Wert hat als für Ausgangsprodukte (*freie Reaktionsenthalpie* $\Delta G = \Sigma v_j G_j$ (Produkte) $- \Sigma v_i G_i$ (Edukte)).

Für *spontane Prozesse* bei konstanter Temperatur und konstantem Druck gilt $dG/dt < O$, wobei im Zustand des *chemischen Gleichgewichts* der Wert für die *freie Enthalpie* als *thermodynamisches Potential* einen Extremal- bzw. Minimalwert annimmt.

Damit ist die Möglichkeit, bei der Wechselwirkung von Stoffen die Größe G zu senken, auch ein Ausdruck der *"chemischen Affinität"* A zwischen diesen Stoffen, wobei eine ungehemmte chemische Reaktion so lange abläuft, bis gilt: $-A = \Sigma v_i \mu_i = 0$ (entsprechend dem Ausgleich des *Potentialunterschieds* bzw. (der Differenz) der *chemischen Potentiale*). Die erreichbare Abnahme ΔG von G wird selbst in Lehrbüchern der theoretischen Physik als Maß der *Affinität* – entsprechend der Terminologie von van't Hoff – benutzt. Die *Affinität* hat dasselbe Vorzeichen wie die *Reaktionslaufzahl* ξ, womit ihre Definition als $-A = (\delta G/\delta \xi)_{T,\,p}$ auch sinnvoll ist (Wert 0 im *Gleichgewicht*). Die *Affinität* ist also die Triebkraft einer chemischen Reaktion.

Als Beispiel sei erwähnt, daß eine Substanz mit der *(Standard-) Bildungsenthalpie* $G < 0$ bei der entsprechenden Temperatur gegenüber dem Zerfall in die Elemente stabil, dagegen eine mit $G > 0$ thermodynamisch instabil gegenüber dem Zerfall in die Elemente ist. Der Zerfall muß jedoch nicht erfolgen, wenn kinetische Aspekte (hohe *Aktivierungsenergie*) dem entgegenstehen: Es besteht die MÖGLICHKEIT des Zerfalls, der jedoch nur unter bestimmten Bedingungen (*Katalysator*anwesenheit) zur WIRKLICHKEIT wird. Damit greift – philosophisch gesprochen – im aristotelischen Sinne eine Beschreibung der Situation mit dem aristotelischen Begriffspaar *dynamis/energeia* bzw. dem Begriff *Potentialität*. (vgl. Hund, F. *Theoretische Physik, Bd. 3: Wärmelehre und Quantentheorie*, 3. Aufl., Teubner, Stuttgart, **1966**; Landau, L. D.; Lifschitz E. M. *Lehrbuch der Theoretischen Physik, Bd. 5: Statistische Physik, Teil 1*, 8. Aufl., Akademie-Verlag, Berlin, **1987**; Kluge, G.; Neugebauer, G. *Grundlagen der Thermodynamik*, Spektrum, Heidelberg, **1994**; Sommerfeld, A. *Vorlesungen über Theoretische Physik, Bd. 5: Thermodynamik und Statistik*, 2. Aufl., Deutsch, Thun, **1988**; Prigogine, I.; Defay, R. *Chemical Thermodynamics*, Longmans, London, **1967**, sowie andere Lehrbücher der Physikalischen Chemie jeweils unter den hier kursiv gedruckten Stichwörtern).

hebung aller durch die lineare Denkweise implizierten theoretischen Restriktionen. So richtet sich das nicht-lineare Weltbild gegen die Reduktion des Geschehens auf eine Summe notwendiger und vorausbestimmter Kausalbeziehungen sowie gegen die Rückführung des komplexen Systemverhaltens auf das Verhalten der isolierten Elemente. Nicht-Linearität umfaßt objektive Zufälle bei der Realisierung von Möglichkeiten, Strukturbildung fernab und in der Nähe von Gleichgewichtszuständen, stochastische und deterministische Verhaltensweisen komplexer Systeme und sie erlaubt auch die Zyklizität eines Entwicklungsgeschehens (Details zum philosophischen und physikalischen Begriff der Nicht-Linearität in 15.4). Für die Chemie sind hier z.B. oszillierende und autokatalytische Reaktionen besonders relevant. Es gibt immer wieder Formen einer neomechanizistischen Weltbildes als Ausdruck des linearen Denkens, der Komplexes auf Einfaches reduziert und damit z.B. die Philosophiefähigkeit der Chemie bezweifelt. Propensitäten oder objektiv zur Verwirklichung anstehende Möglichkeiten sind relative Ziele des Geschehens. Warum, so kann man in diesem Zusammenhang fragen, hat sich die Natur unter den grundsätzlich möglichen für *einen* bestimmten – für uns z.T. nachvollziehbaren – Ablauf der Ereignisse während der Evolution entschieden.

Der Chemiker nützt sein Wissen über die natürliche Entwicklung, über die Art der Qualitätsänderung, über das Wesen und die möglichen Konstruktionsprozesse von Stoffen zur eigenen Konstruktion des Geschehens. Er weckt schlafende Propensitäten.

Als Fazit halten wir fest: die Chemie ist voll philosophisch relevanter Fragen und Einsichten. Philosophie und Chemie ergänzen sich gegenseitig bei der Deutung der Welt und bei der Klärung der Rolle der Chemie in unserem Leben. Wir verstehen unter Chemie die (Natur-) Wissenschaft, die sich mit dem Aufbau und den Eigenschaften sowie der Veränderung von Stoffen beschäftigt. In relevanten Reaktionen werden Strukturen gebildet, verändert oder zerstört. Bei dieser Definition sind chemische Prozesse der Kosmo-, Atmo-, Geo-, Hydro- und natürlich der Biosphäre, aber auch der Anthroposphäre mit ihrer unglaublich großen Vielzahl von Produkten, die aufgrund des chemischen Wissens vom Menschen seit vielen tausend Jahren hergestellt werden, mit einzubeziehen. Hierbei sollten wir zugleich die Universalität der Chemie beachten, denn die philosophisch relevanten chemischen Mechanismen verlangen nicht unbedingt eine genaue Unterscheidung zwischen den gerade angesprochenen Sphären bzw. relevanten Arbeitsgebieten, z.B. denen der Bio- oder Geochemie, welche streng genommen nochmals Differenzierungen unterliegen [*]. Bei der Biochemie kann man z.B. zwischen deskriptiver, funktioneller und angewandter Arbeitsweise unterscheiden.

Einerseits geht es also um die Einheit der Chemie in der Vielfalt ihrer Disziplinen und Arbeitsrichtungen; anderseits ist ihre Universalität nur als Vielfalt existent. Eine Übersicht über die Vielfalt der chemischen Erscheinungsformen in den Sphären läßt sich aus dem Anhang (Kapitel 15.6) entnehmen, wobei bewußt zur Veranschaulichung *einige* relevante Buchtitel aufgeführt worden sind.

Die Natur als "Gewordenes" kann nur unter Einbeziehung chemischen Wissens verstanden werden. Der Philosophie der Chemie gebührt also ein kaum zu unterschätzender Platz in der Natur- bzw. Wissenschaftsphilosophie.

---

[*] Auch der Geologe fragt: *"Welches Prinzip beherrscht die Erdgeschichte: einsinniger Ablauf oder zyklische Wiederkehr, Akzentuierung oder Nivellierung, zeitliche Begrenzung oder endlose Dauer?"* (vgl. Brinkmann, R. *Lehrbuch der Allgemeinen Geologie*, Bd. 1, 2. Aufl., Enke, Stuttgart, **1974**, 38).

15.2 Berechtigte Reduktionen oder Reduktionismus?

"Chemie" ist allgegenwärtig: Sie hat unser tägliches Leben im Zuge der industriellen und wissenschaftlich-technischen Revolution erheblich verändert. Die gesellschaftliche Auswirkung einer von Menschen initiierten chemischen Forschung ist das Resultat ihrer Aufgabenstellung, sich als naturwissenschaftliche Disziplin mit dem Aufbau bzw. der Struktur, den Eigenschaften im weitesten Sinne und den Veränderungen von Stoffen bzw. materiellen Systemen der belebten und unbelebten Natur zu beschäftigen. Allerdings birgt diese mehr geistig-ideelle Seite der Chemie ein nur ihr eigentümliches Faszinosum der im Erkenntnisprozeß zu Tage tretenden Vielfalt der an die Materie gebundenen Erscheinungsformen, die sich im relevanten Möglichkeitsfeld herauskristallisiert haben und in Zukunft weiter herausbilden werden. Man ist beeindruckt von der offensichtlichen Unerschöpflichkeit der Eigenschaften und Reaktionen der Stoffe, ja sogar – bei abstrakter Betrachtungsweise – jedes einzeln herausgegriffenen Stoffes. Die Zahl der verschiedenen in der Natur vorhandenen und der vom Menschen synthetisierten Verbindungen ist unvorstellbar groß: jedes Jahr wird in der Literatur über weit mehr als 100.000 neue Verbindungen berichtet.

Das menschliche Bestreben, Gesetzmäßigkeiten bzw. innere strukturelle Zusammenhänge in der Natur aufzudecken und verstehen zu wollen, wie z.B. den uralten Prozeß ständigen Werdens und Vergehens allen Irdischen und Kosmischen, erinnert an Schellings Plädoyer für die "einzige Aufgabe", die Materie (Welt) zu konstruieren [7]. Betrachten wir den Prozeß ständigen Werdens, dann darf die Basis der Entwicklung materieller Systeme nicht nur in der Entstehung dissipativer Strukturen gesehen werden. Konservativen, spontanen Selbstorganisationsprozessen – aktuellen Untersuchungsobjekten der Chemie – die allerdings in innovativen Arbeiten von Haken [9] und Prigogine [10] nicht direkt erfaßt worden sind, kommt nämlich in diesem Zusammenhang eine mindestens ebensogroße Bedeutung zu [8]. Die genannten Selbstorganisationsprozesse [8] besitzen eine archetypische Signifikanz im Rahmen des genannten Möglichkeitsfeldes, das durch intrinsische Eigenschaften der Materie bzw. der relevanten Stoffe aufgespannt wird. Sie zeigen letztlich, was vom Möglichen real zu werden vermag und was nicht. Vergehensprozesse sind demgegenüber von einem Dualismus gekennzeichnet, der sich im Nebeneinander von zyklischen und linearen Zeitabläufen widerspiegelt: Der "individuelle Tod" irgendeines (materiellen) Strukturgefüges kann auf diese Weise Ausgangspunkt neuer Entwicklung und damit neuen Werdens sein. Dies zeigt sich beispielhaft in den noch zu diskutierenden Elementkreisläufen.

Die Kreation jeder (neuen) chemischen Substanz erzeugt eine neue Propensität oder im aristotelischen Sinne Potentialität, darüberhinaus auch eine feingliedrigere Strukturierung des Möglichkeitsfeldes. Das Problem besteht darin, aus der unglaublichen Vielfalt von Eigenschaften und Reaktionen der Stoffe mittels wissenschaftlich berechtigter Reduktionen bestimmte Eigenschaften und Reaktionen isoliert betrachten zu müssen, um sie beeinflussen oder gestalten zu können, und damit in die Nähe einer philosophischen Haltung zu geraten, die als Reduktionismus bezeichnet werden kann, mit welcher das Suchen nach Zusammenhängen aufgegeben und die Erklärung des Ganzen allein aus seinen Teilen gewonnen wird. Auch der Versuch, eine Gesamtsicht durch Spekulation zu erreichen, hat nur dann wissenschaftliche Bedeutung, wenn die spekulativen Momente empirisch untermauert und vor allem *geprüft werden können* (vgl. hierzu relevante Kritik an Schellings romantischer Naturphilosophie [11], einer Philosophie, die auch die Chemie mit einbezieht [12]).

Chemie als die Gesamtheit der Analyse und Konstruktion aller materiellen Systeme, bei gleichzeitiger theoretischer Reflektion ihrer Eigenschaften und der Reaktionen und Veränderungen der Stoffe, befaßt sich mit komplexen Systemen. Diese geben den Rahmen für das Verhalten der Systemelemente. Heisenberg betonte in seiner Arbeit von 1942 über die "Ordnung der Wirklichkeit", daß die Behauptung, die Bohrsche Theorie führe das chemische Verhalten auf die Bewegung der Elektronen zurück, eigentlich lauten müsse: Die Quantentheorie habe geradezu gezeigt, *"daß die chemischen Gesetze einen selbständigen neuen Zusammenhang darstellen, der nicht durch die mechanischen Bewegungen kleinster Teilchen erklärt werden könne."* [13] Der philosophische Reduktionismus, der chemische Prozesse auf die mechanische Bewegung kleinster Teilchen zurückführen wollte, war mit der Quantentheorie prinzipiell widerlegt. Diese zeigt vor allem an Hand der EPR-Korrelationen (vgl. z.B. [14, 15]), daß sich komplexe Systeme nicht auf "ihre" Elemente reduzieren lassen.

Die Konsequenzen aus dieser Einsicht in die Ordnung der Wirklichkeit sind kaum bis zu Ende gezogen worden. *Wenn sie ernst genommen werden, dann hat auch die philosophische Analyse chemischen Denkens eine kaum zu überschätzende Bedeutung für die des physikalischen und biologischen Denkens.* Durch den Qualitätswandel der Stoffe, der von der Chemie untersucht wird, werden die Rahmenbedingungen für das physikalische Verhalten der Systemelemente verändert. Es kommt zu einer Reduktion der in den von der Physik untersuchten Naturgesetzen enthaltenen Möglichkeiten des Verhaltens, wie etwa die Bedeutung der Chiralität in den Naturprozessen zeigt [16]. Es sind die spezifischen Reaktionen chemischer Systeme zu suchen, die von der Physik als Rahmentheorie chemischen Verhaltens zwar als mögliche Verhaltensweise in den gesetzmäßigen Mechanismen enthalten sind, die aber von der Physik nicht in ihrer vollen Komplexität oder "Historizität" untersucht werden können. Für die Biowissenschaften ist in gleicher Weise eine chemische Rahmentheorie relevant.

Physik und Chemie des Lebendigen haben eine lange Geschichte. Im 19. Jahrhundert untersuchten die "organischen Physiker" Helmholtz, du Bois-Reymond, Ludwig und Brücke mit physikalischen und chemischen Methoden grundlegende Lebensprozesse (vgl. z.B. [15]).

In einem Brief an Helmholtz bemerkte Ludwig: *"Meine anatomische Arbeit über die Niere* [17] *wird eben gedruckt. Mir hat es viel Freude gemacht zu sehen wie die drei Ströme v. Blut, Lymphe und Harn, die durch die Niere gehen sich gegenseitig reguliren. In seiner Art hat die Niere mit den Augen Analogien, freilich ist mit dem Druck und der Geschwindigkeit einer Flüssigkeit nicht so viel zu machen wie mit dem Licht; aber ich hoffe ganz im Stillen, daß es Dich auch ein wenig erfreut zu sehen wo dereinst einmal in diesem merkwürdigen Organ der Physik ein Feld eröffnet ist; in hydraulischer Beziehung kenne ich nichts, selbst nicht die Gärten von Wilhelmshöhe ausgenommen was sich an Feinheit u. Schönheit damit vergleichen ließe."* (vgl. [18]). Die damalige Reduktion komplexer biotischer und physiologischer Phänomene und die Entwicklung von Methoden, welche die physiologischen Mechanismen und Organe der Lebewesen einer Messung zugänglich machen, war erforderlich, um sich mit den Spekulationen der von Schelling stark beeinflußten romantischen Naturphilosophie auseinanderzusetzen und die Aussagen über die materielle Bedingtheit der Lebensprozesse empirisch zu fundieren. Die romantische Naturphilosophie interessierte nur, *was* Natur ist, *nicht wie* Natur ist (Trotz der hier geäußerten Kritik ist die immense Intuition und Kreativität eines Schelling bewundernswert).

Gegen die Arbeitsweise der organischen Physiker gab es prinzipielle Einwände. So stellte der Wiener Anatom Joseph Hyrtl die Frage: *"Ist die Seele das Product des nach unabweichlichen organischen Gesetzen arbeitenden Gehirns, oder ist dieses Gehirn vielmehr nur eine jener Bedingungen, durch welche der Verkehr eines immateriellen Seelenwesens mit der Welt im Raume vermittelt wird?"* [19] Er vertrat die Haltung von der Immaterialität der Seele, stieß damit allerdings auf allgemeinen Protest.

Immerhin hatte sich die "organische Physik" als ein fruchtbares Forschungsprogramm erwiesen, das sich mit wissenschaftlich berechtigten Reduktionen gegen eine andere Form des philosophischen Reduktionismus wandte, der nicht, wie der Mechanizismus, alles auf die Bewegung kleinster unteilbarer Teilchen zurückführte, die sich nach den Gesetzen der klassischen Mechanik bewegen, sondern bestimmte Bereiche aus der Forschung wegen ihres immateriellen und deshalb nicht meßbaren Charakters ausklammerte. Hyrtl meinte: *"Der Naturforscher glaubt sofort nur seinen Beobachtungsresultaten, der Mathematiker seinen Ziffern und ihrer unwiderstehlichen Logik, der Physiker und Chemiker seinen Versuchen, der Physiolog dem anatomischen Messer. Keiner scheint es zu fühlen oder zu beachten, dass, wenn es etwas Uebersinnliches giebt, es nur unter der Bedingung existirt, dass es eben nicht gemessen, nicht gewogen, nicht zergliedert werden kann."* [20].

Hyrtl reduzierte Erforschbares auf Unerkennbares. Er kritisierte die experimentell an Lebewesen arbeitenden Physiologen, wozu ihn auch persönlicher Streit mit Brücke führte.

In seinem berühmten Vortrag über *"Grenzen des Naturerkennens"* aus dem Jahre 1877 äußert sich hierzu auch du Bois-Reymond [15]:

> *"Mag die Wissenschaft fortschreiten, mag der Astronom schließlich den ganzen*
> *Bau und die Bewegungsgesetze des Weltalls durchschauen, mag der Biologe die*
> *Lebensgesetze enträtseln – die letzten Gegebenheiten, nämlich die Materie und*
> *das Bewußtsein, werden unserem Verständnis immer unzugänglich bleiben.*
> *Über sie muß die Wissenschaft nicht nur sagen: Ignoramus (wir wissen es nicht),*
> *sondern auch: Ignorabimus (wir werden es nicht wissen)."*

Im Gegensatz zu Hyrtl anerkannte du Bois-Reymond allerdings die Bedeutung experimenteller Erforschung der Lebensprozesse. Er warnte jedoch davor, prinzipielle Welträtsel durch Detailforschung lösen zu wollen.

Hiergegen setzte Ernst Haeckel seine Thesen in dem bekannten Werk *"Welträtsel"* und später David Hilbert seinen Ausruf [*]:

"Nescimus sed sciemus".

---

[*] Kurz vor seinem Tod – d.h. während des Krieges – hielt Hilbert in Berlin einen Vortrag, den er mit seinem Veto gegen das "Ignorabimus" in lateinischer Sprache schloß (vgl. Meschkowski, H. *Denkweisen großer Mathematiker*, Vieweg, Braunschweig, **1990**, 233). Er wiederholte die Gedanken seines früheren berühmten Königsberger Referates: *"Der wahre Grund, warum es COMTE nicht gelang, ein unlösbares Problem zu finden, besteht meiner Meinung nach darin, daß es ein unlösbares Problem überhaupt nicht gibt. Statt des törichten Ignorabimus heißt im Gegenteil unsere Losung: Wir müssen wissen, Wir werden wissen."* (vgl. Hilbert, D. *Gesammelte Abhandlungen*, Bd. 3, 2. Aufl., Springer, Berlin, **1970**, 387, sowie *Naturwiss.* **1930**, *18*, 959). In einer Biographie Hilberts von Frau Reid wird berichtet, daß er über die grundlegende Arbeit von Gödel, nach der man die Widerspruchsfreiheit der formalisierten Zahlentheorie nicht aus den Mitteln des Systems herausführen könne, verärgert reagiert habe (vgl. Angaben im Werk von Meschkowski, S. 233).

Die stufenweise Lösung der Welträtsel vom Verhältnis von Materie und Bewegung, von Materie und Leben oder gar von Materie und Bewußtsein geht weiter. Die Chemie bzw. Neurochemie wird dazu einen nicht zu überschätzenden Beitrag leisten.

In diesem Zusammenhang sind aktuelle neurochemische Forschungsergebnisse von Interesse, die mit Schlagworten oder Buchtiteln wie *"Symbole, Synapsen und Systeme – Die molekulare Biologie des Geistes"* und *"Chemie der Psyche"* umrissen werden können [21].

Zu den wissenschaftlichen Voraussetzungen und zum Umfeld der "organischen Physiker" gehören natürlich auch die Arbeiten des Chemikers Liebig. Dem entspricht z.B. der Titel seines Berzelius gewidmeten Buches *"Die organische Chemie in ihrer Anwendung auf Physiologie und Pathologie"* [22].

Jede komplexe Erscheinung muß zur Analyse bzw. zur Charakterisierung in bestimmte Komponenten zerlegt werden. Das kann experimentell oder theoretisch geschehen. (Das gilt schon für eine relativ einfache Substanz, z.B. das isolierte Enzym mit seinem aktiven Zentrum und natürlich erst recht für einen komplexen Organismus). Damit wird die Voraussetzung geschaffen, *gewisse Aspekte des Wesens einer Erscheinung*, d.h. Teile der ihren Charakter bestimmenden Seiten, zu erkennen. Philosophisch gesprochen sind Experimente des Naturwissenschaftlers, darunter die des Chemikers, Analysatoren der Wirklichkeit, die nach vorgegebener Anordnung mit konstanten und variablen Faktoren bei der Untersuchung von Eigenschaften und Reaktionen von Stoffen ein Versuchsergebnis liefern, das nachvollziehbar ist. So wird z.B. im Experiment aus der Fülle möglicher Reaktionen bzw. potentiell vorhandener Eigenschaften eines Stoffes ein Ausschnitt bestimmt, der ein Moment des gesuchten Wesens umfaßt. (Beispiele: Untersuchung *eines* speziellen Reaktionsverhaltens oder seiner morphologischen, mechanischen, thermischen, optischen, magnetischen und elektrischen Eigenschaften, die alle natürlich mit den chemischen Qualitäten korrelieren). Die theoretische Synthese der unterschiedlichen Wesensmomente läßt uns weitere wesentliche Seiten erkennen; *vollständig* ist das Wesen eines Stoffes jedoch niemals zu ergründen. Es gibt nämlich unerschöpfliche Kombinations- bzw. Reaktionsmöglichkeiten mit bekannten und darüberhinaus noch unbekannten(!) Stoffen, die zu immer wieder neuen geometrischen Anordnungen der Atome führen.

Die "organischen Physiker" versuchten, die Geheimnisse der viel komplexeren Lebensprozesse mit wissenschaftlich berechtigten Reduktionen in experimentellen Anordnungen, die auch Tierexperimente und Vivisektionen einschlossen, durch die Suche nach bestimmten Wesensmomenten zu entschlüsseln. Sie trugen, z.B. mit der Suche nach wirksamen Narkotika sowie mit der Entdeckung chemischer Mechanismen in physiologischen Regelkreisen, wesentlich zur Entwicklung der Chemie bei.

Die chemische Reduktion physiologischer Prozesse auf meßbare Eigenschaften und Reaktionen von Stoffen liefert zwar keine Theorie für das Wesen von lebenden Organismen und speziell des Menschen als vernunftbegabtem Gestaltungswesen, der bewußt seine Existenzbedingungen mit der Einsicht in die Gesetze der Natur, der Gesellschaft und seines eigenen Erkennens und Handelns in Grenzen gestalten kann, bzw. will. Die Reduktion deckt jedoch Detailaspekte für physiologische Prozesse und – auf einer höheren Abstraktionsebene – Analogien für soziale Mechanismen auf, wie etwa die Forschungen zur Selbstorganisation zeigen [23]. Man kann deshalb in der Hierarchie der Theorien, die auf der Mannigfaltigkeit der Strukturen und Reaktionen in den verschiedenen Struktur- und Entwicklungsniveaus beruht, nicht nur die Physik als Rahmentheorie der Chemie, sondern auch die Chemie als die der Biologie bzw. der Physiologie, die Physiologie als die der Psychologie, die Psychologie als Rahmentheorie des Sozialverhaltens usw. erkennen. Mit dem Problem hat sich schon der Physiker *und* Philosoph E. Mach, der sich bezeichnenderweise als *Naturforscher* betrachtete [24], beschäftigt:

> *"...... Bei unbefangenem Blick wird man es aber eher für möglich halten, dass eine Chemie der Zukunft zugleich auch die Physik umfaßt, als umgekehrt."* [25]

Wir wollen Mach im Kontext des Problems Philosophie und Naturwissenschaft später noch einmal zu Wort kommen lassen.

Wer kreativ sein will, kann im oben angesprochenen Kontext gerade an den Schnittpunkten von Bekanntem und Unbekanntem durch gezielte Reduktionen der Komplexität zu neuen Detailerkenntnissen kommen, da die (unerschöpfliche) Mannigfaltigkeit der Strukturen und Reaktionen viele Geheimnisse enthält, die es noch zu lüften gilt (z.B. hat man die Existenz vieler der heute schon in Lehrbüchern der anorganischen und organischen Chemie aufgeführten Verbindungen noch vor einigen Jahren nicht für möglich gehalten). Die dabei erfaßten Wesensmomente können zu der Entwicklung einer Theorie der komplexen Systeme beitragen.

Wir bemühen uns, mit unseren Begriffen die Mannigfaltigkeit der Erscheinungen so zu analysieren und zu synthetisieren, daß wir die empirisch und theoretisch erfaßten Einzelheiten, d.h. hier speziell alle Detailinformationen über das Stoffverhalten, in Zusammenhänge einordnen, um die Struktur der Wirklichkeit besser erfassen und erklären zu können. Dabei wird die Relativität unseres Begriffsgefüges deutlich. Wir kommen immer wieder in die Situation, mit unscharfen Begriffen komplexe Systeme u.a. solche der Biosphäre, ja sogar der Chemisphäre (als Gesamtheit aller "isoliert" existierenden chemischen Substanzen) erfassen zu wollen. Wir stoßen dabei auf eine interessante Unbestimmtheitsrelation zwischen Inhalt und Umfang der Begriffe. Je präziser wir den Inhalt eines Begriffes bestimmen, wie etwa den des Inertialsystems in der Physik oder die Definition mathematischer Begriffe, desto kleiner wird der Bereich der Objekte, auf die der Begriff zutrifft. Wir erreichen in der Mathematik Idealisierungen, die so in der uns hier interessierenden Wirklichkeit nicht mehr vorkommen. Je unspezifischer dagegen die Inhaltsbestimmung eines Begriffes ist, desto größer ist sein Umfang. Das ist bei philosophischen Begriffen wie Geist, Materie, Raum, Zeit und Bewegung der Fall. (In der Chemie ergäbe sich etwa die Hierarchie Einzel-Substanz, Substanzklasse, anorganische Verbindungen, Chemisphäre). Zur Erkenntnis der Wirklichkeit bedürfen wir jedoch der Präzisierung der Begriffe, um zu meßbaren und empirisch prüfbaren Resultaten unserer Aussagen zu kommen. Wir sind also in der Situation, die Niels Bohr *etwa* so charakterisierte: *Man versucht im Winter auf einer Berghütte mit schmutzigem Wasser und schmutzigen Tüchern schmutzige Gläser sauber zu bekommen. Daß dies gelingt, würde uns kein "Philosoph" glauben.* Die Mehrheit der Chemiker handelt ständig nach diesem Prinzip, jedoch ohne sich immer dessen bewußt zu sein.

Bei der Frage nach der Exaktheit unserer Erkenntnisse und ihrer praktischen Verwertbarkeit ist es sinnvoll, zwischen theoretisch und praktisch orientierter Exaktheit zu unterscheiden. Bedeutet "exakt" die Einordnung in eine mathematisch formulierte Theorie (z.B. zur Charakterisierung des H_2^+-Moleküls durch die Angabe der "exakten" elektronischen Wellenfunktion, da sich in diesem Fall für die Schrödingergleichung im Rahmen der Born-Oppenheimer-Approximation eine geschlossene Funktion als Lösung angeben läßt), dann ist die Definition theoriebezogen, strenger mathematikbezogen. Umfang und Inhalt der Begriffe werden an der Theorie und nicht an ihrer Bedeutung für die Gestaltung der Wirklichkeit gemessen. Dagegen kann eine wirklichkeitsbezogene Definition "unexakt" sein.

Wirklichkeitsbezogene "Exaktheit" von Definitionen und theoretischen Aussagen verlangt, daß die daraus abgeleiteten Orientierungen für das Handeln bei den Konstruktionen komplexer Systeme bzw. für das Experimentieren z.B. im Chemie-Labor den gedachten Zweck erreichen lassen. So werden nicht selten theoriebezogene Aussagen durch Praktiker theoretisch zwar "unexakt" korrigiert, aber wirklichkeitsbezogen erst mit der Korrektur effektiv umgesetzt. Diese Vorgehensweise ist für den Chemiker "lebenswichtig", wobei er den geschilderten Zusammenhang "vor Ort" allerdings im Allgemeinen nicht reflektiert. Beispiele in der Chemie sind: die "Veranschaulichung" von Reaktionsmechanismen in der organischen Chemie durch sogenanntes "Elektronenschieben" und die überall angewandte Methode, komplexe Sachverhalte wie Eigenschaften und Reaktionsverhalten von Molekülen auf der Basis von Inkrementen (z.B. Bindungslängen, Kraftkonstanten, Bindungspolaritäten und -polarisierbarkeiten, Elektronegativitäten) zu verstehen [26]. In diesem Zusammenhang kann man auch die Hierarchie der verschiedenen Methoden der Quantenchemie sehen: Die "exaktere" Methode erlaubt weniger anschauliche Deutungen für den Praktiker als die semiempirische. Entscheidend für die Bewertung der "Exaktheit" einer Formulierung ist also nicht einfach deren mathematische bzw. theoretische Exaktheit.

Mit der genannten Relativität unserer Begriffe müssen wir leben. Für die reale chemische Forschung ist dies besonders relevant. Wir müssen sie zudem immer bewußt beachten, um nicht einem philosophischen Reduktionismus zu verfallen, der entweder einen Teil für das Ganze, das Ganze für einen spezifischen Teil, das Unerklärte als unerklärbar oder den Begriff für die Wirklichkeit nimmt.

H. Primas, der sich mit Fragen bzw. Problemen wie *"Kann Chemie auf Physik reduziert werden?"* und *"Ein Ganzes, das nicht aus Teilen besteht"* [27] intensiv beschäftigt hat, macht darauf aufmerksam, wie kompliziert die "Beherrschung" komplexer Systeme ist. In seiner holistischen Sichtweise werden von ihm grundsätzliche erkenntnistheoretische Aspekte angesprochen [27a]:

"Quarks, Elektronen, Atome oder Moleküle sind keine Bausteine der Materie, sie sind nicht Ge-fundenes, sondern Er-fundenes, das heißt Konstruktionen [.....]."

"Ganz ist, was nur durch eine Vielheit von komplementären Beschreibungen erfaßt werden kann."

"..... in der als fundamental zu betrachtenden Quantentheorie (ist) die Menge der potentiellen Eigenschaften immer viel größer als die Menge der zu einem bestimmten Zeitpunkt aktualisierten Eigenschaften."

Während die Aussage des ersten Satzes für Quarks und Elektronen, denen natürlich keine Individualität zukommt, sicherlich zutrifft, muß dies möglicherweise entsprechend für Atome und vor allem Moleküle, die "extreme" Ausmaße, wie z.B. im Falle von Polymeren oder Viren, annehmen können, nicht unbedingt gelten! Man kann mit modernen physikalischen bzw. spektroskopischen Methoden einzelne Atome und Moleküle "untersuchen", z.B. durch Schalten von Atomen mit dem Rastertunnelmikroskop. Hierbei muß natürlich berücksichtigt werden, daß für das Atom keine "Größe" faßbar ist, was der Chemiker häufig übersieht (vgl. hierzu den relevanten Exponentialausdruck des Radialanteils der Elektronenwellenfunktion!). Erst im Verbund mit anderen Atomen läßt sich eine fiktive(!) Größe durch Bestimmung von Atom-Atom-Abständen *definieren(!)*. Die dritte obige Aussage ist vor allem in Zusammenhang mit den Erläuterungen in Kapitel 15.4 über Potentialität von Interesse. Die zweite Aussage entspricht z.B. dem Sachverhalt, daß Erscheinungen auf der Makroebene, d.h. der der chemischen Substanzen, die der Sinneswahrnehmung zugänglich sind, nicht einfach auf die der Mikroebene zurückgeführt werden können. Beide Beschreibungen kann man allerdings als komplementär auffassen. Primas macht deutlich, daß wissenschaftlich berechtigte Reduktionen oft Konstruktionen sind, die als Grundlage eines philosophischen Reduktionismus mißdeutet werden können, wenn sie nicht in eine komplementäre Betrachtungsweise eingeordnet werden, die Ganzes erfaßt.

Kritisch ist anzumerken, daß sich der forschende Chemiker an den Universitäten über die hier geäußerten erkenntnistheoretischen Probleme kaum Gedanken macht. Dies war sicherlich um die Jahrhundertwende bei anstehenden Paradigmenänderungen in der Physik und Chemie grundsätzlich anders [28]. Relevante Problemstellungen betrafen die Fragen nach der Struktur der Materie, die über einen langen Zeitraum und z.T. in scharfer Auseinandersetzung diskutiert wurden. Beteiligt waren u.a. Boltzmann, Mach, Ostwald, Helmholtz, Planck, Bohr, Born, Einstein und Heisenberg.

Um 1900 war die Atom- und Molekülexistenz noch strittig! Boltzmann standen Mach, Ostwald und Duhem gegenüber, die die Betrachtung von Atomen als ungesunde Metaphysik bezeichneten. Boltzmann diskutierte 1895 eine "H(t)-Kurve" (nach dem H-Theorem entspricht H prinzipiell der Entropie; vgl. [28]). Anhand der Kurve konnte er unmittelbar erklären, daß sich der Widerspruch zwischen den reversiblen Grundgleichungen der Mechanik und den irreversiblen Gleichungen für die Entropie auflösen ließ. Er verstand die Nichtumkehrbarkeit des Geschehens, indem er die in der Figur absteigenden Äste der kleinen Schwankungen den großen irreversiblen *wesentlichen* Änderungen des wirklichen Geschehens gegenüberstellte und als wesensgleich ansah (vgl. hierzu unsere späteren Erläuterungen in 15.3 über Systemeigenschaften). Die Nichtumkehrbarkeit des Geschehens war dann eine Folge extremer Anfangsbedingungen. Dem bekannten "Wiederkehreinwand" von Poincaré und Zermelo widersprach er: bei einer sehr großen Zahl von Freiheitsgraden werden die Wiederkehrzeiten extrem lang (vgl. [28]).

Der Chemiker ist stets gezwungen, komplexe Systeme auf Meßbares zurückzuführen, womit er sie reduziert. Mit seinen Erkenntnissen konstruiert er Modelle, die ihrerseits häufig selbst neue Komplexitäten darstellen. Um nicht einem philosophischen Reduktionismus zu verfallen, muß er das Verhältnis vom Komplexen zum Elementaren reflektieren, also philosophisch denken.

15.3 Qualitätswandel, die relevante Geschichte in der Zeit und schlafende Propensitäten in materiellen Systemen

"πάντα ῥεῖ" [*]
(Aussage der Herakliteer)

---

[*] "Alles fließt" (vgl. Kirk, G. S.; Raven, J. E.; Schofield, M. *Die Vorsokratischen Philosophen: Einführung, Texte und Kommentare*, Metzler, Stuttgart, **1994**, 203).

*"Quod enim nullum universale sit aliqua substantia extra animam exsistens, evidenter
probari potest."* *
W. von Ockham (Summa Logicae)*

Entscheidend für chemische Prozesse ist der konkrete, auf die *reale Welt der einzelnen
Substanzen* bezogene Qualitätswandel. Daher ist die Untersuchung der Genese materieller
Agglomerate und des Verlaufs der Umwandlung eines materiellen Systems in ein anderes
von besonderem Interesse. Dies führt uns auf die Frage nach den Beziehungen zwischen
den Erscheinungsformen der Materie, nach der Art ihrer Strukturiertheit und ihren Eigenschaften, also auf die Frage, wie die großartige Vielfalt der Strukturiertheiten der Stoffe mit
den unterschiedlichen materiellen Erscheinungsformen unter verschiedenen Bedingungen
korreliert.

Bedacht werden sollte in diesem Zusammenhang der Umstand, daß im Laufe der Zeit
materielle Systeme unterschiedlichster Art auf Kosmo-, Atmo-, Geo-, Hydro-, Bio- und
Anthroposphären verteilt worden sind (ein übrigens nie zum Stillstand kommender Prozeß;
vgl. Anhang 15.6), weshalb die Erforschung der Genese und der Veränderungen der Erscheinungsformen der Materie auch die gravierenden Unterschiede der Qualitäten, Quantitäten und Funktionen materieller Aggregate in ihren jeweiligen "Lebenswelten" in Rechnung
zu stellen hat. Wie schwierig eine solche Vorgehensweise sein kann, mag man daran er-
messen, daß noch immer viele Rätsel aus dem Bereich der Biosphäre, ja selbst aus der Geosphäre **, bis zum heutigen Tage ihrer Lösung harren. So sind weder die Entstehung des
Lebens, dessen Existenz an materielle, d.h. chemische Systeme und deren Wechselwirkungen untereinander gebunden ist, noch die Morphogenese der Organismen und
speziell die der molekularen Epigenese bisher hinreichend verstanden.

Seit Beginn der Geschichte der Erde gab es einen ständigen Formenwandel der Materie,
und zwar durchaus im Sinne der eingangs zitierten Beobachtung der Herakliteer (Die Vielfalt der Möglichkeiten der "Veränderung" selbst eines(!) Stoffes ist immens). Hierzu zählen
Änderungen der Qualitäten, Strukturen und Prozeßmechanismen. Quasilinearen, irreversiblen Prozessen (beispielsweise die durch Zufall und Notwendigkeit bedingte Informa-
tionsänderung von Steuerungssystemen im Rahmen der biologischen Evolution)
überlagerten sich zyklische Abläufe, wie z.B. die natürlichen Kreisläufe der Elemente.
Details hierzu sind insbesondere für die Elemente C, N, O, P und S bekannt. Hier handelt es
sich übrigens um Prozesse, die für unser Leben bzw. unsere Ernährung von entscheidender
Bedeutung sind (Stichworte: Fotosynthese und N_2-Fixierung durch Pflanzen bzw. Mikro-
organismen [30]). Während der Evolution der Biosphäre erfolgte eine Differenzierung von
Struktur und Funktion, verbunden mit einer Zunahme an Komplexität, wobei
systemerhaltende und -auflösende Prozesse Hand in Hand gingen. Die Entwicklung der
Biosphäre war und ist gekoppelt bzw. rückgekoppelt an Veränderungen der sich ebenfalls in
Entwicklung befindenden Atmo-, Geo- und Hydrosphäre [31]. Hierzu liegen relevante
Erkenntnisse aus dem Bereich der Geomikrobiologie und Biogeochemie [32] vor (vgl. z.B.
das Auftreten des Sauerstoffs in der präkambrischen Atmosphäre durch die "Entdeckung
der bakteriellen Fotosynthese" und der daraus folgenden Generierung der Bändereisenerze
nach Oxidation von Fe^{2+} zu Fe^{3+}). Die Lebensweise bestimmter Mikroorganismen hängt

---

* "Es kann mit Evidenz aufgewiesen werden, daß kein Universale eine extramentale
  Substanz ist."
** Die Entstehung bestimmter, für die industrielle Produktion wichtiger Lagerstätten – wie
  z.B. von Metallsulfiderzen – konnte bisher nicht geklärt werden [29].

von der Art des Bodens ab; sie beeinflussen aber auch ihrerseits die Geosphäre und Hydrosphäre. Die Veränderung der chemischen Substanzen im Naturhaushalt erfolgt durch Produzenten von Biomasse (grüne Pflanzen), Konsumenten der primären Biomasse (Tiere) und Destruenten (Mikroorganismen und Pilze). Letztere bewirken den Abbau organischer Substanz durch Mineralisation [33]. Der ständige Wandel von Materie durch die ubiquitären Mikroorganismen ist immens. Die Vielfalt der beteiligten elementaren chemischen Redox-Prozesse ist faszinierend, beginnend mit den sehr unterschiedlichen Qualitäten ihrer "Nahrungsquellen", die vom einfachsten Nahrungsmittel, dem gasförmigen Wasserstoff, bis zu "Steinen", wie z.B. Metallsulfiderzen, reichen [33].

Mit dem Auftreten des Menschen ging eine weitere Veränderung der Verteilung materieller Systeme einher: Anthropogene Einflüsse führten zu einer tiefgreifenden Umgestaltung unserer Umwelt und speziell, über die Kenntnis chemischer Prozeßabläufe, auch zur Erzeugung von Stoffen höherer aus solchen von niedrigerer Qualität, wobei allerdings häufig gleichzeitig vormals unbekannte und daher zunächst *nicht* in das planerische Kalkül einbezogene sekundäre Stoffströme generiert wurden. Die mit einer gewissen zeitlichen Verzögerung eingetretene Veränderung der Ökosysteme unserer Tage ist eine Konsequenz eines Denkens in linearen Kausalzusammenhängen. (Grundsätzlich können im Zuge einer Beeinflussung der Biosphäre durch Toxine und der Erzeugung von Resistenzen unsere "natürlichen" Lebensgrundlagen beeinflußt werden.) Der Mensch kann schließlich auch über Eingriffe in das genetische Material von Organismen eine "künstliche Evolution" und damit eine relevante anthropogene Änderung der Verteilung der materiellen Systeme bewirken.

Entwicklung bedeutet generell im Rahmen einer geschichtlich-zeitlichen Auffassung zum einen Kontinuität etwa entsprechend den physikalischen Phasenübergängen 2. Ordnung und zum anderen Diskontinuität entsprechend den Sprüngen zu einer neuen Qualität und nach bestimmten Kriterien von niederer zu höherer Qualität. Beispiele hierfür sind die Entstehung des Lebens aus "unbelebter" Materie (bei Nichtberücksichtigung der möglichen Bedeutung von chemisch interessanten Übergangsformen, z.B. "Mineralwesen" [34]) und zeitlich versetzt, des Bewußtseins, wobei grundsätzlich die Genese aller höher entwickelten Systeme – z.B. auch einfacher supramolekularer Gebilde [35, 36] (vgl. relevante Kapitel dieses Buches) – eine Reduktion sowohl des Möglichkeitsfeldes als auch der Aggregationsvariabilität der Materie beinhaltet. In diesem Zusammenhang ist wichtig, daß nicht alle Möglichkeiten auf der Basis "anorganischer" Bestandteile, die zur Manifestation einer Informationssteuerung geführt haben, auch tatsächlich auf der Erde im frühen Präkambrium ausprobiert bzw. benutzt wurden. So zeigt die irdische Entwicklung einer biologischen DNA/RNA-Basis aus dem Pool der DNA/RNA-Chemie und damit aus grundsätzlich vielen vorhandenen potentiellen Wahlmöglichkeiten, daß schon in frühester Zeit eine bestimmte Entwicklungsrichtung im Hinblick auf die gesteuerte Konstruktion von Strukturen im Rahmen von Selbstorganisationsprozessen durch die natürlichen Umweltbedingungen selektiert worden ist (vgl. hierzu z.B. die Wahl von Pentosen statt möglicher Hexosen in den Polynukleotiden bzw. im genetischen Material! [37]).

Verstehen wir unter der *Struktur* eines abstrakten oder realen Systems, speziell also auch eines durch herausgegriffene materielle Ingredienzien der Chemisphäre gebildeten Systems, die Gesamtheit der wesentlichen und unwesentlichen, allgemeinen und besonderen, notwendigen und zufälligen Beziehungen zwischen den Elementen des Systems in einem bestimmten Zeitintervall, dann können wir das Wesen des Systems als die Gesamtheit der relativ invarianten inneren Beziehungen betrachten [38]. Das System ist eine relativ stabile, geordnete Gesamtheit von Elementen und Beziehungen, die durch die Existenz bestimmter

Gesetze charakterisiert ist. In der im Evolutionsprozeß natürlichen oder durch den Experimentator erzwungenen Wechselwirkung des Systems mit seiner Umgebung werden Momente des Wesens deutlich, die jedoch nie die Gesamtheit aller inneren Beziehungen ausdrücken. Wir unterscheiden dabei unter den Faktoren, die sich auch auf die Wechselwirkung mit anderen Systemen beziehen, zwischen wesentlichen und unwesentlichen. Als wesentlich kann dabei nur das genommen werden, was den Charakter der speziellen Erscheinung bestimmt, Unwesentliches erscheint in *dieser(!)* Beziehung als vernachlässigbar.

Gelegentlich spielen allerdings manche primär als "Schmutzeffekte" abgetane Faktoren in der Entwicklung einer Wissenschaft später eine wichtige Rolle. Die Historie der Entdeckung des ersten Kronenethers als nicht gewünschtes Nebenprodukt durch Pedersen, der mit Lehn und Cram später den Nobelpreis erhielt, liefert dafür ein Beispiel. Der erste Kronenether war nämlich das Ergebnis einer "Zufallsentdeckung": Der Industriechemiker *C. J. Pedersen* (DuPont, Delaware, USA) wollte aus mono-geschütztem Brenzcatechin **1** und einem Bis(2-chlorethyl)ether **2** ein Bis-phenol **3** synthetisieren. Dabei wurde **1** zufällig in unreiner Form eingesetzt (es enthielt ungeschütztes Brenzcatechin). Aus diesem Grund fiel bei der Aufarbeitung des Reaktionsproduktes neben **3** in geringer Ausbeute (0.4%!) der relevante cyclische Hexaether **4**, d.h. ein Kronenether mit an (vgl. [39]).

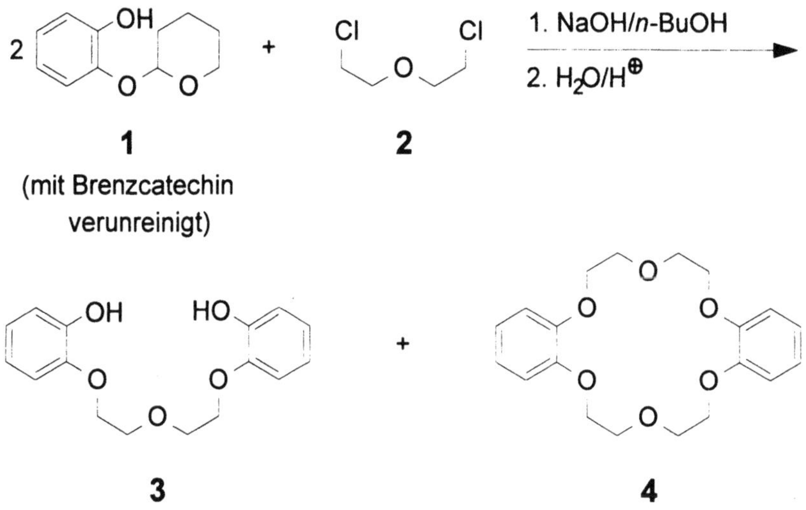

Abbildung 15-1   Entdeckung der Kronenether [Dibenzo[18]krone-6 (**4**)].

Die Gesamtheit der wesentlichen Beziehungen eines Systems, die in einem bestimmten Zusammenhang zwischen seinen Teilsystemen oder auch zu anderen Systemen auftreten, nennen wir Qualität *. Wir folgen damit – auch mit unseren die Chemie betreffenden Aus-

---

*   Qualität ist erscheinendes Wesen. Das Wesen verhält sich zur Qualität wie das Mögliche zum Wirklichen. Qualität ist nicht Eigenschaft schlechthin, sondern wesentliche Eigenschaft, und sie ist nicht nur vom Objekt, dem sie angehört, abhängig, sondern auch vom Bezugssystem. Ein Schritt zur Erfassung des Wesens ist getan, wenn an einem Objekt solche Eigenschaften (Qualitäten) identifiziert werden können, die beim Übergang von einem Bezugssystem zum anderen invariant bleiben. Die Frage, worin das Wesen der

sagen – dem aristotelischen Qualitätsverständnis. In den *"Kategorien"* [40a] hatte Aristoteles auf verschiedene Arten der Qualität verwiesen. Qualität ist für ihn die Art der Beschaffenheit, *"denn Dinge, die sie besitzen, werden deswegen als so und so beschaffen bezeichnet."* Nur auf Grund ihrer Qualität sind sich Dinge ähnlich. Aristoteles betonte in der *"Metaphysik"* [40b]: *"Erste Qualität nämlich ist der Unterschied des Wesens."* Er faßte Qualitäten als *"die Bestimmtheiten der bewegten Wesen [.....], nach welchem man, wenn es wechselt, den Körpern Qualitätsveränderung zuschreibt."*

Eine Qualität eines Systems bestimmen wir also dadurch, daß wir die Beziehungen des Systems hervorheben, die es von anderen unterscheiden. *Die verschiedenen Qualitäten eines Stoffes, den wir in verschiedenen Zusammenhängen in seinen Eigenschaften und Reaktionen, also in verschiedenen qualitativen Äußerungen untersuchen, entsprechen Wesensmomenten.* Dieser Sachverhalt bestimmt z.B. die Arbeitsweise des Chemikers bei der klassischen "Qualitativen Analyse", d.h. beim Nachweis eines Stoffes aufgrund verschiedener charakteristischer Reaktionen, die den verschiedenen Qualitäten entsprechen. Hierzu ist in einem für Nicht-Wissenschaftler geschriebenen Buch, *dem "Schüler Duden – Die Chemie"* folgendes richtig vermerkt: *"Bei der qualitativen Analyse anorganischer Stoffe (Minerale, Salze) geht es darum, durch bestimmte, für die Ionen der einzelnen Elemente charakteristische Nachweisreaktionen die An- oder Abwesenheit dieser Ionen in der Analysensubstanz festzustellen. Da viele Ionen den einwandfreien Nachweis anderer Ionen stören können, wird hier ein Trennungsgang durchgeführt, bei dem bestimmte Gruppen ähnlich reagierender Ionen durch geeignete Reagenzien von der Analysensubstanz abgetrennt werden. Diese Gruppen werden anschließend weiter zerlegt, bis ein einwandfreier Nachweis auf An- oder Abwesenheit der Ionen eines Elements durchführbar ist."* [*]

Die genannten Wesensmomente, die in dem speziellen Fall in Form von verschiedenen charakteristischen Reaktionen zum Tragen kommen, müssen also etwas Gemeinsames haben. Dieses nennen wir Grundqualität oder – im aristotelischen Sinn – das, was die Beschaffenheit ausmacht, die wesentliche Qualität. Die Grundqualität eines Stoffes der Chemisphäre wird durch geometrische und elektronische Strukturelemente in ihren allgemeinen, notwendigen und wesentlichen Beziehungen ausgedrückt. Da wir objektive Gesetze als allgemein-notwendige, d.h. reproduzierbare und wesentliche, den Charakter der Erscheinung bestimmende Beziehungen betrachten können, erkennen wir Wesensmomente der Grundqualität eines Stoffes, indem wir sein gesetzmäßiges Verhalten, etwa sein Reaktionsverhalten, wie z.B. bei der qualitativen Analyse, erforschen (Hierbei kann die Identifikation eines charakteristischen Reaktionsproduktes am Ende durch einfache Sinneswahrnehmung erfolgen und zwar anhand von Farben und Kristallformen). Philosophisch relevant ist, daß wir das Wesen eines Stoffes nicht vollständig erfassen können. Dies würde nämlich *zumindest* die Untersuchung der Wechselwirkung mit "allen" anderen bekannten und *darüberhinaus* noch unbekannten Stoffen voraussetzen.

Chemiker verständigen sich intuitiv durch Qualitätsangaben, die nicht detaillierte Meßergebnisse wiedergeben, sondern auf Ähnlichkeiten mit anderen Substanzen beruhen, wie sie z.B. wie oben beschrieben durch das Studium des Reaktionsverhaltens eruiert worden sind. Auch die gezielte Syntheseplanung erfolgt aufgrund der Kenntnis von Qualitäten von

---

Dinge bestehe, hat in der Geschichte der Philosophie bekanntlich eine zentrale Bedeutung.

[*] In diesem Zusammenhang muß allerdings darauf hingewiesen werden, daß wesentliche Fortschritte der Chemie erst durch die Behandlung quantitativer Aspekte möglich waren.

Stoffen (vgl. hierzu die Einteilung in Stoffgruppen – z.B. Alkohole, Aldehyde, Säuren – die die Basis der Kommunikation bilden).

Wir messen aber immer Quantitäten bestimmter Qualitäten. Die Varianzbreite der Quantität in einer Qualität bezeichnete Hegel als das Maß [40c]. Die *Wahrheit des Seins* ist nach Hegel *das Wesen, das zum Begriff führt*. In diesem Sinne haben wir das Wesen als das gefaßt, was den Charakter der Erscheinung bestimmt, auch damit ein chemisches System von einem anderen unterscheidet. Jedes qualitativ bestimmte System erscheint in der Wechselwirkung mit anderen in bestimmten, wesentlichen Beziehungen. Diese wesentlichen Beziehungen sind relativ. Sie haben zwei Bezugspunkte:

Zum einen ist für ein System, z.B. ein molekulares Gebilde, jede innere Beziehung wesentlich, die seine Funktion und damit seine Existenz garantiert. Chemierelevant sind insbesondere Beziehungen, die thermodynamische oder kinetische Stabilität bewirken. Erkenntnistheoretische Schwierigkeiten ergeben sich aus der hier immer vorhandenen Multifunktionalität von Elementen und Beziehungen (Eine Substanz kann sich in ihrem Verhalten zu verschiedenen anderen Stoffen als "Chamäleon" erweisen; vgl. obige Aussage zur Qualitativen Analyse sowie spätere Ausführungen hierzu). Wesentliche Beziehungen entsprechen Realisierungen der in objektiven Gesetzen enthaltenen Möglichkeiten. Jedes konkrete Objekt oder Ereignis ist gesetzmäßig, weil es einem System von Gesetzen unterliegt, die sein Verhalten hierarchisch beeinflussen und damit zu einer Hierarchie der wesentlichen Beziehungen führen.

Zum anderen zeigen sich wesentliche Beziehungen in der Wechselwirkung mit anderen Systemen. Dem entspricht das intuitive Vorgehen des Chemikers, wenn er die Eigenschaften einer Substanz aus dem Reaktionsverhalten mit einer gezielt als "sinnvoll" herausgegriffenen Menge von anderen Substanzen ermittelt. Die Chemie muß daran interessiert sein, durch Reaktionen Stoffe mit neuen, vielleicht emergenten oder auch maßgeschneiderten Eigenschaften zu finden. Dazu werden die Stoffe mit anderen "in Beziehung bzw. zur Reaktion gebracht". Hier ist wesentlich all das, was die Erklärungs- oder Konstruktionsziele der erklärenden und konstruierenden Subjekte konkret betrifft.

Die bisherigen Betrachtungen zu den Erscheinungsformen des Wesens eines Stoffes in verschiedenen Zusammenhängen führen zu einer Differenzierung der Qualitäten. Es existiert immer ein Bezugssystem für die Qualität. Die Grundqualität eines Stoffes, ausgedrückt in wesentlichen, inneren geometrischen und elektronischen Strukturelementen, äußert sich in verschiedenen Reaktionen unterschiedlich [*]. Wir können dabei von verschiedenen Ausprägungen der gleichen Grundqualität sprechen. Die konzentrierte Schwefelsäure kann z.B. wasserentziehend wirken, sie kann Metalle oxidieren bzw. auflösen, und sie kann Kohlenwasserstoffverbindungen in Sulfonsäuren (eine grundlegende Reaktion der chemischen Industrie) überführen. Das in anderem Zusammenhang genannte Aspirin® hat zahlreiche pharmakologische Wirkungen von der analgetischen bis zur antipyretischen [41]. Die Existenz *unterschiedlicher Ausprägungen* der gleichen Grundqualität in einfachen klassischen chemischen Reaktionen ist von einem Qualitätswandel bei einem "evolvierenden" Prozeß zu unterscheiden. Qualitätswandel basiert auf einer neuen Gesamtheit von relativ beständigen inneren Beziehungen des Systems. Das bei der Definition der Struktur genannte Zeitintervall ist dabei durch die Grundqualität bestimmt.

---

[*] Hier wird bewußt noch nicht zwischen den komplementären Mikro- und Makroebenen, d.h. zwischen der Ebene der Moleküle und derjenigen der Substanzen, unterschieden, da dies Gegenstand einer weiteren Arbeit sein soll.

Ändert sich die Grundqualität, dann existiert eine neue wesentliche Struktur mit *neuen Qualitäten*.

Neue Qualitäten sind gemessen an Kriterien, die für die Endqualität gegenüber der Ausgangsqualität eine qualitativ umfangreichere und bessere Funktionserfüllung konstatieren, *höhere Qualitäten*. Das ist deshalb ein relativer Begriff, weil immer der Bezug zu bestimmten Kriterien herzustellen ist, gemäß denen das "Höhere" einer Endqualität, z.B. in einem Entwicklungszyklus, im Vergleich zu der Ausgangsqualität, die als die niedrigere bezeichnet wird, festgestellt werden kann. Von herausragender Bedeutung ist die natürliche Entwicklung, die über die Entstehung das Lebens bis hin zum sozial organisierten Menschen führte. Die natürliche Entwicklung unterscheidet sich hierbei grundsätzlich von der menschlichen Konstruktion artifizieller Systeme, in der andere, neue und möglicherweise höhere Qualitäten im Rahmen der Möglichkeitsfelder, die durch die Systemgesetze aufgespannt werden, erzeugt werden.

Hier kann auch beispielhaft auf Fitnessbetrachtungen hinsichtlich der Produktion verschiedener Polynukleotide im Rahmen des Anderson-Modells der präbiologischen Evolution [42] hingewiesen werden. Auch auf "der höheren Proteinebene" läßt sich mittels vergleichender Sequenzanalyse – bei phylogenetischer Betrachtungsweise – das Entstehen immer neuer (verbesserter) Funktionalitäten nachweisen. Diese Überlegungen sind leicht nachvollziehbar und im übertragenen Sinne auch von realem praktischen Interesse, insbesondere wenn sich die chemischen Grundqualitäten (also die innere Geometrie und die elektronischen Strukturelemente) bei der "Reaktion", d.h. bei den zu neuen Funktionalitäten führenden Prozessen, nicht wesentlich ändern. Dies gilt z.B. für organisch-chemische Reaktionen, bei denen lediglich funktionelle Gruppen verändert werden. Hier können Qualitätsverbesserungen pharmakologisch relevant sein. So lassen sich z.B. unterschiedliche Wirkungen und Nebenwirkungen von der in Weiderinden vorkommenden Salicylsäure und der Acetylsalicylsäure (ASS®, Aspirin®), dem bekannten Produkt der Fa. Bayer, vergleichen. Die schädlichen Nebenwirkungen – z.B. auf die Magenschleimhaut – sind beim Naturstoff signifikant größer [41].

Qualitäts-Kriterien sind in eine Hierarchie von Entwicklungsniveaus einzuordnen. Im groben Raster, bezogen auf die Existenz und Entwicklung komplexerer Organismen, sind diese an die Herausbildung chemischer Vorformen für die Entstehung des Lebens, der Vielfalt der Arten und der menschlichen Sozietät gebunden. Bei Konstruktionen, also in der artifiziellen Entwicklung, sind die Kriterien – wie bereits erwähnt – an den Gestaltungszielen orientiert. So besitzt etwa die elektronische Datenverarbeitung eine höhere Qualität als das Rechenbrett, und zwar bezogen auf die Komplexität von Programmen und den Umfang der verarbeitbaren Daten. Die Feststellung, ob etwas eine höhere Qualität darstellt, verlangt die Kenntnis des Entwicklungsablaufs, Beachtung der Kriterien und Vergleichbarkeit der jeweiligen Qualitäten.

Das Wesen einer chemischen Substanz kann, wie betont, nicht vollständig erfaßt werden. Es geht auf in der Multidimensionalität seiner Potentialität, d.h. des Reaktionsvermögens mit allen bekannten *und* noch unbekannten Stoffen. Als Beispiele seien genannt die ungewöhnlich große Zahl von pharmakologischen Wirkungen der Acetylsalicylsäure, die Einsatzmöglichkeiten des Porphyrinringes in verschiedenen biochemischen Prozessen und die zahlreichen Reaktionen, die selbst mit einfachen Substanzen wie der Schwefelsäure durchgeführt werden können. Aspekte der Potentialität einer seit langer Zeit bekannten Substanz erfährt man häufig erst ziemlich spät. Ein aktuelles Beispiel ist Eisensulfid, das mit Schwefelwasserstoff eine für die präbiotische Evolution möglicherweise wichtige Reaktion zeigt [43]. Generell wird die Potentialität des Eisenschwefelsystems als immens wichtig für die Evolution angesehen (vgl. z.B. die Überschrift des ersten Kapitels *"Die Zeugen des Anfangs: Eisen und Schwefel"* eines für Laien geschriebenen, neueste Erkenntnisse zusammentragenden Buches [44]). Spannend in dieser Hinsicht ist z.B. sicherlich die vielschichtige Geschichte eines Eisenatoms im Zeitraum von etwa 4 Mrd. Jahren (vgl. Anhang 15.6), aber auch die große Zahl der "Einsatzmöglichkeiten" eines Eisen(II/III)-kations nach der Resorption in einen Organismus, übrigens ein interessanter Fall für nichtlineares Geschehen. Bemerkenswerterweise verläuft häufig weder der anthropogen initiierte Reaktionsweg einer Substanz in der Umwelt, wie z.B. der eines Pestizides, noch der eines Pharmakons im menschlichen Organismus wegen der immer vorhandenen Vielfalt der Qualitäten und des komplexen Basissystems mit gekoppelten Reaktionsabläufen linear im Sinne von vorhersehbar.

Entwicklung ist ein komplizierter Prozeß, der andere, neue (gelegentlich vielleicht gar "höhere"?) Qualitäten hervorbringt, was wir häufig auch als Emergenz * bezeichnen. Meist betrachten wir nur bestimmte Phasen von Entwicklungszyklen und denken kaum über den Platz einer Reaktion in unterschiedlichen Zeithorizonten nach [45a]. So kann Entwicklung erst hinreichend begriffen werden, wenn Ausgangs- und Endqualität über einen, meist historisch langen, Zeitraum betrachtet worden sind. Qualitätswandel bei einfachen chemischen Reaktionen kann grundsätzlich Teil größerer Entwicklungszusammenhänge sein. Die Einbeziehung der Zyklizität bei der Entwicklung zeigt sich darin, daß höhere Qualitäten in einem Prozeß der Stagnation und Regression sowie der Ausbildung von Elementen einer Entwicklungsphase entstehen können.

Die Unerschöpflichkeit der Materieformen läßt die Annahme (schlafender) Propensitäten zu. Die gesamte Vielfalt der heutigen irdischen Erscheinungsformen, die den menschlichen Geist zum Prinzip der Unerschöpflichkeit der Materie geführt hat, wurde durch (spontane) konservative (z.B. in der Geosphäre) sowie dissipative Strukturbildung, speziell durch Einbeziehung informationsgesteuerter Abläufe (z.B. bei der Genexpression), aber auch durch anthropogene Prozesse erzeugt. Diese Vielfalt stellt aber nur einen kleinen Ausschnitt aus dem Möglichkeitsfeld dar, das durch die Ingredienzien der Materie, wie z.B. die chemischen Elemente, aufgespannt wird. Der Begriff Unerschöpflichkeit der Materie läßt sich in diesem Zusammenhang eindrucksvoll zu dem eingangs erwähnten "Statement" von R. P. Feynman in Beziehung setzen. Die Realisierung eines geringen Teils der in diesem Möglichkeitsfeld verborgenen und schlummernden Optionen, *vor allem auch durch nichtgesteuerte Abläufe*, vollzieht sich in den Spannungsfeldern von Möglichkeit und Wirklichkeit, von Zufall und Notwendigkeit und bezüglich erkenntnistheoretischer Aspekte von Emergenz und Reduktion. Die Rolle des Zufalls im Evolutionsprozeß oder überhaupt im Naturgeschehen ist fundamental. Selbst für die Erzeugung der Vielfalt der Millionen Antikörper hat nach der klonalen Selektionstheorie der Zufall Pate gestanden [46]. Wir wären ohne die angesprochene Vielfalt den Angriffen der feindlichen Welt pathogener Keime wehrlos ausgesetzt. Zufälle *und* die intrinsischen Eigenschaften materieller Systeme bestimmten/bestimmen das Geschehen. Stellen wir die sibyllinische Frage: Hat der Zufall einen Willen? [47] Sicher ist er kein bewußt handelndes Wesen, wohl aber Entwicklungs- und Konstruktionsprinzip für die Entstehung von Neuem durch das in ihm "enthaltene" Mögliche, nicht Präformierte.

Der Begriff des Möglichkeitsfeldes wird zum zentralen Terminus der Naturphilosophie. In jeder materiellen Einheit, d.h. speziell jeder chemischen Substanz, sind schlafende Pro-

---

* Bedingungen für Emergenz sind Unvorhersehbarkeit, Neuartigkeit und Irreduzibilität (relevante Definitionen sowie Literatur in [8b]).

pensitäten enthalten, wobei es gegebenenfalls vom Zufall als einer Erscheinungsform der Notwendigkeit * und der Gesetze abhängt, welche Propensitäten unter den jeweiligen "Umweltbedingungen" zum Tragen kommen und ob und auf welche Weise hierdurch Veränderungen im chemisch-materiellen Reaktionsgeschehen verursacht werden können. Damit stellt sich die Frage nach den Voraussetzungen für die Entstehung *unserer Welt*, nach den Faktoren, welche für ein Verständnis der Genese, Interaktion und Mutation materieller Systeme im Rahmen chemischer Prozesse eine außerordentliche Relevanz besitzen; es stellt sich die Frage nach den intrinsischen Propensitäten materieller (chemischer) Systeme. Hier kann der Chemiker auf der Ebene der unbelebten materiellen Systeme ausloten, welche Grundprozeßarten und Grundphänomene relevant sind, wie z.B. Wachstum und dessen Begrenzung, artifizielle Replikation, Steuerung der Verknüpfung einfacher chemischer Fragmente, Ablesefehler bei der informationsgesteuerten Verknüpfung, Irreversibilität, Singularitäten in Systemen mit Fehlordnungen und katalytisch bzw. steuernd wirkende Oberflächenstrukturen.

15.4 Determinismus, Linearität und Nicht-Linearität des anthropogenen und evolutionären Geschehens

Philosophie hat sich als Kritik überholter Denkweisen zu bewähren. Unsere bisherigen Überlegungen zeigen die Mängel einer mechanistischen Denkweise beim Versuch, Naturprozesse zu verstehen. In sich konsistent und übersichtlich bei der Welterklärung war die mechanistische Denkweise des 18. und 19. Jahrhunderts, z.B. basierend auf der bekannten Laplaceschen Annahme: Wenn die Wirklichkeit aus letzten unteilbaren Teilchen besteht, die schwer und träge sind sowie konzentriert den Raum erfüllen, dann kann man diese als Massenpunkte behandeln, deren Bewegungsgleichungen durch die klassische Mechanik gegeben sind. Dieses einfache Weltbild unterlag aus verschiedenen Gründen der Kritik. Es konnte die biotische Evolution und generell die Entstehung von Neuem nicht erklären, negierte den Zufall und wurde der wirklichen Nicht-Linearität des Geschehens nicht gerecht. Seither dominiert die Auseinandersetzung mit dem mechanistischen Weltbild, ohne daß schon eine einheitliche Welterklärung auf einer neuen Basis existiert.

Die Linearität der mechanistischen Denkweise drückte sich in verschiedenen Positionen aus. Der kausale Ablauf des Geschehens, den man allgemein als die konkrete inhaltlich und zeitlich gerichtete Vermittlung des Zusammenhangs von Elementen eines Systems betrachten kann, wurde im mechanischen Determinismus auf die *notwendige* Verwirklichung *einer(!)* Möglichkeit unter gegebenen Bedingungen eingeschränkt. Damit war theoretisch klar, daß nichts Neues im Sinne von Emergenz entstehen konnte, es sei denn, es wäre, bezogen auf den Prozeß der Evolution, vorher schon präformiert in der Ursuppe vorhanden und nur noch nicht bekannt gewesen, oder es wurde durch immaterielle, also wissenschaftlich nicht erfaßbare "Kräfte" geschaffen. Das ist eine Position, die auch schon einem Helmholtz, der von der prinzipiellen Begreifbarkeit der Natur ausging, nicht paßte. Führt man jedoch etwa mit der Quantentheorie den "objektiven Zufall" [45b] in die Naturerklärung ein, dann wird die Kausalität anders zu fassen sein. Die Nicht-Linearität faßt Kausalität als die Verursachung von Wirkungen, wobei ein vorhandener Bedingungs-

---

* Hegel spricht von der Erkenntnis der unter dem Schein der Zufälligkeit verborgenen Notwendigkeit (vgl. Hegel, G. W. F. *Enzyklopädie der philosophischen Wissenschaften im Grundrisse (1830)*, Teil 1, Suhrkamp, Frankfurt am Main, **1986**, 286).

komplex der materiellen Ingredienzien ein Möglichkeitsfeld des (evolvierenden) Geschehens konstituiert, von dem, durch eine Anfangsursache ausgelöst, sich mit einer bestimmten Wahrscheinlichkeit eine Möglichkeit verwirklicht, die Endwirkung des Prozesses ist [*]. Die Vereinfachung der Beziehungen zwischen Anfangsursache und Endwirkung als

---

[*] *"Die heutige Wissenschaft zeigt, daß die Natur erbarmungslos nicht-linear ist. [.....]. Die Linearität ist eine Falle. [......]. Der lineare Habitus war so tief verwurzelt, daß viele Wissenschaftler und Ingenieure in den vierziger und fünfziger Jahren kaum etwas anderes kannten."*
(*I. Stewart*)

(a) *Linearität ist eine Struktureigenschaft mathematischer Ausdrücke.* (Die Elemente der Struktur sind mit den Elementen eines zugeordneten (Zahlen-)Körpers (reelle und komplexe Koeffizienten) multiplikativ verknüpft und treten untereinander in additiver Verknüpfung auf). Wichtige Gleichungen der Physik – wie die Hamilton-kanonischen und die Schrödinger-Gleichung – sind linear. Die Welten der klassischen Mechanik und der Quantenmechanik sind konservativ, d.h. nicht dissipativ und auch invariant gegenüber einer Zeitumkehr. Charakteristisch für lineare Systeme ist das Superpositionsprinzip. (Während bei Gültigkeit des Prinzips der Effekt der gemeinsamen Wirkung zweier Ursachen der Überlagerung der Wirkungen der Einzelursachen entspricht, kann in nicht-linearen Systemen eine zusätzliche Ursache dramatische Effekte zur Folge haben). Die relevante Physik linearer Prozesse kann – wie im Text angesprochen – allerdings nicht die Kreativität des Kosmos erklären.

Streng betrachtet sind nun alle realen Prozesse – speziell auch chemische Reaktionen – irreversibel, da die Ausgangssituation nicht ohne bleibende Änderungen in der Umgebung wieder hergestellt werden kann. Approximativ lassen sich einige spezielle irreversible Prozesse auf der Basis der Gibbsschen Fundamentalgleichung (lokales Gleichgewicht) mit einem linearen Ansatz beschreiben. Bei geringen Abweichungen vom Gleichgewicht bestehen lineare Beziehungen zwischen (allgemeinen) Strömen und Kräften entsprechend den Onsagerschen Reziprozitätsbeziehungen. Der zeitliche Ablauf chemischer Reaktionen, die so lange ablaufen, bis sich der Gleichgewichtszustand eingestellt hat, kann aber im allgemeinen nicht mit einem linearen Ansatz beschrieben werden, sondern nur durch ein kompliziertes System von Differentialgleichungen. Es gibt Ansätze zur Beschreibung von komplizierten gekoppelten Reaktionen in komplexen Multikomponentensystemen. Von besonderem Interesse sind für uns – vor dem Hintergrund der Betrachtung von Strukturbildung bzw. von Emergenz – Situationen, die mit dem Auftreten von Instabilitäten verbunden sind. So kann die autokatalytische Reaktion $A_1 + A_2 \leftrightarrow 2A_1$ (Koeffizienten der entsprechenden Markow-Gleichungen werden nicht-linear) zu Instabilitäten und dissipativen Strukturen führen, wenn sie zusammen mit einem anderen stationären chemischen Umwandlungsprozeß abläuft. Dies gilt z.B. für die im Text erwähnte oszillierende B.Z.-Reaktion.

Man kann mit Recht sagen, daß einige in den Nicht-Linearitäten verborgenen Möglichkeiten in oder nahe dem Gleichgewicht nur durch das Nichtgleichgewicht verwirklicht werden. Die Komplexität unserer Welt – speziell von dissipativen Strukturen – kann erst unter Berücksichtigung nicht-linearer Gleichungen zutreffend beschrieben werden, wobei Nicht-Linearität und Abstand vom thermodynamischen Gleichgewicht zu Mehrfachlösungen bzw. Bifurkationen als Folge von Instabilitäten führen (weitere Folgen sind: Ausbildung und Aufrechterhaltung von Korrelationen makroskopischer Reichweite). Dabei ist Nicht-Linearität eine notwendige, aber nicht hinreichende Bedingung.

In physikalisch-chemischen Systemen sind wegen der Wirkung thermischer Fluktuationen die im Text angesprochenen zufälligen Elemente vorhanden. Dies ist wichtig, da stochastische Effekte eine entscheidende Rolle in der Nähe von Bifurkationspunkten spielen. Für dissipative Systeme ist

linear und notwendig mußte aufgegeben werden, da die Wirklichkeit sich als komplizierter erwies, als es die in bestimmtem Rahmen fruchtbaren Denkkonstruktionen der mechanischen Deterministen ausweisen. Nicht-Linearität ist mit der Existenz des "objektiven Zufalls" – mit den wirklichen bei evolutionären Prozessen nicht vorhersehbaren Übergängen von einem Zustand in den anderen – verbunden. Das analytische Herauslösen bestimmter Beziehungen aus dem Beziehungsgeflecht durch das Denken ist gestattet, wenn

---

daher das fundamentale Niveau das einer statistischen Behandlungsweise, so daß auch die Entschlüsselung der im Text angesprochenen wichtigen Beziehungen zwischen konservativen und dissipativen dynamischen Systemen durch die Aufklärung des Zusammenhangs zwischen deterministischer Dynamik und Wahrscheinlichkeit zugängig ist. Die Propensität der Materie äußert sich nämlich sowohl bei konservativen als auch bei dissipativen Selbstorganisationsprozessen, bei letzteren nur wenn – wie schon erwähnt – nicht-lineare Dynamik und Nichtgleichgewichtsbedingungen zusammenwirken. In diesem Zusammenhang ist auch interessant, daß Neurophysiologen, die sich mit Signalwandlung und Informationsverarbeitung befassen, zu der Aussage kommen, daß sich auf der zellulären Ebene das Neuron als stochastisches, nicht-lineares dynamisches System darstellt.

Prozesse der Selbstorganisation in der Natur basieren auf bestimmten Elementarprozessen und -eigenschaften (wie z.B. Selbstreproduktion (Replikation), Bi-/Multistabilität, Konkurrenz, Mutation, Selektion, Speicherung, Informationsverarbeitung), von denen hier nur die ersten beiden kurz charakterisiert werden sollen.

Zuerst zur der Selbstreproduktion, die für die Evolution von grundlegender Bedeutung ist: Obwohl es für bestimmte Prozesse der Selbstreproduktion lineare Wachstumsgesetze gibt (d.h. die Geschwindigkeit ist der Anzahl der vorhandenen Moleküle proportional), verläuft das Wachstum bei einigen Prozessen schneller (Selbstbeschleunigung) oder langsamer (Selbsthemmung). Ein Beispiel für den ersteren Fall bzw. für die Selbstreproduktion stellt die autokatalytische Schlögl-Reaktion dar ($A + 2X \leftrightarrow 3X$; $X \leftrightarrow F$).

Nun zur Multistabilität: Liegt bei einem nicht-linearen Prozeß Multistabilität vor, so können in Abhängigkeit von der Vorgeschichte verschiedene Zustände angenommen werden (Gedächtniseffekt!). Hierdurch ist die Möglichkeit der Informationsspeicherung gegeben. Die oben angegebene Reaktion weist z.B. zwei stabile stationäre Zustände für die Konzentrationen von X auf. Offene chemische Systeme, die mehrere stabile Zustände besitzen, sind wegen der Bedeutung von einfachen Informationsspeichersystemen für die frühe Evolution auch als Bioide in der Literatur bezeichnet worden (zu Details vgl. Kapitel 5 von [10], verschiedene Kapitel von [42], Fußnote S. 195 und zum obigen Zitat: Stewart, I. *Spielt Gott Roulette? Chaos in der Mathematik*, Birkhäuser, Basel, **1990**, 89).

(b) *Linearität im philosophischen Sinne* umfaßt gerichtete Prozesse, deren Zusammenhang zwischen Ursache und Wirkung direkt und zeitlich unmittelbar ist. Nicht-Linearität ist verbunden mit Komplexität. Komplexität umfaßt Systeme mit mindestens drei wechselwirkenden Elementen, unterschiedlichen Kopplungen und Verhaltensweisen, die das Entstehen und die Existenz anderer, neuer und höherer Qualitäten unter bestimmten Bedingungen ermöglichen (vgl. auch Mainzer, K. *Thinking in Complexity*, Springer, Berlin, **1994**; Philosophical Foundations of Nonlinear Complex Systems, in: Haken, H.; Mikhailov, A. (Hrsg.), *Interdisciplinary Approaches to Nonlinear Complex Systems*, Springer Series in Synergetics, Vol. 62, Springer, Berlin, **1993**; Götschl, J. *Zur philosophischen Bedeutung des Paradigmas der Selbstorganisation für den Zusammenhang von Naturverständnis und Selbstverständnis*, in: Krohn, W.; Küppers, G. (Hrsg.), *Selbstorganisation: Aspekte einer wissenschaftlichen Revolution*, Vieweg, Braunschweig, **1990**, 181)

man sich der möglichen Folgen der Einschränkung der Kompliziertheit der Ursache-Wirkungs-Relationen auf einen Typ bewußt ist. Historisch interessant ist es, an den genialen Forscher Boltzmann zu erinnern, der probabilistischen Vorgängen und Fluktuationen im Naturgeschehen eine zentrale Bedeutung zuschrieb [48] (vgl. auch obige Ausführung zur H(t)-Kurve). Die mechanistische Kausalität tritt dort wirklich auf, wo der Bedingungs-komplex nur eine wesentliche Möglichkeit zuläßt, die sich mit Notwendigkeit, d.h. der Wahrscheinlichkeit 1, realisiert. (Dies wäre z.B. bei der geplanten Synthese des Chemikers mit der hypothetischen Ausbeute 100% der Fall, was für jedes Molekül *einen* eindeutigen, voraussehbaren Reaktionsweg zur Folge hätte). Das ist aber die am wenigsten in der Wirk-lichkeit auftretende Variante der Kausalität, die von der umfassenderen dialektischen Formulierung von der zeitlich und inhaltlich gerichteten Vermittlung des Zusammenhangs zwischen Anfangsursache und Endwirkung umfaßt wird.

Linearität des Geschehens wird auch angenommen, wenn komplexe Systeme auf die Summe ihrer Elemente reduziert werden. Schon die Thermodynamik zeigt, daß Temperatur nicht auf die kinetische Energie von *einzelnen* Teilchen zurückzuführen ist. Komplexe Systeme, auch biologische oder allgemein dissipative, bedürfen ihrer eigenen Erklärungs-komponenten. Es geht um die Einsicht in die relevanten Systemgesetze, die das Verhalten des Systems bestimmen. Dabei zeigt sich in vielen Fällen erst mit der statistischen Struktur der Gesetze der Zusammenhang zwischen System und Elementen. Wenn sich unter System-bedingungen für das System eine Möglichkeit notwendig realisiert, wie etwa die Schrödinger-Gleichung (eine allerdings lineare Differentialgleichung 2. Ordnung) die Verteilung von "Teilchen" auf einem Schirm nach dem Durchgang durch einen Spalt angibt, dann ergibt sich daraus für das Verhalten der Elemente ein Möglichkeitsfeld, das durch die Gesamtverteilung bestimmt ist, von dem die Elemente mit Wahrscheinlichkeit bestimmte Möglichkeiten so realisieren, daß die gesetzmäßige "beobachtbare" Verteilung heraus-kommt. *Das Systemgesetz bestimmt also keineswegs eindeutig das Elementverhalten, bildet aber einen Rahmen* [49], in dem wahrscheinliche Übergänge auftreten. Dem entspricht formal auch die Unmöglichkeit, das Reaktionsverhalten eines speziell herausgegriffenen Moleküls eines Ausgangsstoffes bei einer Reaktion, die zu mehreren Endprodukten führt, vorauszusagen, obwohl das Massenverhältnis der Reaktionsprodukte durch das Massen-wirkungsgesetz von Guldberg und Waage, das für die Chemie von eminenter Bedeutung ist, vorgegeben ist. Das alles führt zur Frage nach der theoretischen Durchdringung der Komplexität unter dem Aspekt der Nicht-Linearität, die Teil des Geschehens ist. Forschungen zur Selbstorganisation dissipativer Systeme erfassen dies immer besser, und es ist vor allem interessant, bei hoch rückgekoppelten *klassischen* chemischen Reaktionen nach möglichen Bifurkationspunkten zu suchen [50]. (Die bekannte oszillierende (chemische) B.Z.-Reaktion erfolgt dagegen weit ab vom Gleichgewicht [51]). Chemisches Gleichgewicht selbst erscheint als eine Idealisierung, die wesentliche Seiten der Prozesse hervorhebt, und die die vorhandenen Schwankungen um das Gleichgewicht als unwesent-lich vernachlässigt. Hier kann jedoch grundsätzlich im Sinne von [50] unter bestimmten Be-dingungen Unwesentliches in Form von Fluktuationen, die verstärkt werden, wesentlich werden.

Ein weiterer Aspekt linearen Denkens ist das Negieren des Qualitätswandels und der Entwicklung oder der Emergenz. Jede gegenwärtige Struktur ist aber geronnene Ent-wicklung, denn sonst gäbe es Strukturwunder, deren Entstehen nicht erklärt werden könnte.

Damit hat jedes materielle System in seiner Strukturierung eine Geschichte *. Zitieren wir in diesem Zusammenhang Kekulé [52]:

*"Die Chemie ist also die Lehre von den stofflichen Metamorphosen der
Materie. Ihr wesentlicher Gegenstand ist nicht die existirende Substanz,
sondern vielmehr ihre Vergangenheit und ihre Zukunft. Die Beziehungen
eines Körpers zu dem was er früher war und zu dem was er werden kann
bilden den eigentlichen Gegenstand der Chemie."*

Hier sei nochmals an die Multidimensionalität der Potentialität einer einzigen chemischen Substanz, die allerdings erst aus der hypothetischen Wechselwirkung mit allen anderen bekannten und noch unbekannten resultiert, erinnert.

Die Geschichte wird immer unwesentlicher, je isolierter das Objekt ist. Ein Elementarteilchen, völlig isoliert von jeder Umgebung, ist eine nicht realisierbare Gedankenkonstruktion, da immer Wechselwirkung vorhanden ist und selbst ohne Wechselwirkung Systeme immer verschränkt sind [53] (Wechselwirkung besteht natürlich auch im Experiment mit dem Experimentator). Das Elementarteilchen hätte keine Geschichte. Dies ist eine Tatsache, die der Chemiker bei vielen Modellbetrachtungen meist nicht reflektiert, z.B. bei der Annahme der Individualität von Elektronen.

Geschichte hinterläßt Spuren, eben Strukturen. Strukturen verweisen auf Wechselwirkung mit anderen Systemen, enthalten also potentielle Informationen, die dann abgerufen werden, wenn nach einer Art Schlüssel/Schlüsselloch-Prinzip ein Empfänger die in der Struktur codierte Information aufnimmt. Information ist damit widerspiegelnde und steuernde Struktur (auch entsprechend Monods Aussage bezüglich der Epigenese [54]). Information kann als ein wesentlicher Ausdruck der Nicht-Linearität des Geschehens angesehen werden, denn potentielle Informationen können unter bestimmten verschiedenen Bedingungen verschieden zum Tragen kommen, den Reaktionsablauf in Abhängigkeit vom Gegenüber beeinflussen, indem sie als Strukturen und damit als Repräsentanten geronnener Entwicklung bzw. als Widerspiegelung der Geschichte des Systems und Spuren vorheriger Wechselwirkung weiteres Geschehen regulieren.

Hierzu sei ein signifikantes Beispiel von Monod angeführt, ein Beispiel, das sowohl die Bedeutung der Information in der obigen Sichtweise veranschaulicht, als auch der elektromagnetischen Kraft, also einer der vier Kräfte der Natur, in Form von nichtkovalenten Wechselwirkungen in sehr unterschiedlichen Bereichen der Natur eine dominante Funktion zuweist [54]:

*"Und wenn man bei den Ribosomen oder den Bakteriophagen auch nicht mehr von
Kristallisation sprechen kann, da diese Teilchen einen sehr viel höheren Komplexitäts-,
d.h. Ordnungsgrad aufweisen als die uns bekannten Kristalle, so sind doch letzten Endes die
ins Spiel kommenden chemischen Wechselwirkungen von der gleichen Art wie jene, die
einen molekularen Kristall aufbauen. Wie bei einem Kristall bildet eigentlich die Struktur
der aggregierten Moleküle die "Informations"quelle für den Aufbau des Ganzen. Das Wesen
dieser epigenetischen Prozesse besteht folglich darin, daß die Gesamtorganisation eines*

---

* Dies gilt leicht nachvollziehbar für nichtkristalline Stoffe, z.B. Gläser, deren Entropie auch beim absoluten Nullpunkt nicht verschwindet und deren "Struktur" von ihrer Vorgeschichte abhängt.

*komplexen multimolekularen Gebildes potentiell in der Struktur seiner Bestandteile ent-
halten ist, sich aber erst offenbart und damit wirklich wird durch ihren Zusammenschluß."*

Der *Informationsbegriff*, z.B. in der Definition von C. F. von Weizsäcker (*"Information
ist das Maß einer Menge von Form ... Materie hat Form ..."* [55]), ist in diesem Zusammenhang für die Chemie von entscheidender Bedeutung, speziell auch für die Klassifikation von
konservativen Selbstorganisationsprozessen, die zu mesoskopischen Produkten führen [56]
oder zur Faltung von Proteinen [57]. Die genannten Prozesse beruhen nämlich auf der Information, die in den Edukten gespeichert ist, und auf der Art der Wechselwirkung
zwischen den relevanten Informationsträgern. Dabei erzeugt vorhandene Information neue
Information: Im Entstehungsprozeß komplexer, molekularer Systeme kann beispielsweise
nach einer Nukleations- eine Wachstumsphase, wie z.B. beim Tabakmosaikvirus [57],
ablaufen. (Die Natur versucht mit einem Minimum an Genen, die im vorliegenden Fall
lediglich den elementaren Baustein kodieren, auszukommen). Auf die Synthese-Chemie
übertragen bedeutet dies: es besteht grundsätzlich die Möglichkeit, große Intermediate, die
in der Reaktionslösung gebildet worden sind, durch synthetische Kniffe entsprechend der in
ihnen enthaltenen Information in einer "Wachstumsphase" zu Riesensystemen weiter zu
verknüpfen [36] *. Wir erleben dann simultan Entstehung von Information, (molekularer)
Komplexität und Differenziertheit auf nanodimensionaler Ebene. Im Sinne des Prinzips "aus
Kleinheit werde Größe, aus Einfachheit werde Komplexität" entspräche dies evolutionären
Naturprozessen.

Konservative Selbstorganisationsprozesse spielen bei der Morphogenese, speziell bei der
Epigenese, als nichtgengesteuerte Reaktionen eine entscheidende Rolle. Man kann sich dies
anhand eines Vergleiches der relativ kleinen Zahl der menschlichen Gene mit der extrem
großen Zahl der Zellen des Adulten verdeutlichen. Nur durch das ausgeklügelte Wechselspiel von dissipativer und konservativer Selbstorganisation konnte im Möglichkeitsfeld
materieller Systeme die Evolution in Gang gesetzt werden. Die realen Möglichkeiten
resultieren jeweils aus der vorhergehenden Entwicklung. Die immense Bedeutung der
konservativen Prozesse, die zur Vielfalt der Erscheinungsformen wesentlich beitragen,
wurde bisher wegen des problematischen Zugangs zur relevanten Komplexität offensicht-
lich nur unzureichend gewürdigt. So vermißt man in den ausgezeichneten Werken von
Haken [9] und Prigogine [10] (zur Synergetik, zur Theorie dissipativer Strukturen, bzw.
zum Problem der Selbststrukturierung der Materie) detaillierte Aussagen zum Wechselspiel
von konservativen *und* dissipativen Selbstorganisationsprozessen (vgl. [36]). Die Theorie
dissipativer Strukturen liefert, um es deutlich zu sagen, lediglich den Rahmen für die
biotische Evolution.

Mit den Forschungen zur Information, Selbstorganisation und Emergenz zeigen sich
wichtige Tendenzen, die eine beziehungsreiche, neue Sicht auf die Ordnung der Wirklichkeit ermöglichen. Information entspricht der steuernden Struktur und kann sowohl Qualitätserhaltung, als auch Qualitätsänderungen, Qualitätswandel und Emergenz bewirken. Sie
ist eine in allen Struktur- und Entwicklungsebenen der Materie vorhandene Existenzform
mit qualitativ verschiedener Ausprägung in bestimmten Niveaus. Der Mechanismus, der zu
Qualitätserhaltung, Qualitätsänderungen, Qualitätswandel und Emergenz führt, wird durch
die Forschungen zur Selbstorganisation mittels eines Wechselspiels von konservativen und

---

* Der Übergang von der Mikro- zur Makroebene ist vom physikalischen aber auch vom
erkenntnistheoretischen Standpunkt von großem Interesse [8b, 62].

dissipativen Prozessen aufgedeckt und das Ergebnis, die Existenz und Entstehung von anderen, neuen und höheren Qualitäten, erfassen wir als emergent.

Der Übergang von der Analyse einfacher zu der komplexerer Systeme, verbunden mit der Kritik der linearen Denkweise und dem Verständnis der Nicht-Linearität von Prozessen in komplexen Systemen, macht eine Erweiterung der theoretischen Erkenntnismittel erforderlich. So gibt es seit einigen Jahrzehnten Ansätze zur Entwicklung einer allgemeinen Systemtheorie, die die theoretische Durchdringung der Komplexität besser ermöglichen soll. Dabei können sich Systemtheorie und philosophische Erörterungen gegenseitig ergänzen. Besondere Bedeutung hat die Untersuchung komplexer sozialer Systeme. Im Mittelpunkt stehen die Beziehungen zwischen den Systemstrukturen und den kommunikativ handelnden Personen und Gruppen [58]. Eine wichtige Variante zum systemtheoretischen Verständnis sozialer Systeme hat N. Luhmann entwickelt (vgl. Kapitel 16 dieses Buches). Sie richtet sich gegen eine statische Erklärung und berücksichtigt die Zeitlichkeit der strukturellen Elemente in den Operationen der Systeme in einer funktionalen Theorie. Gesellschaft wird als ein sich selbst beobachtendes und beschreibendes System verstanden, wobei im Verhältnis von System und Umwelt Systemdifferenzen zu untersuchen sind.

Die Nicht-Linearität des Geschehens umfaßt die Wechselwirkung als Reaktion und Gegenreaktion und die daraus resultierende Entstehung von Neuem, die Nichtreduzierbarkeit komplexer Systeme auf das Verhältnis der Elemente, die nicht vorhersehbare Strukturbildung in allen Bereichen des Geschehens und die Existenz von Entwicklungszyklen mit Stagnation und Regression. Nicht-Linearität ist dann verbunden mit dem Unerwarteten, das die Fortschreibung des Gewesenen durchbricht, mit der Existenz von Zufällen, mit der Entstehung neuer Qualitäten und dem Auftreten emergenter Prozesse, Strukturen und Systeme.

15.5 Fazit

Die Chemie befaßt sich mit Prozessen im Spannungsfeld zwischen Möglichkeit (dynamis) und Wirklichkeit (energeia) [8b].

Mit den Bemerkungen zum Verhältnis von berechtigten Reduktionen und Reduktionismus, zum Qualitätswandel und zu schlafenden Propensitäten, zur adäquaten Erkenntnis der Nicht-Linearität allen prozessualen realen Geschehens soll angeregt werden, kritisch-konstruktiv über philosophische Aspekte der Chemie nachzudenken. Philosophie hat als Ideengenerator auch die Aufgabe, zum Nachdenken über neue Zusammenhänge zu provozieren. Auch dies ist hier durchaus beabsichtigt.

Sicher ist die Einschätzung berechtigt, daß die Philosophie der Chemie unterentwickelt ist. Die heuristische Funktion der Philosophie für die Chemie, die nur im wechselseitigen Dialog von Philosophen und Chemikern zu realisieren ist, wird ungenügend wahrgenommen. Daß die *Chemie philosophiefähig* ist, kann nur von empiristischen Positionen her bestritten werden, oder von der Auffassung her, daß die Elemente eines Systems das System erklären; denn wer das vertritt, dem reicht die "Philosophie der Physik". Eine Philosophie der Chemie anzuerkennen, das bedeutet, sich auf die (relative) Selbständigkeit chemischer Prozesse und ihre Universalität zu beziehen. Die Philosophie der Chemie könnte verschiedene Aspekte aufweisen: Beiträge der Chemie zum Verständnis der Struktur komplexer Systeme mit eigenen Systemgesetzen; spezifische Erkenntnisprobleme im Verhältnis von Experiment, Modell, Theorie; das Verhältnis von theorieorientierter und praxisgebundener "Exaktheit" beim Übergang vom Laborversuch in die industrielle Fertigung; Begriffsgeschichte. Eine Philosophie der Chemie verlangt Einsicht in theoretische Zusammenhänge,

Abweisung des Reduktionismus und Erkenntnis der Rolle der Chemie an den Schnittpunkten von "anorganischen" komplexen Prozessen und Leben.

Ihrer heuristischen Funktion durch die Aufstellung philosophischer Hypothesen über
den möglichen zukünftigen Beitrag der Chemie zur Präzisierung des Weltbilds, zu allgemeinen Aussagen über die Materiestruktur in ihren Formen und Arten, kann eine künftige
Philosophie nur gerecht werden, wenn neue Erkenntnisse durch konstruktiv-kritische Zu-
sammenarbeit von Philosophen und Chemikern gewonnen worden sind. Die interdiszi-
plinäre Diskussion grundlegender philosophischer Positionen im Lichte chemischer Erkenntnisse kann neue Ideen initiieren. Das sollte im Interesse von Chemie und Philosophie
ausprobiert werden.

Es wurde versucht, solche philosophischen Aspekte, die das Wesen der Chemie betreffen, anzusprechen. Die Chemie stellt sich uns dar mit den Charakteristika der Allumfassend-
heit, der (unerschöpflichen) Potentialität und der Beständigkeit des Wandels (vgl. Anhang
15.6). Kenntnisse der Chemie sind von fundamentaler Bedeutung für das Verständnis der
Komplexität der Welt [59], vielleicht im Augenblick eines der aktuellsten Themen der
Wissenschaft. Dies wird beispielhaft durch die Titel von Büchern wie *"Inseln im Chaos"*
[59], *"Das Quark und der Jaguar, vom Einfachen zum Komplexen – die Suche nach einer
neuen Erklärung der Welt"* [60] und *"Das schöpferische Teilchen, der Grundbaustein des
Universums"* [61] sowie darüber hinaus aus relevanten Berichten über Arbeiten der berühmten Santa-Fe-Gruppe deutlich (vgl. z.B. den Band *"Complexity, Entropy and the Physics of
Information"* [62]). Komplexität als physikalisches Problem wird auch von Eigen in seinem
bemerkenswerten Buch *"Stufen zum Leben"* angesprochen [63].

Es ist nicht immer einfach, zwischen den chemischen Prozessen, wie sie "normale"
Chemiker im Labor oder der Industrie durchführen, und denen, die in hochkomplexen
Systemen ablaufen, zu unterscheiden. Bei Prozessen der noch nicht verstandenen biologischen Entwicklung, die zu hoher Komplexität führen, sind aber auf jeden Fall gesteuerte
Reaktionen mit ausschlaggebend. Zitieren wir hierzu aus einem Buch über Evolutionsbiologie [64] und nochmals Monod [54]: *"Die Entwicklung eines Organismus aus einem
befruchteten Ei beinhaltet zwei Prozesse: die Zytodifferenzierung – wobei Zellen verschiedene biochemische und strukturelle Merkmale erwerben – und die Morphogenese –
dem Erwerb der dreidimensionalen Form von Geweben, Organen und Strukturen. Fast alle
Mechanismen der Zytodifferenzierung und der Morphogenese sind im Detail noch unver-
standen. In dieser Wissenslücke – zwischen primärer Genwirkung und der Entwicklung
komplexer Phänotypen – steckt ein Großteil dessen, was wir noch nicht von der Evolution
verstehen."*

*"Diese Analyse reduziert offenbar den alten Streit zwischen Präformatisten und
Epigenetisten auf ein Wortgefecht, das jeglichen Interesses entbehrt. Die vollendete Struktur
war nirgendwo als solche präformiert. Aber der Strukturplan war schon in seinen
Bestandteilen vorhanden".*

Zum Verständnis der angesprochenen komplexen Vorgänge ist die Kooperation aller
Naturforscher im Sinne von Mach [65] gefragt: *"Das wissenschaftliche Denken tritt uns in
zwei anscheinend recht verschiedenen Typen entgegen: dem Denken des Philosophen und
dem Denken des Spezialforschers. Der erstere sucht eine möglichst vollständige, weltum-
fassende Orientierung über die Gesamtheit der Tatsachen, wobei er nicht umhin kann,
seinen Bau auf Grund fachwissenschaftlicher Anleihen auszuführen. Dem anderen ist es zunächst um Orientierung und Übersicht in einem kleineren Tatsachengebiet zu tun. Da aber*

die Tatsachen immer etwas willkürlich und gewaltsam, mit Rücksicht auf den ins Auge ge-
faßten augenblicklichen intellektuellen Zweck, gegeneinander abgegrenzt werden, so ver-
schieben sich diese Grenzen beim Fortschritt des forschenden Denkens immer weiter und
weiter. Der Spezialforscher kommt schließlich auch zur Einsicht, daß die Ergebnisse aller
übrigen Spezialforscher zur Orientierung in seinem Gebiet berücksichtigt werden müssen.
So strebt also auch die Gesamtheit der Spezialforscher ersichtlich nach einer Weltorien-
tierung durch Zusammenschluß der Spezialgebiete."

Abbildung 15-2   Der Spezialist und der Generalist mit archetypischen Verhaltensmustern.

Die Aussage von Mach entspricht leider heute nicht der Wirklichkeit, d.h. den Hand-
lungsweisen der "scientific community". Die Akteure der Abbildung 15-2 haben wenig
gemeinsam!

Die philosophisch motivierte Analyse der Chemie regt an, tiefer über die Rolle
chemischer Prozesse im Weltgeschehen nachzudenken, was die Beschäftigung mit philosophischen Problemen der Chemie zu einem intellektuellen Vergnügen machen kann. Philoso-

phisches Denken soll dazu anregen, ungelöste Rätsel zu stellen, Bekanntes in umfassendere Zusammenhänge einzuordnen, um es zu verstehen und Neues zu finden. Auch das Erklimmen philosophischer Gipfel macht Mühe, aber es lohnt sich.

Lassen wir Boltzmann, den Widersacher Machs, in dem Zusammenhang zum Schluß zu Wort kommen: *"Ich bin der* .... *festen Hoffnung, daß ein einmütiges Zusammenwirken der Philosophie und Naturwissenschaft jeder dieser Wissenschaften neue Nahrung zuführen wird, ja daß man nur auf diesem Weg zu einem wahrhaft konsequenten Gedankenaustausch gelangen kann"* (zitiert nach [48], S. 201).

Prof. Dr. R. Ahlswede, Prof. Dr. Th. Dorfmüller, Prof. Dr. A. Dress, Dr. B. Redeker und Prof. Dr. W. Saltzer danken wir für wichtige Hinweise sowie dem Fonds der Chemischen Industrie (A. M.) für finanzielle Unterstützung.

15.6 Anhang

unter Mitarbeit von E. Diemann

Die Verteilung materieller Systeme auf die Sphären und deren Wechselwirkungen sowie die Geschichte eines Eisenatoms

Die Entstehung einiger hundert verschiedener Kerne und später der Atome im kosmologischen Prozeß hat zur Gestalt des uns "zugänglichen" Teils der Welt bzw. des Kosmos geführt. Speziell auf der Erde waren diese Ingredienzien Ausgangsmaterial für die unglaubliche Vielfalt der Erscheinungsformen. Die materiellen Systeme, insbesondere die Elemente, sind im Laufe der Zeit auf die Kosmo-, Atmo-, Geo-, Hydro-, Bio- und Anthroposphären verteilt worden (vgl. Abbildung 15-3 sowie Literaturauswahl). Die Wechselwirkung der Sphären miteinander wird über die verschiedenartigsten Mechanismen bewirkt, wobei die Verteilung eine Geschichte in der Zeit hat und nie zum Stillstand kommt. Die Untersuchung dieses höchst interessanten Befundes ist Gegenstand verschiedener Forschungsrichtungen. Basis des Geschehens ist einerseits die Wechselwirkung, andererseits die relative Beständigkeit der meisten Atome.

Wir haben versucht, diese Zusammenhänge durch eine sich über mehrere Milliarden Jahre erstreckende "Autobiographie" *eines* Eisenatoms, durch eine Auswahl relevanter Buchtitel sowie durch eine Graphik (Abbildung 15-3) zu verdeutlichen.

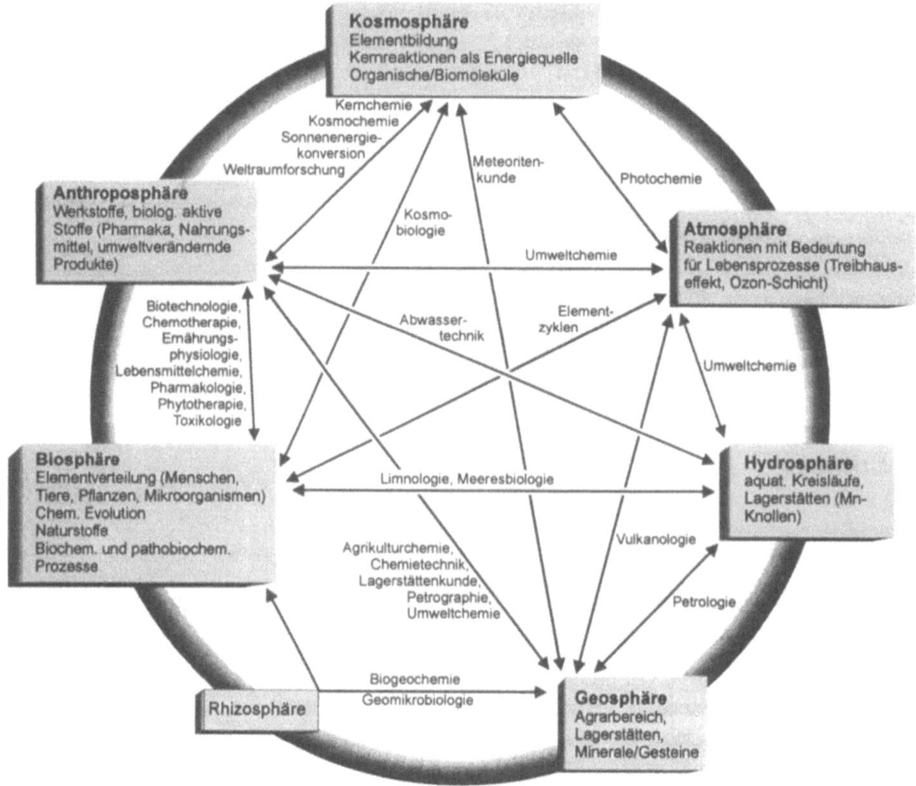

Abbildung 15-3

Lassen wir ein "kleines" Eisenatom, das sich wegen seines Beitrages zur "Eisenspitze"
seiner Bedeutung bewußt ist, seine "Lebensgeschichte" über einen Zeitraum von einigen
Milliarden Jahren einem interessierten Kind erzählen:

Ich weiß nicht mehr ganz genau, wann, wo und wie ich entstanden bin und wie ich zum
Planeten Erde kam. Aber ich erinnere mich noch gut an die Zeit, als ich als Bestandteil eines
Eisensulfids, heute nennt man es Pyrit, Versuche der Natur zur Kenntnis nahm, Vorlebens-
formen zu kreieren. Später fand ich mich dann mit vielen gleichartigen Atomen im Meerwasser wieder, wurde von einem Bakterium aufgenommen und in einem Protein dazu benutzt, gasförmigen Sauerstoff aus Wasser zu erzeugen. Die Photosynthese war entdeckt.
Das hat meine Eisengenossen im Meerwasser sehr erschreckt, da sie sich nach kurzer Zeit
wegen unseres Wertigkeitswechsels nicht mehr im Wasser tummeln konnten und – in ihrer
Beweglichkeit eingeschränkt – zu Bändereisenerzen ausgefällt wurden. Diese werden inter-
essanterweise heute dazu benutzt, den Zeitpunkt der Sauerstoffbildung in der Atmosphäre
festzulegen.

Mein Bakterium starb und ich fand mich nach einiger Zeit in einem anderen wieder,
dem ich als Bestandteil des Magnetits half, sich im magnetischen Feld der Erde zu
"orientieren". Auch die Lebenszeit dieses Mikroorganismus war kurz und ich war dann
glücklich, in einem Artgenossen als Bestandteil eines Metalloproteins etwas für höhere
Lebewesen tun zu können. Man nennt sie heterotroph, da sie auf externe Nahrungsquellen

angewiesen sind. Ich half nämlich dabei, aus dem Stickstoff der Luft Aminosäuren und damit Proteine zu machen. Beinahe hätte ich vergessen, daß ich zwischendurch noch einem anderen Bakterium bei der Eisenatmung geholfen hatte.

Nun wollte ich etwas in höheren Regionen leisten und hatte auch Glück. Einem Kaninchen half ich, sich vor einem Fuchs zu retten: Seine Muskeln benötigten große Mengen Sauerstoff, um Schnelligkeit zu erzielen, und ich war Bestandteil eines sauerstofftransportierenden Proteins, das man Hämoglobin nennt. Es ging damit weiter, daß das Kaninchen von einem Jäger erlegt und verzehrt wurde und so fand ich mich bald bei ihm im Eisenspeicherprotein – dem Ferritin – wieder. Meinem Besitzer half ich dann, als er sich einmal im Hungerstreik befand, zu überleben, da der Eisenmetabolismus bei ihm weiter lief. Hier begegnete ich später auch einem Bekannten aus meiner Magnetitzeit, der als Bestandteil eines Dolches meinen "Inhaber" verletzte. Viele andere dieser alten Bekannten tragen heute auf Tonträgern zum Musikgenuß eines "Klassikfans" bei.

Nun verläßt mich mein Gedächtnis für einige Zeit. Doch erinnere ich mich genau, daß ich einem Menschen einmal das Leben gerettet habe, der eine Krankheit durch mich frühzeitig genug entdeckte: Ich half ihm durch die richtige Deutung seines *Teerstuhls*, der ihn offenbar so sehr erschreckte, daß er einen Arzt aufsuchte. Irgendwann, glaube ich, war ich dagegen direkt als Auslöser in einen Krankheitsprozeß involviert und habe dem armen Menschen das Leben verkürzt. Er litt nämlich an Hämochromatose, und ich "war daran beteiligt", seine Lunge zu zerstören.

Literaturauswahl zum Anhang

I.   Kosmosphäre
II.   Geosphäre (Litho-)
III.   Atmosphäre
IV.   Hydro- und Kryosphäre
V.   Biosphäre
    (a)   Menschen/Tiere/Pflanzen (Biochemie/Pathobiochemie/Neurochemie/Immunologie/ Molekularbiologie)
    (b)   Mikroorganismen (Biochemie)
    (c)   Elementverteilung (Bioanorganische Chemie)
    (d)   Chemische Evolution
VI.   Anthroposphäre
    (a)   Materialien/Werkstoffe/Produkte/Verfahren
    (b)   Biologisch wirksame Substanzen
        (i)   Isolierte Naturstoffe
        (ii)   Nahrungsmittel (Stoffwechsel und Ernährung)
        (iii)   Pharmaka/Toxine (Pharmakologie und Toxikologie)
        (iiii) Umweltrelevante Substanzen
    (c)   Rohstoffe und Energieträger
VII.   Wechselwirkungen
    (a)   Kosmo-/Geosphäre
    (b)   Kosmo-/Atmosphäre
    (c)   Kosmo-/Biosphäre
    (d)   Kosmo-/Hydrosphäre
    (e)   Kosmo-/Anthroposphäre
    (f)   Geo-/Atmosphäre
    (g)   Geo-/Hydrosphäre

(h)   Geo-/Biosphäre (auch Rhizosphäre)

(i)   Geo-/Anthroposphäre

(j)   Atmo-/Hydrosphäre

(k)   Atmo-/Biosphäre

(l)   Atmo-/Anthroposphäre

(m) Hydro-/Biosphäre

(n)   Hydro-/Anthroposphäre

(o)   Bio-/Anthroposphäre

(p)   Kosmo-/Geo-/Hydro-/Bio-/Anthroposphäre (Ökologie)

VIII. Kulturbezüge

IX.   Geschichte

I.   Kosmosphäre

Bernhard, H.; Lindner, K.; Schukowski, M. *Wissensspeicher Astronomie*, Deutsch, Thun, **1987**.
Bühler, R. W. *Meteorite: Urmaterie aus dem interplanetaren Raum*, Birkhäuser, Basel, **1988**.
Dewar, M. J. S. et al. (Hrsg.), Topics in Current Chemistry 44: Cosmochemistry, Springer, Berlin, **1974**.
Dewar, M. J. S. et al. (Hrsg.), Topics in Current Chemistry 99: Cosmo- and Geochemistry (Anders, E.; Hayatsu, R. Organic Compounds in Meteorites and Their Origins; Winnewisser, G. The Chemistry of Interstellar Molecules; Lüst, R. Chemistry in Comets), Springer, Berlin, **1981**.
Goenner, H. *Einführung in die Kosmologie*, Spektrum, Heidelberg, **1994**.
Hoyle, F. *Das intelligente Universum*, Umschau, Frankfurt am Main, **1984**.
Karttunen, H.; Kröger, P.; Oja, H.; Poutanen, M.; Donner, K. J. (Hrsg.), *Astronomie*, Springer, Berlin, **1990**.
Köhler, H. W. *Die Planeten*, Vieweg, Braunschweig, **1983**.
Krautter, J. (Hrsg.), *Die Entstehung der Sterne: Interstellare Materie*, Spektrum, Heidelberg, **1986**.
Kuroda, P. K. *The Origin of the Chemical Elements and the Oklo Phenomenon*, Springer, Berlin, **1982**.
Müller, B. *Grundzüge der Astronomie*, 5. Aufl., Teubner, Leipzig, **1984**.
Reeves, H. *Woher nährt der Himmel seine Sterne? Die Entwicklung des Kosmos und die Zukunft der Menschen*, Birkhäuser, Basel, **1983**.
Sciama, D. W. *Modern Cosmology*, Cambridge University Press, Cambridge, **1971**.
Spiering, Ch. *Auf der Suche nach der Urkraft*, 2. Aufl., Teubner, Leipzig, **1989**.

II.   Geosphäre

Bancroft, P. *Die schönsten Minerale und Kristalle aus aller Welt*, Franckh, Stuttgart, **1975**.
Barnes, C. W. *Earth, Time, and Life: An Introduction to Geology*, Wiley, New York, **1980**.
Baumann, L.; Nikolskij, I. L.; Wolf, M. *Einführung in die Geologie und Erkundung von Lagerstätten*, VEB Dt. Verlag für Grundstoffindustrie, Leipzig, **1979**.
Baumgärtel, R.; Quellmalz, W.; Schneider, H. *Schmuck- und Edelsteine*, VEB Dt. Verlag für Grundstoffindustrie, Leipzig, **1988**.
Brownlow, A. H. *Geochemistry*, Prentice-Hall, Englewood Cliffs, **1979**.
Correns, C. W. *Einführung in die Mineralogie*, 2. Aufl., Springer, Berlin, **1968**.
Bruhns, W.; Ramdohr, P. *Petrographie*, de Gruyter, Berlin, **1972**.
Ernst, W. G. *Bausteine der Erde*, Enke, Stuttgart, **1977**.
Fabian, E. *Die Entdeckung der Kristalle*, VEB Dt. Verlag für Grundstoffindustrie, Leipzig, **1986**.
Fergusson, J. E. *Inorganic Chemistry and the Earth: Chemical Resources, Their Extraction, Use and Environmental Impact*, Pergamon, Oxford, **1982**.
Fotoatlas der Mineralien und Gesteine, Gräfe und Unzer, München, **1980**.
Georgi, K.-H. *Kreislauf der Gesteine*, Rowohlt, Reinbek, **1983**.
Henderson, P. *Inorganic Geochemistry*, Pergamon, Oxford, **1982**.
Herder Lexikon, *Geologie und Mineralogie*, 6. Aufl., Herder, Freiburg, **1990**.
Hollerbach, A. *Grundlagen der organischen Geochemie*, Springer, Berlin, **1985**.
Hunter, D. R. (Hrsg.), *Developments in Precambrian Geology 2: Precambrian of the Southern Hemisphere*, Elsevier, Amsterdam, **1981**.
Huttenlocher, H. *Mineral- und Erzlagerstättenkunde*, Vol. I, II, de Gruyter, Berlin, **1954**.
Jubelt, R. *Mineralien*, 2. Aufl., dtv, Stuttgart, **1976**.
Kleber, W. *Einführung in die Kristallographie*, 14. Aufl., VEB Verlag Technik, Berlin, **1979**.

Lagerstätten der Steine, Erden und Industrieminerale: Untersuchung und Bewertung, Heft 38 der Schriftenreihe der GDMB, Verlag Chemie, Weinheim, **1981**.
Lotze, F. *Geologie*, 5. Aufl., de Gruyter, Berlin, **1973**.
Mason, B.; Moore, C. B. *Principles of Geochemistry*, 4th Ed., Wiley, New York, **1982**.
Matthes, S. *Mineralogie: Eine Einführung in die spezielle Mineralogie, Petrologie und Lagerstättenkunde*, Springer, Berlin, **1983**.
McElhinny, M. W. (Hrsg.), *The Earth: Its Origin, Structure and Evolution*, Academic Press, London, **1979**.
Metz, R. *Antlitz Edler Steine: Mineralien · Kristalle*, Prisma, Gütersloh, **1985**.
Möller, P. *Anorganische Geochemie*, Springer, Berlin, **1986**.
Mueller, R. F.; Saxena, S. K. *Chemical Petrology with Applications to The Terrestrial Planets and Meteorites*, Springer, New York, **1977**.
Neef, E. (Hrsg.), *Das Gesicht der Erde*, 5. Aufl., Deutsch, Thun, **1981**.
O'Donoghue, M. (Hrsg.), *Enzyklopädie der Minerale und Edelsteine*, Herder, Freiburg, **1977**.
Offermann, E.; Schreiber, W. *Von Bergwerken und Kristallschätzen*, Bode, Haltern, **1989**.
Pape, H. *Leitfaden zur Bestimmung von Erzen und mineralischen Rohstoffen*, Enke, Stuttgart, **1977**.
Parker, R. L. *Mineralienkunde*, 5. Aufl., Ott, Thun, **1975**.
Perrin, M. B. *An Introduction to the Chemistry of Rocks and Minerals*, Arnold, London, Reprint with corrections, **1979**.
Petrascheck, W. E.; Pohl, W. *Lagerstättenlehre: Eine Einführung in die Wissenschaft von den mineralischen Bodenschätzen*, 3. Aufl., Schweizerbart, Stuttgart, **1982**.
Richter, D. *Allgemeine Geologie*, 2. Aufl., de Gruyter, Berlin, **1980**.
Rösler, H. J. *Lehrbuch der Mineralogie*, 3. Aufl., VEB Dt. Verlag für Grundstoffindustrie, Leipzig, **1984**.
Sawkins, F. J.; Chase, C. G.; Darby, D. G.; Rapp, Jr., G. *The Evolving Earth: A Text in Physical Geology*, 2nd Ed., Macmillan, New York, **1978**.
Schenck, P. A.; de Leeuw, J. W.; Lijmbach, G. W. M. (Hrsg.), *Advances in Organic Geochemistry 1983*, Pergamon, Oxford, **1984**.
Schmidt, K. *Erdgeschichte*, 3. Aufl., de Gruyter, Berlin, **1978**.
Schöne und seltene Minerale, Herder, Freiburg, **1981**.
Schröcke, H.; Weiner, K.-L. *Mineralogie*, de Gruyter, Berlin, **1981**.
Schulthess, E. *Landschaft der Urzeit*, Meridan, Zürich, **1988**.
Schumann, W. *Steine und Mineralien*, BLV Verlagsges., München, **1982**.
Schumann, W. *Edelsteine und Schmucksteine*, BLV Verlagsges., München, **1976**.
Seim, R.; Tischendorf, G. (Hrsg.), *Grundlagen der Geochemie*, VEB Dt. Verlag für Grundstoffindustrie, Leipzig, **1990**.
Seim, R. *Minerale*, Neumann, Leipzig, **1981**.
Strübel, G.; Zimmer, S. H. *Lexikon der Mineralogie*, Enke, Stuttgart, **1982**.
Tröger, K.-A.; Kozur, H.; Ruchholz, K.; Watznauer, A.; Kahlke, H.-D. (Hrsg.), *Abriß der Historischen Geologie*, Akademie-Verlag, Berlin, **1984**.
U.N. Dept. for Development (Hrsg.), *The Development Potential of Precambrian Mineral Deposits*, Pergamon, New York, **1982**.
Vojtkevič, G. V. *Die stoffliche Geschichte der Erde*, VEB Dt. Verlag für Grundstoffindustrie, Leipzig, **1986**.
Vollstädt, H.; Baumgärtel, R. *Edelsteine*, 2. Aufl., Enke, Stuttgart, **1982**.
Wagenbreth, O.; Steiner, W. *Geologische Streifzüge*, VEB Dt. Verlag für Grundstoffindustrie, Leipzig, **1982**.
Wedepohl, K. H. *Geochemie*, de Gruyter, Berlin, **1967**.

III. Atmosphäre

Crutzen, P. J. (Hrsg.), *Atmosphäre, Klima, Umwelt*, Spektrum, Heidelberg, **1990**.
Finlayson-Pitts, B. J.; Pitts, Jr., J. N. *Atmospheric Chemistry*, Wiley, New York, **1986**.
Fortak, H. *Meteorologie*, 2. Aufl., Reimer, Berlin, **1982**.
Goldberg, E. D. (Hrsg.), *Atmospheric Chemistry*, Springer, Berlin, **1982**.
Graedel, T. E.; Crutzen, P. J.; *Chemie der Atmosphäre*, Spektrum, Heidelberg, **1994**.

IV. Hydro- und Kryosphäre

Lanius, K. *Die Erde im Wandel*, Spektrum, Heidelberg, **1995**.
Melchior, D. C.; Bassett, R. L. (Hrsg.), *Chemical Modeling of Aqueous Systems II*, American Chemical Society, Washington, **1990**.
Pantenburg, V. *Rettet das Wasser! Vom weltweiten Feldzug der Hydrologen*, Schwann, Düsseldorf, **1969**.

Stumm, W.; Morgan, J. J. *Aquatic Chemistry: An Introduction Emphasizing Chemical Equilibria in Natural Waters*, Wiley, New York, **1981**.

V.   Biosphäre

(a)   Menschen/Tiere/Pflanzen (Biochemie/Pathobiochemie/Neurochemie/Immunologie/Molekularbiologie)

Alberts, B.; Bray, D.; Lewis, J.; Raff, M.; Roberts, K.; Watson, J. D. *Molekularbiologie der Zelle*, VCH, Weinheim, **1986**.
Beier, W.; Laue, R.; Leutert, G.; Rotzsch, W.; Schmidt, U. J. (Hrsg.), *Prozesse des Alterns*, 2. Aufl., Akademie-Verlag, Berlin, **1989**.
Burger, K. (Hrsg.), *Biocoordination Chemistry: Coordination Equilibria in Biologically Active Systems*, Horwood, New York, **1990**.
Gregory, R. P. F. *Biochemistry of Photosynthesis*, 3rd Ed., Wiley, Chichester, **1989**.
Hucho, F. *Neurochemistry*, VCH, Weinheim, **1986**.
Karlson, P.; Gerok, W.; Groß, W. *Pathobiochemie*, Thieme, Stuttgart, **1978**.
Kindl, H. *Biochemie der Pflanzen*, 2. Aufl., Springer, Berlin, **1987**.
Lehninger, A. L.; Nelson, D. L.; Cox, M. M. *Prinzipien der Biochemie*, 2. Aufl., Spektrum, Heidelberg, **1994**.
Lewin, B. *Gene: Lehrbuch der molekularen Genetik*, VCH, Weinheim, **1988**.
Libbert, E. *Lehrbuch der Pflanzenphysiologie*, 3. Aufl., Fischer, Stuttgart, **1979**.
Mengel, K. *Ernährung und Stoffwechsel der Pflanze*, 5. Aufl., Fischer, Stuttgart, **1979**.
Müller, A.; Ratajczak, H.; Junge, W.; Diemann, E. *Electron and Proton Transfer in Chemistry and Biology*, Elsevier, Amsterdam, **1991**.
Müntz, K. *Stoffwechsel der Pflanzen: Ausgewählte Gebiete der Physiologie*, Aulis, Köln, **1976**.
Müntz, K. *Stickstoffmetabolismus der Pflanzen*, Fischer, Stuttgart, **1984**.
Peter, G. *Lehrbuch der Klinischen Chemie*, 3. Aufl., Verlag Chemie, Weinheim, **1982**.
Roitt, I. M.; Brostoff, J.; Male, D. K. *Kurzes Lehrbuch der Immunologie*, Thieme, Stuttgart, **1987**.
Strasburger, E.; Noll, F.; Schenck, H.; Schimper, A. F. W. (Begr.), *Lehrbuch der Botanik für Hochschulen*, 31. Aufl., Fischer, Stuttgart, **1978**.
Zech, R.; Domagk, G. F. *Enzyme: Biochemie, Pathobiochemie, Klinik, Therapie*, VCH, Weinheim, **1986**.

(b)   Mikroorganismen (Biochemie)

Brock, T. D. *Thermophilic Microorganisms and Life at High Temperatures*, Springer, New York, **1978**.
Davis, I.; David, B. *Microbiology*, 3rd Ed., Harper, Philadelphia, **1980**.
Doelle, H. W. *Bacterial Metabolism*, 2nd Ed., Academic Press, New York, **1975**.
Gottschalk, G. *Bacterial Metabolism*, Springer, New York, **1979**.
McKane, L.; Kandel, J. *Microbiology*, McGraw-Hill, New York, **1986**.
Schlegel, H. G. *Allgemeine Mikrobiologie*, 6. Aufl., Thieme, Stuttgart, **1985**.
Stanier, R. Y.; Adelberg, E. A.; Ingraham, J. L. *The Microbial World*, 4th Ed., Prentice-Hall, Englewood Cliffs, **1976**.
Weide, H.; Aurich, H. *Allgemeine Mikrobiologie*, Fischer, Stuttgart, **1979**.

(c)   Elementverteilung (Bioanorganische Chemie)

Bertini, I.; Drago, R. S.; Luchinat, C. (Hrsg.), *The Coordination Chemistry of Metalloenzymes: The Role of Metals in Reactions Involving Water, Dioxygen, and Related Species*, Reidel, Dordrecht, **1983**.
Bezkorovainy, A. *Biochemistry of Nonheme Iron*, Plenum, New York, **1980**.
Biological Roles of Copper, Ciba Foundation Symposium 79 (new series), Excerpta Medica, Amsterdam, **1980**.
Bothe, H.; Trebst, A. (Hrsg.), *Biology of Inorganic Nitrogen and Sulfur*, Springer, Berlin, **1981**.
Bothwell, T. H.; Charlton, R. W.; Cook, J. D.; Finch, C. A. *Iron Metabolism in Man*, Blackwell, Oxford, **1979**.
Bowen, H. J. M. *Environmental Chemistry of the Elements*, Academic Press, London, **1979**.
Brätter, P.; Schramel, P. (Hrsg.), *Trace Element Analytical Chemistry in Medicine and Biology*, de Gruyter, Berlin, **1980**.
Brill, A. S. *Transition Metals in Biochemistry*, Springer, Berlin, **1977**.
Chappell, W. R.; Petersen, K. K. (Hrsg.), *Molybdenum in the Environment, Vol. 1: The Biology of Molybdenum*, Dekker, New York, **1976**; *Vol. 2: The Geochemistry, Cycling, and Industrial Uses of Molybdenum*, Dekker, New York, **1977**.

Chasteen, N. D. (Hrsg.), *Vanadium in Biological Systems*, Kluwer, Dordrecht, **1990**.
Chatt, J.; da Câmara Pina, L. M.; Richards, R. L. (Hrsg.), *New Trends in the Chemistry of Nitrogen Fixation*, Academic Press, London, **1980**.
Coughlan, M. P. (Hrsg.), *Molybdenum and Molybdenum-containing Enzymes*, Pergamon, Oxford, **1980**.
Cowan, J. A. *Inorganic Biochemistry*, VCH, New York, **1993**.
Crichton, R. R. (Hrsg.), *Proteins of Iron Storage and Transport in Biochemistry and Medicine*, North-Holland, Amsterdam, and American Elsevier, New York, **1975**.
Crichton, R. R. *Inorganic Biochemistry of Iron Metabolism*, Horwood, New York, **1991**.
Dilworth, M. J.; Glenn, A. R. (Hrsg.), *Biology and Biochemistry of Nitrogen Fixation*, Elsevier, Amsterdam, **1991**.
Dunford, H. B.; Dolphin, D.; Raymond, K. N.; Sieker, L. (Hrsg.), *The Biological Chemistry of Iron: A Look at the Metabolism of Iron and Its Subsequent Uses in Living Organisms*, Reidel, Dordrecht, **1982**.
Engels, S.; Nowak, A. *Auf der Spur der Elemente*, VEB Dt. Verlag für Grundstoffindustrie, Leipzig, **1983**.
Evans, C. H. *Biochemistry of the Lanthanides*, Plenum, New York, **1990**.
Fiedler, H. J.; Rösler, H. J. (Hrsg.), *Spurenelemente in der Umwelt*, Enke, Stuttgart, **1988**.
Filby, R. H.; Branthaver, J. F. (Hrsg.), *Metal Complexes in Fossil Fuels: Geochemistry, Characterization, and Processing*, ACS Symposium Series 344,
American Chemical Society, Washington, **1987**.
Frieden, E. (Hrsg.), *Biochemistry of the Essential Ultratrace Elements*, Plenum, New York, **1984**.
Genesis of Precambrian Iron and Manganese Deposits, Proc. Kiev Symp., 1970, Unesco, Paris, **1973**.
Harrison, P. M. (Hrsg.), *Metalloproteins, Part 1: Metal Proteins with Redox Roles; Part 2: Metal Proteins with Non-redox Roles*, Verlag Chemie, Weinheim, **1985**.
Harrison, P. M.; Hoare, R. J. *Metals in Biochemistry*, Chapman and Hall, London, **1980**.
Hay, R. W. *Bio-Inorganic Chemistry*, Horwood, New York, **1984**.
Hughes, M. N.; Poole, R. K. *Metals and Micro-organisms*, Chapman and Hall, London, **1989**.
Hughes, M. N. *The Inorganic Chemistry of Biological Processes*, Wiley, London, **1972**.
Ingraham, L. L.; Meyer, D. L. *Biochemistry of Dioxygen*, Plenum, New York, **1985**.
Iron Metabolism, Ciba Foundation Symposium 51 (new series), Elsevier, Amsterdam, **1977**.
Jacobs, A.; Worwood, M. (Hrsg.), *Iron in Biochemistry and Medicine, II*, Academic Press, London, **1980**.
Karlin, K. D.; Zubieta, J. (Hrsg.), *Biological & Inorganic Copper Chemistry, Vols. I, II*, Adenine Press, New York, **1986**.
Kendrick, M. J.; May, M. T.; Plishka, M. J.; Robinson, K. D. *Metals in Biological Systems*, Horwood, New York, **1992**.
Kharasch, N. (Hrsg.), *Trace Metals in Health and Disease: New Roles of Metals in Biochemistry, the Environment, and Clinical/Nutritional Studies*, Raven Press, New York, **1979**.
Lancaster, Jr., J. R. (Hrsg.), *The Bioinorganic Chemistry of Nickel*, VCH, New York, **1988**.
Lang, K. *Wasser, Mineralstoffe, Spurenelemente*, Steinkopff, Darmstadt, **1974**.
Lippard, S. J.; Berg, J. M. *Principles of Bioinorganic Chemistry*, University Science Books, Mill Valley, **1994**.
Loehr, T. M. (Hrsg.), *Iron Carriers and Iron Proteins*, VCH, New York, **1989**.
Loneragan, J. F.; Robson, A. D.; Graham, R. D. (Hrsg.), *Copper in Soils and Plants*, Academic Press, Sydney, **1981**.
Lovenberg, W. (Hrsg.), *Iron-Sulfur Proteins, Vol. I: Biological Properties; Vol. II: Molecular Properties*, Academic Press, New York, **1973**; *Vol. III: Structure and Metabolic Mechanisms*, Academic Press, New York, **1977**.
Martell, A. E. (Hrsg.), *Inorganic Chemistry in Biology and Medicine*, ACS Symposium Series 140, American Chemical Society, Washington, **1980**.
Merian, E. (Hrsg.), *Metals and Their Compounds in the Environment: Occurrence, Analysis, and Biological Relevance*, VCH, Weinheim, **1991**.
Meyer, B. *Sulfur, Energy, and Environment*, Elsevier, Amsterdam, **1977**.
Müller, A.; Newton, W. E. (Hrsg.), *Nitrogen Fixation: The Chemical-Biochemical-Genetic Interface*, Plenum, New York, **1983**.
Müller, A.; Krebs, B. (Hrsg.), *Sulfur: Its Significance for Chemistry, for the Geo-, Bio- and Cosmosphere and Technology*, Elsevier, Amsterdam, **1992**.
Neilands, J. B. (Hrsg.), *Microbial Iron Metabolism*, Academic Press, New York, **1974**.
Newton, W. E.; Otsuka, S. (Hrsg.), *Molybdenum Chemistry of Biological Significance*, Plenum, New York, **1980**.
Nriagu, J. O. (Hrsg.), *Zinc in the Environment, Part I: Ecological Cycling*, Wiley, New York, **1980**.
Nriagu, J. O. (Hrsg.), *Sulfur in the Environment, Part II: Ecological Impacts*, Wiley, New York, **1978**.

Nriagu, J. O.; Nieboer, E. (Hrsg.), *Chromium in the Natural and Human Environments*, Wiley, New York, **1988**.
Ochiai, E. *Bioinorganic Chemistry*, Allyn and Bacon, Boston, **1977**.
Ochiai, E. *General Principles of Biochemistry of the Elements*, Plenum, New York, **1987**.
Otsuka, S.; Yamanaka, T. (Hrsg.), *Metalloproteins: Chemical Properties and Biological Effects*, Elsevier, Amsterdam, **1988**.
Prasad, A. S. *Trace Elements and Iron in Human Metabolism*, Wiley, Chichester, **1978**.
Saltzman, E. S.; Cooper, W. J. (Hrsg.), *Biogenic Sulfur in the Environment*, ACS
Symposium Series 393, American Chemical Society, Washington, **1989**.
Shamberger, R. J. *Biochemistry of Selenium*, Plenum, New York, **1983**.
Sigel, H.; Sigel, A. (Hrsg.), *Metal Ions in Biological Systems, Vol. 23: Nickel and Its Role in Biology*, Dekker, New York, **1988**; *Vol. 25: Interrelations among Metal Ions, Enzymes, and Gene Expression*, Dekker, New York, **1989**.
de Sousa, M.; Brock, J. H. (Hrsg.), *Iron in Immunity, Cancer and Inflammation*, Wiley, Chichester, **1989**.
Spiro, T. G. (Hrsg.), *Molybdenum Enzymes*, Wiley, New York, **1985**.
Spiro, T. G. (Hrsg.), *Iron-Sulfur Proteins*, Wiley, New York, **1982**.
Sulphur in Biology, Ciba Foundation Symposium 72 (new series), Excerpta Medica, Amsterdam, **1980**.
Toy, A. D. F.; Walsh, E. N. *Phosphorus Chemistry in Everyday Living*, 2nd Ed., American Chemical Society, Washington, **1987**.
Underwood, E. J. *Trace Elements in Human and Animal Nutrition*, 4th Ed., Academic Press, New York, **1977**.
Wendel, A. (Hrsg.), *Selenium in Biology and Medicine*, Springer, Berlin, **1989**.
Westbroek, P.; De Jong, E. W. (Hrsg.), *Biomineralization and Biological Metal Accumulation*, Reidel, Dordrecht, **1983**.
Williams, R. J. P.; Da Silva, J. R. R. F. (Hrsg.), *New Trends in Bio-inorganic Chemistry*, Academic Press, London, **1978**.
Winkelmann, G.; v. d. Helm, D.; Neilands, J. B. (Hrsg.), *Iron Transport in Microbes, Plants and Animals*, VCH, Weinheim, **1987**.
Zumkley, H. (Hrsg.), *Spurenelemente: Grundlagen - Ätiologie - Diagnose - Therapie*, Thieme, Stuttgart, **1983**.

(d)   Chemische Evolution

Cairns-Smith, A. G.; Hartman, H. (Hrsg.), *Clay Minerals and the Origin of Life*, Cambridge University Press, Cambridge, **1986**.
Deamer, D. W.; Fleischaker, G. R. *Origins of Life*, Jones and Bartlett, Boston, **1994**.
Dose, K.; Rauchfuss, H. *Chemische Evolution und der Ursprung lebender Systeme*, Wiss. Verlagsges., Stuttgart, **1975**.
de Duve, C. *Ursprung des Lebens: Präbiotische Evolution und die Entstehung der Zelle*, Spektrum, Heidelberg, **1994**.
Ebeling, W.; Engel, A.; Feistel, R. *Physik der Evolutionsprozesse*, Akademie-Verlag, Berlin, **1990**.
Follmann, H. *Chemie und Biochemie der Evolution: Wie und wo entstand das Leben?*, Quelle & Meyer, Heidelberg, **1981**.
Gutfreund, H. (Hrsg.), *Biochemical Evolution*, Cambridge University Press, Cambridge, **1981**.
Kaplan, R. W. *Der Ursprung des Lebens*, dtv/Thieme, Stuttgart, **1972**.
Miller, S. L.; Orgel, L. E. *The Origins of Life on the Earth*, Prentice-Hall, Englewood Cliffs, **1974**.
Mortlock, R. P. (Hrsg.), *Microorganisms as Model Systems for Studying Evolution*, Plenum, New York, **1984**.
Oparin, A. I. *The Origin of Life*, 2nd Ed., Dover Publ., New York, **1953**.
Reinbothe, H.; Krauß, G.-J. *Entstehung und molekulare Evolution des Lebens*, Fischer, Jena, **1982**.
Rutten, M. G. *The Origin of Life by Natural Causes*, Elsevier, Amsterdam, **1971**.
Schopf, J. W. (Hrsg.), *Earth's Earliest Biosphere: Its Origin and Evolution*, Princeton University Press, Princeton, **1983**.
Siewing, R. (Hrsg.), *Evolution*, 2. Aufl., Fischer, Stuttgart, **1982**.
Unsöld, A. *Evolution kosmischer, biologischer und geistiger Strukturen*, Wiss. Verlagsges., Stuttgart, **1981**.
Vollmert, B. *Das Molekül und das Leben: Vom makromolekularen Ursprung des Lebens und der Arten: Was Darwin nicht wissen konnte und Darwinisten nicht wissen wollen*, Rowohlt, Reinbek, **1985**.

VI. Anthroposphäre

(a) Materialien/Werkstoffe/Produkte/Verfahren

Autorenkollektiv, *Technische Anorganische Chemie*, VEB Dt. Verlag für Grundstoffindustrie, Leipzig, **1978**.

Baumgartl, E. *Neue Werkstoffkunde kurz und einprägsam*, Archimedes, Kreuzlingen, **1980**.

Böttger, H.; Bryksin, V. V. *Hopping Conduction in Solids*, VCH, Weinheim, **1985**.

Bradley, S. A.; Gattuso, M. J.; Bertolacini, R. J. (Hrsg.), *Characterization and Catalyst Development*, ACS Symposium Series 411, American Chemical Society, Washington, **1989**.

Brostow, W. *Einstieg in die moderne Werkstoffwissenschaft*, Hanser, München, **1984**.

Büchner, W.; Schliebs, R.; Winter, G.; Büchel, K. H. *Industrielle Anorganische Chemie*, Verlag Chemie, Weinheim, **1984**.

Cardon, F.; Gomes, W. P.; Dekeyser, W. (Hrsg.), *Photovoltaic and Photoelectrochemical Solar Energy Conversion*, Plenum, New York, **1981**.

Delmon, B.; Jacobs, P.; Poncelet, G. (Hrsg.), *Preparation of Catalysts I*, Elsevier, Amsterdam, **1976**.

Eckert, M.; Schubert, H. *Kristalle, Elektronen, Transistoren*, Rowohlt, Reinbek, **1989**.

Emons, H.-H.; Bräutigam, G.; Hellmold, P.; Holldorf, H.; Kümmel, R.; Martens, H. *Grundlagen der technischen anorganischen Chemie*, 3. Aufl., VEB Dt. Verlag für Grundstoffindustrie, Leipzig, **1983**.

Feltz, A. *Amorphe und glasartige anorganische Festkörper*, Akademie-Verlag, Berlin, **1983**.

Gates, B. C. *Catalytic Chemistry*, Wiley, New York, **1992**.

Gellings, P. J. *Korrosion und Korrosionsschutz von Metallen*, Hanser, München, **1981**.

Glasser, F. P.; Potter, P. E. (Hrsg.), *High Temperature Chemistry of Inorganic and Ceramic Materials*, The Chemical Society, London, **1977**.

Grellert, R.; Hauschild, G.; Ziemann, J. *Die Chemie des Erdöls*, 2. Aufl., Volk und Wissen, Berlin, **1982**.

Hadamovsky, H.-F. (Hrsg.), *Werkstoffe der Halbleitertechnik*, VEB Dt. Verlag für Grundstoffindustrie, Leipzig, **1985**.

Hammond, G. S.; Kuck, V. J. (Hrsg.), *Fullerenes: Synthesis, Properties, and Chemistry of Large Carbon Clusters*, ACS Symposium Series 481, American Chemical Society, Washington, **1992**.

Hecht, F. *Grundzüge der Radio- und Reaktorchemie*, Akad. Verlagsges., Frankfurt am Main, **1968**.

Hiemenz, P. C. *Principles of Colloid and Surface Chemistry*, 2nd Ed., Dekker, New York, **1986**.

Ibelgaufts, H. *Gentechnologie von A bis Z*, VCH, Weinheim, **1990**.

Iler, R. K. *The Chemistry of Silica: Solubility, Polymerization, Colloid and Surface Properties, and Biochemistry*, Wiley, New York, **1979**.

Kassatkin, A. G. *Chemische Verfahrenstechnik, Bd. II*, 4. Aufl., VEB Dt. Verlag für Grundstoffindustrie, Leipzig, **1961**.

Keller, C. *Radiochemie*, 2. Aufl., Diesterweg, Frankfurt am Main, **1981**.

Klöppel, E.; Klöppel, B. *Chemische Technologie: Theoretische Grundlagen und Grundverfahren*, 2. Aufl., VEB Dt. Verlag für Grundstoffindustrie, Leipzig, **1983**.

Kolthoff, I. M.; Elving, P. J. (Hrsg.), *Treatise on Analytical Chemistry*, Wiley, New York, **1961-1977**.

Kung, H. H. *Transition Metal Oxides: Surface Chemistry and Catalysis*, Elsevier, Amsterdam, **1989**.

Lieser, K. H. *Einführung in die Kernchemie*, Verlag Chemie, Weinheim, **1969**.

Marr, I. L.; Cresser, M. S.; Ottendorfer, L. J. *Umweltanalytik*, Thieme, Stuttgart, **1988**.

Matkovich, V. I. (Hrsg.), *Boron and Refractory Borides*, Springer, Berlin, **1977**.

Merkel, M.; Thomas, K.-H. *Technische Stoffe*, 2. Aufl., VEB Fachbuchverlag, Leipzig, **1984**.

Mikhail, R. Sh.; Robens, E. *Microstructure and Thermal Analysis of Solid Surfaces*, Wiley, Chichester, **1983**.

Neubauer, A. *Chemie? Chemie! Entwicklungen - Ergebnisse - Wirkungen*, Aulis, Köln, **1981**.

Nitzsche, K.; Ullrich, H.-J. (Hrsg.), *Funktionswerkstoffe der Elektrotechnik und Elektronik*, VEB Dt. Verlag für Grundstoffindustrie, Leipzig, **1985**.

Oppermann, K.; Blessing, K. *Ökonomie der Metallurgie*, VEB Dt. Verlag für Grundstoffindustrie, Leipzig, **1983**.

Peacocke, T. A. H. *Radiochemistry: Theory and Experiment*, Wykeham Publ., London, **1978**.

Petzold, A. *Anorganisch-nichtmetallische Werkstoffe*, 2. Aufl., VEB Dt. Verlag für Grundstoffindustrie, Leipzig, **1986**.

Philipp, B.; Stevens, P. *Grundzüge der Industriellen Chemie*, VCH, Weinheim, **1987**.

Poncelet, G.; Grange, P.; Jacobs, P. A. (Hrsg.), *Preparation of Catalysts III: Scientific Bases for the Preparation of Heterogeneous Catalysts*, Elsevier, Amsterdam, **1983**.

Schade, W. *Einführung in die chemische Technologie*, 2. Aufl., VEB Dt. Verlag der Wissenschaften, Berlin, **1980**.

Schatt, W. (Hrsg.), *Einführung in die Werkstoffwissenschaft*, 5. Aufl., VEB Dt. Verlag für Grundstoffindustrie, **1983**.

Schmalzried, H. *Festkörperreaktionen: Chemie des festen Zustandes*, Verlag Chemie, Weinheim, **1971**.
Scholze, H. *Glas: Natur, Struktur und Eigenschaften*, 2. Aufl., Springer, Berlin, **1977**.
Schultze, D. *Differentialthermoanalyse*, 2. Aufl., VEB Dt. Verlag der Wissenschaften, Berlin, **1972**.
Shriver, D. F.; Atkins, P. W.; Langford, C. H. *Inorganic Chemistry*, Oxford University Press, Oxford, **1990**.
Sonntag, H. *Lehrbuch der Kolloidwissenschaft*, VEB Dt. Verlag der Wissenschaften, Berlin, **1977**.
Stöcklin, G. *Chemie heißer Atome*, Verlag Chemie, Weinheim, **1969**.
Thomas, J. M.; Lambert, R. M. *Characterisation of Catalysts*, Wiley, Chichester, **1980**.
Thompson, R. (Hrsg.), *The Modern Inorganic Chemicals Industry*, Special Publ. No. 31, The Chemical Society, London, **1977**.
Trimm, D. L. *Design of Industrial Catalysts*, Elsevier, Amsterdam, **1980**.
Vogel, R. *Einführung in die Metallurgie*, Musterschmidt, Göttingen, **1955**.
Vogel, W. *Glaschemie*, 2. Aufl., VEB Dt. Verlag für Grundstoffindustrie, Leipzig, **1983**.
Wendlandt, W. W. *Thermal Methods of Analysis*, Interscience Publ., New York, **1964**.
White, M. G. *Heterogeneous Catalysis*, Prentice-Hall, Englewood Cliffs, **1990**.
Wynne, M. D. *Chemical Processing in Industry*, The Royal Institute of Chemistry, London, **1970**.

(b)   Biologisch wirksame Substanzen

(i)   Isolierte Naturstoffe

Dugas, H.; Penney, C. *Bioorganic Chemistry: A Chemical Approach to Enzyme Action*, Springer, New York, **1981**.
Hollenberg, M. D. (Hrsg.), *Insulin: Its Receptor and Diabetes*, Dekker, New York, **1985**.
Nuhn, P. *Chemie der Naturstoffe - Bioorganische Chemie*, Akademie-Verlag, Berlin, **1981**.
Pfaender, P. *Enzyme: Bezugsquellen, Daten, Eigenschaften*, Fischer, Stuttgart, **1982**.
Wikström, M.; Krab, K.; Saraste, M. *Cytochrome Oxidase*, Academic Press, London, **1981**.

(ii)   Nahrungsmittel (Stoffwechsel und Ernährung)

Anemueller, H. *Das Grunddiät-System. Leitfaden der Ernährungstherapie*, 2. Aufl., Hippokrates, Stuttgart, **1983**.
BASF (Hrsg.), *Chemie und Ernährung*, Verlag Wissenschaft und Politik, Köln, **1983**.
Belitz, H.-D.; Grosch, W. *Lehrbuch der Lebensmittelchemie*, 3. Aufl., Springer, Berlin, **1984**.
Cherry, J. P. (Hrsg.), *Food Protein Deterioration: Mechanisms and Functionality*, ACS Symposium Series 206, American Chemical Society, Washington, **1982**.
Franzke, C. *Lehrbuch der Lebensmittelchemie, Bd. 1: Lebensmittelinhaltsstoffe*, Akademie-Verlag, Berlin, **1981**; *Bd. 2: Die Lebensmittel*, Akademie-Verlag, Berlin, **1982**.
Johnston, J. F. M. *Die Chemie des täglichen Lebens, Bd. 1, 2*, Duncker, Berlin, **1869**.
Ketz, H.-A.; Baum, F. (Hrsg.), *Ernährungslexikon*, VEB Fachbuchverlag, Leipzig, **1986**.
Kies, C. (Hrsg.), *Nutritional Bioavailability of Manganese*, ACS Symposium Series 354, American Chemical Society, Washington, **1987**.
Muntwyler, E. *Elektrolytstoffwechsel und Säure-Basen-Gleichgewicht*, de Gruyter, Berlin, **1973**.
Täufel, A.; Tunger, L.; Zobel, M. (Hrsg.), *Lebensmittel-Lexikon*, 2. Aufl., VEB Fachbuchverlag, Leipzig, **1981**.
Welzl, E. *Biochemie der Ernährung*, de Gruyter, Berlin, **1985**.

(iii)   Pharmaka/Toxine (Pharmakologie und Toxikologie)

Auterhoff, H.; Knabe, J. *Lehrbuch der pharmazeutischen Chemie*, 6. Aufl., Wiss. Verlagsges., Stuttgart, **1971**.
Birgersson, B.; Sterner, O.; Zimerson, E. *Chemie und Gesundheit: Eine verständliche Einführung in die Toxikologie*, VCH, Weinheim, **1988**.
Czygan, F.-C. (Hrsg.), *Biogene Arzneistoffe: Entwicklungen auf dem Gebiet der Pharmazeutischen Biologie, Phytochemie und Phytotherapie*, Vieweg, Braunschweig, **1984**.
Forth, W.; Henschler, D.; Rummel, W. (Hrsg.), *Allgemeine und spezielle Pharmakologie und Toxikologie*, 5. Aufl., BI, Mannheim, **1987**.
Hathway, D. E. *Molecular Aspects of Toxicology*, The Royal Society of Chemistry, London, **1984**.
Holland, W. C.; Klein, R. L.; Briggs, A. H. *Molekulare Pharmakologie*, Thieme, Stuttgart, **1967**.
Lewin, L. *Die Gifte in der Weltgeschichte: Toxikologische allgemeinverständliche Untersuchungen der historischen Quellen*, reprograph. Nachdr. d. Ausg. von 1920, 2. Aufl., Gerstenberg, Hildesheim, **1983**.
Markwardt, F. (Hrsg.), *Allgemeine und spezielle Pharmakologie*, 4. Aufl., VEB Verlag Volk und Gesundheit, Berlin, **1983**.

Marquardt, H.; Schäfer, S. G. (Hrsg.), *Lehrbuch der Toxikologie*, BI, Mannheim, **1994**.
Müller, W. E. G. *Chemotherapie von Tumoren: Biochemische Grundlagen*, Verlag Chemie, Weinheim, **1975**.
Mutschler, E. *Arzneimittelwirkungen: Lehrbuch der Pharmakologie und Toxikologie*, 5. Aufl., Wiss. Verlagsges., Stuttgart, **1986**.
Nriagu, J. O. (Hrsg.), *Changing Metal Cycles and Human Health*, Springer, Berlin, **1984**.
Prestayko, A. W.; Crooke, S. T.; Carter, S. K. (Hrsg.), *Cisplatin: Current Status and New Developments*, Academic Press, New York, **1980**.
Rapoport, S. M. *Medizinische Biochemie*, 8. Aufl., VEB Verlag Volk und Gesundheit, Berlin, **1983**.
Schneider, G. *Pharmazeutische Biologie*, 2. Aufl., BI, Mannheim, **1985**.
Snyder, S. H. *Chemie der Psyche: Drogenwirkungen im Gehirn*, Spektrum, Heidelberg, **1988**.
Taylor, J. B.; Kennewell, P. D. *Introductory Medicinal Chemistry*, Horwood, Chichester, **1981**.
Wirth, W.; Gloxhuber, C. *Toxikologie*, 3. Aufl., Thieme, Stuttgart, **1981**.

(iiii) Umweltrelevante Substanzen

Bailey, R. A.; Clarke, H. M.; Ferris, J. P.; Krause, S.; Strong, R. L. *Chemistry of the Environment*, Academic Press, New York, **1978**.
Cairns, V. W.; Hodson, P. V.; Nriagu, J. O. (Hrsg.), *Contaminant Effects on Fisheries*, Wiley, New York, **1984**.
Calvert, S.; Englund, H. M. (Hrsg.,) *Handbook of Air Pollution Technology*, Wiley, New York, **1984**.
Duedall, I. W.; Ketchum, B. H.; Park, P. K.; Kester, D. R. (Hrsg.), *Wastes in the Ocean, Vol. 1: Industrial and Sewage Wastes in the Ocean*, Wiley, New York, **1983**.
Duffus, J. H. *Environmental Toxicology*, Arnold, London, **1980**.
Förstner, U.; Wittmann, G. T. W. *Metal Pollution in the Aquatic Environment*, Springer, Berlin, **1979**.
Haug, G.; Hoffmann, H. (Hrsg.), *Chemistry of Plant Protection, Vol. 6: Controlled Release, Biochemical Effects of Pesticides, Inhibition of Plant Pathogenic Fungi*, Springer, Berlin, **1990**.
Hedin, P. A. (Hrsg.), *Bioregulators for Pest Control*, ACS Symposium Series 276, American Chemical Society, Washington, **1985**.
Hock, B.; Elstner, E. F. (Hrsg.), *Schadwirkungen auf Pflanzen: Lehrbuch der Pflanzentoxikologie*, 2. Aufl., BI, Mannheim, **1988**.
Irgolic, K. J.; Martell, A. E. (Hrsg.), *Environmental Inorganic Chemistry*, VCH, New York, **1985**.
Kester, D. R.; Ketchum, B. H.; Duedall, I. W.; Park, P. K. (Hrsg.), *Wastes in the Ocean, Vol 2: Dredged-Material Disposal in the Ocean*, Wiley, New York, **1983**.
Koch, R. *Umweltchemikalien: Physikalisch-chemische Daten, Toxizitäten, Grenz- und Richtwerte, Umweltverhalten*, VCH, Weinheim, **1989**.
Kümmel, R.; Papp, S. *Umweltchemie*, VEB Dt. Verlag für Grundstoffindustrie, Leipzig, **1988**.
Laws, E. A. *Aquatic Pollution*, Wiley, New York, **1981**.
Legge, A. H.; Krupa, S. V. (Hrsg.), *Air Pollutants and Their Effects on the Terrestrial Ecosystem*, Wiley, New York, **1986**.
Lepp, N. W. (Hrsg.), *Effect of Heavy Metal Pollution on Plants, Vol. 1: Effects of Trace Metals on Plant Function; Vol. 2: Metals in the Environment*, Applied Science Publ., London, **1981**.
Martin, A. *Einführung in den Umweltschutz*, 3. Aufl., VEB Dt. Verlag für Grundstoffindustrie, Leipzig, **1986**.
Nriagu, J. O. *Lead and Lead Poisoning in Antiquity*, Wiley, New York, **1983**.
Nriagu, J. O. (Hrsg.), *Aquatic Toxicology*, Wiley, New York, **1983**.
Nriagu, J. O. (Hrsg.), *Environmental Impacts of Smelters*, Wiley, New York, **1984**.
Nriagu, J. O.; Davidson, C. I. (Hrsg.), *Toxic Metals in the Atmosphere*, Wiley, New York, **1986**.
Nriagu, J. O.; Simmons, M. S. (Hrsg.), *Toxic Contaminants in the Great Lakes*, Wiley, New York, **1984**.
Nriagu, J. O.; Sprague, J. B. (Hrsg.), *Cadmium in the Aquatic Environment*, Wiley, New York, **1987**.
Pitts, Jr., J. N.; Metcalf, R. L.; Grosjean, D. (Hrsg.), *Advances in Environmental Science and Technology*, Wiley, New York, **1980**.
Ragsdale, N. N.; Menzer, R. E. (Hrsg.), *Carcinogenicity and Pesticides*, ACS Symposium Series 414, American Chemical Society, Washington, **1989**.
Raiswell, R. W.; Brimblecombe, P.; Dent, D. L.; Liss, P. S. *Environmental Chemistry*, Arnold, London, **1980**.
Robinson, F. A. (Hrsg.), *Environmental Effects of Utilising More Coal*, The Royal Society of Chemistry, London, **1980**.
Schubert, R. (Hrsg.), *Lehrbuch der Ökologie*, Fischer, Jena, **1984**.
Schwartz, S. E. (Hrsg.), *Trace Atmospheric Constituents: Properties, Transformations, and Fates*, Wiley, New York, **1983**.
Strubelt, O. *Gifte in unserer Umwelt: Toxische Gefahren von Arsen bis Zyankali*, DVA, Stuttgart, **1989**.
Stugren, B. *Grundlagen der Allgemeinen Ökologie*, 4. Aufl., Fischer, Stuttgart, **1986**.

Umweltbundesamt (Hrsg.), *Umweltchemikalien: Prüfung und Bewertung von Stoffen auf ihre Umweltgefährlichkeit im Sinne des neuen Chemikaliengesetzes*, Berlin, **1980**.
Wadden, R. A.; Scheff, P. A. *Indoor Air Pollution: Characterization, Prediction, and Control*, Wiley, New York, **1983**.

(c)   Rohstoffe und Energieträger

Osteroth, D. *Von der Kohle zur Biomasse: Chemierohstoffe und Energieträger im Wandel der Zeit*, Springer, Berlin, **1989**.
Roschlau, H.; Heintze, W. *Wissensspeicher Bergbau: Erzbergbau und Kalibergbau*, 3. Aufl., VEB Dt. Verlag für Grundstoffindustrie, Leipzig, **1986**.
Schobert, H. H. *Coal: The Energy Source of the Past and Future*, American Chemical Society, Washington, **1987**.
Skinner, B. J. *Earth Resources*, 2nd Ed., Prentice-Hall, Englewood Cliffs, **1976**.
(vgl. auch Literatur über Erzlagerstätten in II.).

VII.   Wechselwirkungen

(a)   Kosmo-/Geosphäre

(Veränderung der Erdkruste durch extraterrestrische Objekte; vgl. Lehrbücher der Geochemie in II.).

(b)   Kosmo-/Atmosphäre

(Strahlenwirkungen; Fotochemie; vgl. Lehrbücher der Chemie der Atmosphäre in III.).

(c)   Kosmo-/Biosphäre

Feinberg, G.; Shapiro, R. *Life Beyond Earth: The Intelligent Earthling's Guide to Life in the Universe*, Morrow, New York, **1980**.
Goldsmith, D.; Owen, T. *The Search for Life in the Universe*, Benjamin/Cummings, California, **1980**.
Hoyle, F.; Wickramasinghe, C. *Evolution aus dem Weltraum*, Ullstein, Berlin, **1981**.
Ponnamperuma, C. (Hrsg.), *Comets and the Origin of Life*, Reidel, Dordrecht, **1981**.
Ponnamperuma, C. (Hrsg.), *Cosmochemistry and the Origin of Life*, Reidel, Dordrecht, **1983**.

(d)   Kosmo-/Hydrosphäre

(vgl. VII(a)).

(e)   Kosmo-/Anthroposphäre

Boschke, F. L. *Kernenergie: Eine Herausforderung unserer Zeit*, Birkhäuser, Basel, **1988**.
Maas, K. (Hrsg.), *Themen zur Kosmo-, Geo- und Rohstoffchemie*, Hüthig, Heidelberg **1975**.
(UV-Strahlung und Hautschäden; vgl. Lehrbücher der Dermatologie, sowie III.).

(f)   Geo-/Atmosphäre

Pichler, H. (Hrsg.), *Vulkanismus*, Spektrum, Heidelberg, **1988**.
(Verwitterungsprozesse; vgl. Lehrbücher der Geochemie in II.).

(g)   Geo-/Hydrosphäre

Barnes, H. L. (Hrsg.), *Geochemistry of Hydrothermal Ore Deposits*, 2nd Ed., Wiley, New York, **1979**.
Cronan, D. S. *Underwater Minerals*, Academic Press, London, **1980**.
Kennett, J. P. *Marine Geology*, Prentice-Hall, Englewood Cliffs, **1982**.
(vgl. Lehrbücher in II.).

(h)   Geo-/Biosphäre (auch Rhizosphäre)

Alexander, M. *Introduction to Soil Microbiology*, 2nd Ed., Wiley, New York, **1977**.
Brasier, M. D. *Microfossils*, Allen & Unwin, London, **1980**.
Day, W. *Genesis on Planet Earth: The Search for Life's Beginning*, 2nd Ed., Yale University Press, New Haven, **1984**.
Diels, L.; Mattick, F. *Pflanzengeographie*, 5. Aufl., de Gruyter, Berlin, **1958**.
Ehrlich, H. L. *Geomicrobiology*, Dekker, New York, **1981**.
Fenchel, T.; Blackburn, T. H. *Bacteria and Mineral Cycling*, Academic Press, London, **1979**.
Fraas, E. *Der Petrefaktensammler: Ein Leitfaden zum Bestimmen von Versteinerungen*, 4. Aufl., Franckh, Stuttgart, **1976**.

Holland, H. D.; Schidlowski, M. (Hrsg.), *Mineral Deposits and the Evolution of the Biosphere*, Springer, Berlin, **1982**.
Kinzel, H. *Pflanzenökologie und Mineralstoffwechsel*, Ulmer, Stuttgart, **1982**.
Krumbein, W. E. (Hrsg.), *Microbial Geochemistry*, Blackwell, Oxford, **1983**.
Krumbiegel, G.; Krumbiegel, B. *Fossilien der Erdgeschichte*, Enke, Stuttgart, **1981**.
Lynch, J. M. (Hrsg.), *The Rhizosphere*, Wiley, Chichester, **1990**.
Orr, W. L.; White, C. M. (Hrsg.), *Geochemistry of Sulfur in Fossil Fuels*, ACS Symposium Series 429, American Chemical Society, Washington, **1990**.
Trudinger, P. A.; Walter, M. R.; Ralph, B. J. (Hrsg.), *Biogeochemistry of Ancient and Modern Environments*, Springer, New York, **1980**.

(i) Geo-/Anthroposphäre

Chemistry and Agriculture, Special Publ. No. 36, The Chemical Society, London, **1979**.
Davies, B. E. (Hrsg.), *Applied Soil Trace Elements*, Wiley, Chichester, **1980**.
Fauser, O. *Kulturtechnische Bodenverbesserungen. Bd. I: Allgemeines, Entwässerung*, 5. Aufl., de Gruyter, Berlin, **1959**; *Bd. II: Bewässerung, Ödlandkultur Flurbereinigung*, 5. Aufl., de Gruyter, Berlin, **1961**.
Foth, H. D. *Fundamentals of Soil Science*, 7th Ed., Wiley, New York, **1984**.
Hayes, M. H. B.; MacCarthy, P.; Malcolm, R. L.; Swift, R. S. (Hrsg.), *Humic Substances II: In Search of Structure*, Wiley, Chichester, **1989**.
Kuntze, H.; Niemann, J.; Roeschmann, G.; Schwerdtfeger, G. *Bodenkunde*, 3. Aufl., Ulmer, Stuttgart, **1983**.
Lange, H. *Agrarchemie*, 3. Aufl., VEB Dt. Landwirtschaftsverlag, Berlin, **1986**.
LeBaron, H. M.; Mumma, R. O.; Honeycutt, R. C.; Duesing, J. H. (Hrsg.), *Biotechnology in Agricultural Chemistry*, ACS Symposium Series 334, American Chemical Society, Washington, **1987**.
Müller, G. (Hrsg.), *Bodenkunde*, 2. Aufl., VEB Dt. Landwirtschaftsverlag, Berlin, **1985**.
Roschlau, H. *ABC Erzbergbau*, VEB Dt. Verlag für Grundstoffindustrie, Leipzig, **1984**.
Scharrer, K. *Agrikulturchemie, Bd. I: Pflanzenernährung*, de Gruyter, Berlin **1953**; *Bd. II: Futtermittelkunde*, de Gruyter, Berlin, **1956**.
Tivy, J. *Landwirtschaft und Umwelt*, Spektrum, Heidelberg, **1993**.
Wild, A. *Umweltorientierte Bodenkunde*, Spektrum, Heidelberg, **1995**.
Ziechmann, W.; Müller-Wegener, U. *Bodenchemie*, BI, Mannheim, **1990**.
(vgl. Lehrbücher der Bodensphäre in II., der Agrikulturchemie, der Ökologie und Lagerstättenkunde).

(j) Atmo-/Hydrosphäre

Holland, H. D. *The Chemical Evolution of the Atmosphere and Oceans*, Princeton University Press, Princeton, **1984**.
(vgl. Lehrbücher der Ökologie sowie der Geo- und Biochemie).

(k) Atmo-/Biosphäre

Nisbet, E. G. *Globale Umweltveränderungen*, Spektrum, Heidelberg, **1994**.
(vgl. Lehrbücher der Ökologie und Biochemie).

(l) Atmo-/Anthroposphäre

(vgl. Angaben in VI(b)(iiii), Lehrbücher der Ökologie, der Umweltchemie sowie der Medizin).

(m) Hydro-/Biosphäre

Dewar, M. J. S. et al. (Hrsg.), *Topics in Current Chemistry 69: Interactions between Metal Ions and Living Organisms in Sea Water*, Springer, Berlin, **1977**.
Hempel, G. (Hrsg.), *Biologie der Meere*, Spektrum, Heidelberg, **1991**.
Rheinheimer, G. *Mikrobiologie der Gewässer*, Fischer, Stuttgart, **1981**.
Tardent, P. *Meeresbiologie*, Thieme, Stuttgart, **1979**.

(n) Hydro-/Anthroposphäre

Clark, R. B. *Kranke Meere?*, Spektrum, Heidelberg, **1992**.
(vgl. Angaben in VI(b)(iiii), Lehrbücher der Ökologie, der Hygiene sowie Medizinischen Mikrobiologie).

(o) Bio-/Anthroposphäre

Dörfler, H.-P.; Roselt, G. *Heilpflanzen*, Enke, Stuttgart, **1984**.

Franklin, T. J.; Snow, G. A.; Barrett-Bee, K. J.; Nolan, R. D. (Hrsg.), *Biochemistry of Antimicrobial Action*, 4th Ed., Chapman and Hall, London, **1989**.

Hiller, K.; Bickerich, G. *Giftpflanzen*, Enke, Stuttgart, **1988**.

Koch, H. P.; Hahn, G. *Knoblauch: Grundlagen der therapeutischen Anwendung von Allium sativum L.*, Urban & Schwarzenberg, München, **1988**.

Präve, P.; Faust, U.; Sittig, W.; Sukatsch, D. A. (Hrsg.), *Handbuch der Biotechnologie*, 3. Aufl., Oldenbourg, München, **1987**.

Prince, L. M.; Sears, D. F. (Hrsg.), *Biological Horizons in Surface Science*, Academic Press, New York, **1973**.

Reinhard, E. *Pharmazeutische Biologie, Bd. 1: Cytologie, Genetik, Physiologie, Viren, Bakterien, Pilze, Algen*, 3. Aufl., Wiss. Verlagsges., Stuttgart, **1986**.

Wagner, H. *Pharmazeutische Biologie: Drogen und ihre Inhaltsstoffe*, 2. Aufl., Fischer, Stuttgart, **1982**.

Walker, J. M.; Gingold, E. B. (Hrsg.), *Molecular Biology and Biotechnology*, 2nd Ed., Royal Society of Chemistry, London, **1988**.

Weiss, R. F. *Lehrbuch der Phytotherapie*, 5. Aufl., Hippokrates, Stuttgart, **1982**.

Zepernick, B.; Langhammer, L.; Lüdcke, J. B. P. *Lexikon der offizinellen Arzneipflanzen*, de Gruyter, Berlin, **1984**.

(vgl. auch Lehrbücher der Medizinischen Mikrobiologie, der Forst- und Landwirtschaft und des Fischereiwesens sowie VI(b)(iii)).

(p)   Kosmo-/Geo-/Hydro-/Bio-/Anthroposphäre (Ökologie)

Heinrich, D.; Hergt, M. *Atlas zur Ökologie*, dtv, München, **1990**.

VIII. Kulturbezüge

Benesch, K. *Archäologie: Eine Einführung*, Bertelsmann, Gütersloh, **1983**.

Leute, U. *Archeometry: An Introduction to Physical Methods in Archeology and the History of Art*, VCH, Weinheim, **1987**.

Moesta, H. *Erze und Metalle: Ihre Kulturgeschichte im Experiment*, Springer, Berlin, **1983**.

Riederer, J. *Kunstwerke chemisch betrachtet: Materialien, Analysen, Alters-bestimmung*, Springer, Berlin, **1981**.

IX.   Geschichte

Amberger-Lahrmann, M.; Schmähl, D. *Gifte: Geschichte der Toxikologie*, Fourier, Wiesbaden.

Bax, K. *Schätze aus der Erde: Die Geschichte des Bergbaus*, Econ, Düsseldorf, **1981**.

Engels, S.; Stolz, R. *ABC Geschichte der Chemie*, VEB Dt. Verlag für Grundstoffindustrie, Leipzig, **1989**.

Figurowski, N. *Die Entdeckung der chemischen Elemente und der Ursprung ihrer Namen*, MIR, Moskau, **1981**.

Ihde, A. J.; Kieffer, W. F. (Hrsg.), *Selected Readings in the History of Chemistry*, Journal of Chemical Education, Easton, **1965**.

Latz, G. *Die Alchemie: Das ist die Lehre von den großen Geheim-Mitteln der Alchemisten und den Spekulationen, welche man an sie knüpfte* (Originalausgabe: Bonn, **1869**), Fourier, Wiesbaden.

Lockemann, G. *Geschichte der Chemie*, Band I, *Vom Altertum bis zur Entdeckung des Sauerstoffs*, de Gruyter (Göschen), Berlin, **1950**.

Mierzecki, R. *The Historical Development of Chemical Concepts*, Kluwer, Dordrecht, **1991**.

Neuburger, A. *Die Technik des Altertums*, R. Voigtländers Verlag, Leipzig, **1919**; Reprint: **1994**.

Russell, C. A. (Hrsg.), *Recent Developments in the History of Chemistry*, The Royal Society of Chemistry, London, **1985**.

Strube, I.; Stolz, R.; Remane, H. *Geschichte der Chemie: Ein Überblick von den Anfängen bis zur Gegenwart*, VEB Dt. Verlag der Wissenschaften, Berlin, **1986**.

Strube, W. *Der historische Weg der Chemie*, 3. Aufl., VEB Dt. Verlag für Grundstoffindustrie, Leipzig, *Bd. I*, **1976**; *Bd. II*, **1981**.

Walden, P. *Geschichte der Chemie*, 2. Aufl., Athenäum, Bonn, **1950**.

Zimmermann, P. A. *Patentwesen in der Chemie: Ursprünge, Anfänge, Entwicklung*, BASF, Ludwigshafen, **1965**.

16 Über systemtheoretische Grundlagen der Gesellschaftstheorie

Niklas Luhmann

Abstract

At present, sociology does not offer an adequate theory for comprehending what is actually happening in our society today. In this note, new developments in general systems theory are discussed as they offer new perspectives for setting up an appropriate research program in theoretical sociology. Based on a fundamental distinction between system and environment, these developments have led to a reevaluation of characteristic traits of "non-trivial" or "autopoietic" systems: they consist of events (nerve impulses, thought, communications, ...) rather than matter; they mainly refer to and operate on themselves; they react to environmental pressures, perceived as perturbations, through highly selective modes of structural coupling, yet their evolution is driven by the internal dynamics of the system experimenting with its own complexity.

Theoretical sociology will profit considerably from viewing society (and itself as being part of this society), stating precisely which system it is analyzing, which operations – including the modes of self-observation and external observation – the system performs, and how it reproduces itself in this way.

Gegen Ende unseres Jahrhunderts scheinen die Möglichkeiten der klassischen Soziologie, die moderne Gesellschaft zu beschreiben, ausgeschöpft zu sein. Nach wie vor beherrschen freilich "Klassiker" die Theoriediskussionen der Soziologie, und man findet sogar die Auffassung, daß damit zwar nicht die Einzelheiten, wohl aber die grundlegenden Perspektiven fixiert seien [1]. Ein schwacher Trost ist der Verzicht auf jede Art von "Orthodoxie", aber die Diskrepanz zu den aktuellen Erfahrungen, die wir mit der heutigen Gesellschaft machen, fällt zunehmend auf. Neue Themen wie Probleme der Ökologie und Probleme des internationalen Geldflusses, Probleme eines die staatliche Ordnung unterlaufenden Regionalismus und Internationalismus, Probleme der Entwicklungsungleichgewichte und Probleme mit neuartig-eigensinnigem Individualismus bleiben einer soziologischen Schriftstellerei überlassen oder auch den sozialen Bewegungen. Intellektuelle Moden – gegenwärtig zum Beispiel Risikoforschung oder Technologiefolgenabschätzung – werden mitvollzogen, hinterlassen aber keine Spuren in der soziologischen Theorieentwicklung. Es fehlt an einer auch nur einigermaßen adäquaten Gesellschaftstheorie.

Die folgenden Überlegungen haben das Ziel zu prüfen, ob und in welchen Hinsichten neuere Entwicklungen in der allgemeinen Systemtheorie hierfür Anregungen bieten können. Diese Quellen liegen außerhalb der fachspezifischen soziologischen Forschung, aber auch außerhalb der Bemühungen, die als "Philosophie" bezeichnet werden. Ihre Relevanz ist daher nicht ohne weiteres zu erkennen. Und es bedarf eines scharf abstrahierenden Zugriffs, um sie für eine soziologische Auswertung aufzubereiten. Wir haben also keine Analogie im Sinn und auch keine nur metaphorische Übertragung von Maschinenmodellen oder Organismusmodellen. Das wird mit den folgenden Ausführungen hoffentlich deutlich werden.

16.1

Von "Systemtheorie" spricht man in nennenswertem Umfang erst seit den 50er Jahren. Damals lag das Ausgangsproblem im Entropiesatz der Thermodynamik und in Forschungen, die zeigen konnten, daß und wie "umweltoffene Systeme" sich dem Wärmetod entziehen, Negentropie aufbauen, Ungleichgewichte stabilisieren, morphogenetische Prozesse einleiten, Inputs in Outputs transformieren und "order from noise" produzieren können. Aus rückblickender Sicht lag der entscheidende Durchbruch darin, daß das traditionelle Schema des Ganzen und seiner Teile abgelöst wurde durch die Differenz von System und Umwelt. Die alte Unterscheidung des Ganzen und seiner Teile konnte verlustlos übernommen, und als Theorie der Systemdifferenzierung reformuliert werden, und zugleich konnte man (vor allem in Organisationsanalysen) zeigen, daß die Form der internen Differenzierung mit den Außenbeziehungen des Systems variiert. Alle Ordnung hängt demnach von Grenzerhaltung ab, aber Grenzen sind durchlässig – sei es für Energie, sei es für Information.

Bei allem Bewußtsein theoretischer Innovation blieb jedoch die System-Umwelt-Differenz als solche unreflektiert. Die Theorie schwankte zwischen einem konkreten und einem nur analytischen Begriff des Systems und seiner Grenzen. Systeme wurden als eine besondere Art von Objekten angesehen und die Systemtheorie als eine Beschreibung dieser Objekte verstanden. Daraus ergaben sich endlose Kontroversen über die Reichweite der Systemtheorie sowie eine teils humanistische, teils politische Polemik gegen überzogene Theorieansprüche und gegen die Übertragbarkeit der Systemtheorie auf menschliche und gesellschaftliche Belange [2].

Man wird zugeben können, daß es der Systemtheorie bis dahin nicht gelungen war, die Differenz von System und Umwelt mit ausreichender Radikalität zu formulieren. Heute stehen dafür bessere Möglichkeiten zur Verfügung, die aber außerhalb der Systemtheorie entwickelt worden sind. Das gilt vor allem für den von George Spencer Brown entwickelten Formenkalkül [3]. G. S. Brown beginnt mit der Feststellung, daß etwas nur bezeichnet werden kann, wenn es unterschieden werden kann. "Draw a distinction" ist daher die erste Anweisung seines Kalküls. Das Unterscheiden führt zur Markierung einer Form, die folglich immer zwei Seiten hat: das Bezeichnete und das, wovon es unterschieden wird. In Anwendung auf die Systemtheorie müßte man entsprechend sagen: das System und seine Umwelt [4]. Die Form ist demnach nicht die schöne (oder weniger schöne) Gestalt. Sie ist die Differenz, sie ist eine Zwei-Seiten-Form, sie ist eine Grenze, deren Überschreiten Zeit kostet.

Als Form markiert, macht die Unterscheidung den Zugriff auf den *unmarked space* unmöglich. Sie läßt statt dessen etwas sehen, nämlich das, was im Unterschied zu anderem bezeichnet werden kann. Jeder Versuch, im Ausgang von der Form zum *unmarked space* zurückzukehren, müßte die Einheit des Unterschiedenen zu erfassen versuchen und würde also die Form einer Paradoxie annehmen.

Wenn man diesen differenztheoretischen Ausgangspunkt akzeptiert, kann man sehen, daß das Forschungsprogramm der Systemtheorie nicht von besonderen Objekten handelt, sondern von einem Einschnitt, der durch eine zeitliche Verkettung von Operationen produziert wird und die Welt als Einheit des Differenten (als *unmarked space* im Sinne von G. S. Brown) voraussetzt, aber nie wieder erreicht, da eben alles Bezeichnen ein Unterscheiden und alles Unterscheiden einen Unterscheider voraussetzt. Und erst damit wird es möglich, zu fragen, was mit der Welt geschieht, wenn sich in ihr und innerhalb selbstgezogener Grenzen ein Gesellschaftssystem ausdifferenziert, das die Möglichkeit hat, sich selbst von anderem zu unterscheiden und die innere bzw. die äußere Seite der dadurch produzierten Form zu bezeichnen.

16.2

Alle wichtigen Innovationen der Systemtheorie beziehen sich auf diese Differenz von System und Umwelt, und der Trend geht in Richtung auf eine Theorie selbstreferentieller, operativ geschlossener Systeme. Dies soll an Hand einiger Beispiele erläutert werden.

1. Empirisch Überzeugendes hat vor allem die Neurobiologie und speziell die Gehirnforschung beigetragen. Das hierzu nötige Wissen stammt in wesentlichen Zügen bereits aus dem 19. Jh., zeigt aber in systemtheoretischer Interpretation neue Facetten. Gehirne benutzen eine eigene "Sprache", eine eigene elektrische Form der Datenverarbeitung, die nicht in die Umwelt hinein verlängert werden kann. Man hat von indifferenter (oder undifferenzierter) Codierung der Informationsverarbeitung des Gehirns gesprochen [5] und meint damit, daß das Gehirn nur mit eigenen Unterscheidungen arbeiten kann. Optische und akustische Wahrnehmungen werden zum Beispiel mit den gleichen Mitteln prozessiert, und erst das Gehirn selbst gibt ihnen eine unterschiedliche Qualität. Außenkontakte müssen daher in einem anderen Medium (etwa photochemisch) hergestellt werden; aber darüber weiß das Gehirn nichts. Auch quantitativ läßt sich diese Differenz von außen und innen eindrucksvoll nachweisen. Man schätzt, daß auf eine Außenkontaktreizung einhunderttausend interne Verarbeitungsvorgänge entfallen.

Im Unterschied zu klassischen erkenntnistheoretischen Problemen führen diese Einsichten keineswegs zu der Folgerung, daß ein Gehirn nur als Idee oder nur subjektiv existiere. Es ist selbstverständlich ein reales System, das von zahllosen hochkomplexen Umweltbedingungen abhängt, vermittelt durch das Leben des entsprechenden Organismus, aber auch durch eine hinreichend unruhige Umwelt. Um so bemerkenswerter dann die Einsicht, daß operative Geschlossenheit eine Realbedingung jeder Erkenntnis ist. Man kann die Umwelt nur erkennen, *weil* (der Idealismus hätte gesagt: obwohl) man keinen operativen Kontakt zu ihr unterhalten kann. Die Bedingung der Kontaktlosigkeit ermöglicht und wird ausgeglichen durch intern aufgebaute Eigenkomplexität.

Wenn dies aber schon eine Bedingung für Gehirnarbeit ist, gilt Entsprechendes erst recht für alle darauf aufbauenden Systeme, also für psychische Systeme (Bewußtseinssysteme) und soziale Systeme (Kommunikationssysteme). Diese sind dann aber auf eigene Operationen angewiesen und auf dieser Grundlage operativ geschlossen. Das Bewußtsein kann von seinem Gehirn im Moment des Gebrauchs nichts wissen; und dasselbe gilt für das Verhältnis der Kommunikation zu ihren neurobiologischen bzw. psychischen Bedingungen.

2. In engem Zusammenhang mit diesen Überlegungen unterscheidet Heinz von Foerster triviale Maschinen und nichttriviale Maschinen. "Maschine" heißt in diesem kybernetischen Zusammenhang einfach: Transformationsregel. Triviale Maschinen transformieren einen

Input in einen Output auf eine bestimmte, vorgesehene und wiederholbare Weise. Triviale Maschinen sind zuverlässige Maschinen. Wenn die Maschine nicht oder anders als vorgesehen funktioniert, ist sie kaputt und muß zur Reparatur oder ersetzt werden. Nichttriviale Maschinen sind selbstreferentielle Maschinen. Sie reagieren immer auch auf ihren jeweiligen eigenen Zustand, der seinerseits das Ergebnis vorheriger eigener Operationen ist. Es handelt sich, anders gesagt, um historische Maschinen, die sich mit jeder Operation in eine andere Maschine verwandeln. Solche Maschinen sind notorisch unzuverlässig, da sie auf die gleichen Anreize je nach ihrer Befindlichkeit verschiedene Reaktionen produzieren können [6].

Bei nichttrivialen Maschinen ist die Transformationsfunktion, die das Verhalten reguliert, ein adäquater Ausdruck ihrer Komplexität. Nichttriviale Maschinen werden schon bei wenigen Möglichkeiten des Inputs und des Outputs aus mathematisch faßbaren Gründen derart komplex, daß sie nicht mehr adäquat beschrieben und vor allem nicht prognostiziert werden können. Der Umgang mit ihnen ist wegen dieser unaufhebbaren Intransparenz angewiesen auf die Emergenz eines neuartigen Kontaktsystems, das eigenen Strukturen gehorcht. Im Falle psychischer Systeme, die selbstverständlich zur Klasse der nichttrivialen Maschinen gehören, entstehen anstelle von wechselseitiger Transparenz der psychischen Systeme soziale Systeme, an die man sich *statt dessen* halten kann.

3. Häufig werden die hier behandelten Theorien auch als Kybernetik zweiter Ordnung bezeichnet und damit von den älteren Formen einer Regelungskybernetik unterschieden. Auch diese Bezeichnung geht auf das Problem der Zirkularität zurück. Wir zeigen dies am klassischen Fall der Regelung der Temperatur eines Raumes durch einen Thermostaten. Der Thermostat ist konstruiert, um die Temperatur des Raumes zu kontrollieren und mit Hilfe von Heiz- oder Kühlapparaten konstant zu halten. Genausogut kann man aber auch sagen, daß der Raum den Thermostaten kontrolliert; denn immer und nur dann, wenn der Raum seine Temperatur ändert, tritt der Thermostat in Funktion. Der Thermostat kontrolliert die Raumtemperatur, die Raumtemperatur kontrolliert den Thermostaten: es handelt sich um ein symmetrisches, zirkulares Verhältnis. Am Objekt selbst läßt sich keine Kontrollasymmetrie feststellen. Nur ein Beobachter wird sich selbst auf eine asymmetrisierende Perspektive festlegen können, also als Techniker die eine, als Wohnzimmerbewohner die andere Kausalrichtung bevorzugen. Wenn jemand wissen will, welche Kausalität "gilt", muß er also nicht das Thermostat-Raum-System beobachten, er muß Beobachter beobachten (*second order cybernetics*). Und avancierte Theorien schließen daraus, daß es überhaupt genügt, Beobachter zu beobachten im Hinblick auf das, was sie beobachten, und im Hinblick auf das, was sie *nicht* beobachten können. Die Weltkonstruktion wird auf der Ebene des Beobachtens von Beobachtern errechnet, und auch dies in einem operativ geschlossenen selbstreferentiellen System. Zusätzlich zu und in engem Verbund mit der unter 1. behandelten Theorie operativer Schließung als Bedingung jeglicher Kognition gibt es also noch eine zweite epistemologische Revolution: die des Beobachtens zweiter Ordnung in einer nicht als gemeinsam vorausgesetzten, unterschiedlich perspektivierten Welt.

4. Im engen wissenschaftlichen Zusammenhang mit den erwähnten Theorieinnovationen hat der chilenische Neurobiologe Humberto Maturana vorgeschlagen, lebende Systeme als "autopoietische" Systeme zu bezeichnen [7]. Die Formulierung ist bewußt gewählt. Sie betont "poiesis" (nicht "praxis") im griechischen Verständnis, also nicht sich selbst befriedigendes Handeln, sondern Produktion; aber das produzierte Werk ist nicht ein Resultat außerhalb des produzierten Systems, sondern das produzierte System selbst. Die Zelle produziert ihre eigenen Elemente durch das Netzwerk ihrer eigenen Elemente.

Es gibt inzwischen viel unfruchtbare Diskussion darüber, ob dieser Begriff, als Definition des Lebens eingeführt, auf die Biologie beschränkt bleiben müsse [8]. Ohne weitere Theoriefestlegungen ist eine solche Frage jedoch nicht zu entscheiden. Wichtiger ist es daher zunächst, sich die theoretische Innovation klarzumachen. Sie besteht in der Übertragung des Konzepts der Selbstorganisation von der Ebene der Strukturen auf die Ebene der Elemente, ja der für das System selbst unauflösbaren Letztelemente eines Systems. Scharf formuliert heißt dies: Alles, was für ein System als Einheit fungiert (Strukturen, Elemente, aber auch das System selbst und die Umwelt des Systems), muß durch das System selbst produziert werden. Es gibt keinen Import von Einheit (also auch keinen Import von Information) in ein System und ebensowenig einen Export. Natürlich kann das System selbst die Welt unter einer solchen Voraussetzung beobachten und beschreiben; aber das ist und bleibt dann seine Eigenleistung.

Es gibt keine Merkmale dieser Begrifflichkeit, die dazu zwängen, sie auf Biologie zu beschränken. Allerdings dürfen gewichtige Unterschiede zwischen biologischem und sozialem System nicht verwischt werden. Die wohl wichtigste Differenz besteht darin, daß bereits Gehirne, erst recht aber psychische und soziale Systeme voll temporalisierte Systeme sind. Sie bestehen nicht aus Substanzen, sondern aus Ereignissen (Nervenimpulsen, Gedanken, Kommunikationen), die mit ihrem Entstehen schon wieder verschwinden. Das hat weittragende, noch keineswegs ausgelotete Konsequenzen für die Funktion von Strukturen. Denn Strukturen müssen jetzt nicht etwa Replikation sichern, sondern die Produktion ständig neuer, ständig anderer Ereignisse ermöglichen. Wer denkt, denkt nicht immer nur denselben Gedanken. Wer redet, darf sich nicht wiederholen. Die Zeit ist also schon auf der Ebene der Elemente in solche Systeme eingebaut (und nicht etwa nur im klassischen Diskussionskontext des "Strukturwandels" relevant). Es handelt sich in diesem Sinne um dynamisch stabile Systeme. Und das mag auch erklären, daß die soziokulturelle Evolution im Vergleich zur biologischen Evolution sehr viel schneller läuft, weil sie schon aus Gründen der Form von Autopoiesis zur ständigen Produktion von Diskontinuitäten gezwungen ist.

5. Zu den Mißlichkeiten einer inzwischen weitläufigen Diskussion über "Autopoiesis" gehört, daß einem Einzelbegriff (besonders von seinen Kritikern) zu viel zugemutet wird. Es versteht sich von selbst, daß ein Begriff noch keine Theorie ist und daß das Konzept der autopoietischen Systeme das gesamte Theoriedesign noch keineswegs determiniert. Es geht – wie unter 16.1 gesagt – um eine Formulierung der Differenz von System und Umwelt, und zahllose weitere Theorieentscheidungen müssen dann erst noch getroffen werden. Sie sind durch diesen Ausgangspunkt nicht festgelegt, stehen aber selbstverständlich unter Bedingungen der Kompatibilität, die beim Auftreffen auf klassische Theorien in einem faszinierenden Maße innovativ, aber auch verwirrend und irritierend wirken können. So muß zum Beispiel die Erkenntnistheorie von repräsentationalen Annahmen befreit und auf konstruktivistischer Basis neu formuliert werden (ohne daß man dadurch in die Falle des Solipsismus/Idealismus/Subjektivismus geriete). Aus genau parallel liegenden Gründen muß in der Evolutionstheorie der Adaptionismus aufgegeben werden [9]. Die Evolution führt nicht qua strenger Auslese zu immer besserer Anpassung der Systeme an ihre Umwelt, sondern sie experimentiert mit Möglichkeiten des Aufbaus von Eigenkomplexität in autopoietischen Systemen, die freilich immer schon umweltangepaßt sein müssen, wie ein Beobachter feststellen kann; denn andernfalls würden sie schlicht aufhören, ihre Autopoiesis fortzusetzen.

Um die Notwendigkeit von Zusatzkonzepten zu belegen, wähle ich als Beispiel die Begriffe strukturelle Kopplung und Irritation. Der Begriff der strukturellen Kopplung bezeichnet eine Form für reguläre Interdependenzen zwischen System und Umweltverhältnissen, die nicht operativ disponibel sind, sondern vorausgesetzt werden müssen. Wichtig ist die hohe Selektivität dieser Formen. Sie betreffen keineswegs pauschal die Gesamtumwelt eines Systems. So sind alle Kommunikationssysteme selbstverständlich an Bewußtseinsvorgänge gekoppelt. Ohne Bewußtsein keine Kommunikation. Aber das heißt gerade nicht, daß Bewußtseinsereignisse (Wahrnehmungen, Gedanken) als solche schon Elemente eines Kommunikationsprozesses sein könnten. Das Kommunikationssystem bleibt, mit anderen Worten, ein operativ geschlossenes selbstreferentielles System. Strukturelle Kopplung bedeutet andererseits und vor allem, daß die Umweltkopplung der Kommunikation auf Bewußtseinssysteme beschränkt ist und daß es keinen direkten (nicht über Bewußtsein vermittelten) physikalischen, chemischen oder biologischen Einwirkungen ausgesetzt wird. Damit werden nicht zuletzt die ökologischen Risiken einer solchen Abschließung verständlich.

Die hochselektive (und damit evolutionär unwahrscheinliche) Distinktheit von Formen struktureller Kopplung wird deutlich, wenn man bedenkt, daß es in diesem Fall um Sprache geht [10]. Daß die entsprechenden Geräusche bzw. (bei Schrift) optischen Figuren von selbst entstehen, ist extrem unwahrscheinlich. Die Distinktheit der Sprache erklärt sich aus den Anforderungen struktureller Kopplung, die ihrerseits erst entstehen, wenn es die entsprechenden autopoietischen Systeme bereits gibt. Abstrahiert man von Zeit, handelt es sich mithin um ein zirkuläres Phänomen der wechselseitigen Bedingtheit von Bewußtsein, Sprache und Gesellschaft. In Wirklichkeit liegt dem natürlich ein evolutionärer Prozeß zugrunde.

Was strukturelle Kopplungen für den Systemaufbau bedeuten, kann man mit dem Begriff der *Irritation* (Maturana sagt "perturbation") beschreiben. Auch Irritation ist eine voll systeminterne Angelegenheit. Es gibt keinen Transfer von Irritationen aus der Umwelt in das System [11]. Denn Irritationen ergeben sich immer erst aus Unterscheidungen und aus Vergleichen mit systeminternen Strukturen (Erwartungen), sind also – ebenso wie Information – notwendig systemeigenes Produkt. Trotzdem kann ein Beobachter erkennen, daß hochselektive Formen struktureller Kopplung auch Irritationsanlässe kanalisieren und damit den historischen Selbststrukturierungsprozeß autopoietischer Systeme beeinflussen. Darin liegt ein wichtiger Beitrag zur Erklärung des Tempos der Evolution. Auch die Theorie der Sozialisation wird von diesem begrifflichen Apparat profitieren können, denn Sozialisation ist einerseits immer Selbstsozialisation im geschlossenen System des Einzelbewußtseins, aber andererseits im Gesamteffekt natürlich nicht unabhängig von der Umwelt, die sich zwar nicht operativ, aber über strukturelle Kopplungen geltend macht.

16.3

Die Soziologie hat sich in den letzten Jahrzehnten in vielen Forschungsbereichen spektakulär entwickelt. Eine Gesellschaftstheorie, d.h. eine Theorie der Gesamtheit sozialer Verhältnisse (um nicht gleich zu sagen: eine Theorie des umfassenden Sozialsystems), hat sie jedoch nicht zustande gebracht. Alle Forschungen, etwa im Bereich Schichtung und Mobilität, Organisation, Jugend, Familienbildung, Geldwirtschaft und ihre Folgen usw., können natürlich auf die Gesamtheit moderner Lebensbedingungen bezogen werden; aber dieser Bezug bleibt unklar, solange dafür keine passende Begrifflichkeit ausgearbeitet ist.

Man kann den Grund für dieses Defizit in der Komplexität des Gesamtsystems vermuten, aber das würde nur auf Abstraktionsnotwendigkeiten verweisen, die der Wissenschaft vertraut sind. Erschwerend kommen Erkenntnishindernisse hinzu – mit Gaston Bachelard könnte man von "obstacles épistémologiques" sprechen [12] –, die sich aus einem Überhang traditioneller Begrifflichkeiten, Semantiken, Erkenntnisinstrumente ergeben, die durch den Stand der Forschung eigentlich überholt sind. Solche Traditionsüberhänge, die bis in die Anthropologie und die politische Philosophie der alteuropäischen Semantik zurückreichen, lassen sich leicht identifizieren. Ich nenne nur die drei wichtigsten:

1. Die Gesellschaftslehre müsse vom Menschen handeln, sei es von der Gattung Mensch, sei es von der Gesamtheit existierender Menschen und ihren sozialen Beziehungen. Das würde bedeuten, daß konkrete Menschen "Teile" des Gesellschaftssystems sind – mit Haut und Haaren, mit Chromosomen und Neuronen, mit Bewußtsein und mit Unterbewußtsein, und dies multipliziert mit fünf Milliarden.

2. Die Gesellschaftslehre behandle, um diesem Mengenproblem und auch gewichtigen regionalen Unterschieden Rechnung tragen zu können, territoriale Einheiten. Die USA und die UdSSR, Paraguay und Uruguay, die DDR und die Bundesrepublik seien verschiedene Gesellschaften (vielleicht sogar: Katalonien und Kastilien, Nordostbrasilien und Südbrasilien etc.). Die Gesellschaftsgrenzen seien territoriale Grenzen, vielleicht sogar politische Grenzen.

3. Die Gesellschaft sei ein Objekt, das durch ein Subjekt "objektiv" beschrieben werden könne. Gesellschaftstheorie sei in Durkheims Sinne "positive Wissenschaft". Eine Wissenschaft, die dieses Objekt behandle, könne (ungeachtet der wirkungsvollen Kritik der politischen Ökonomie durch Karl Marx) ihre eigenen gesellschaftlichen Bedingungen vernachlässigen oder durch methodische Tricks neutralisieren. Methodisch gesicherter Konsens könne als funktionales Äquivalent für einen archimedischen Punkt, für einen externen Standort dienen.

Es fällt – mehr oder weniger – schwer, diese Erkenntnisvoraussetzungen aufzugeben und ins Museum für soziologische Altertumskunde abzustellen. Man muß – und dies ist ein Problem, das typisch die Fortdauer solcher "obstacles épistémologiques" sichert – wissen, was denn statt dieser Annahmen gelten soll. Meine These ist, daß die skizzierten Entwicklungen der Systemtheorie diesen Sprung ermöglichen. Denn sie können zeigen, daß die klassischen Prämissen unbrauchbar sind und warum. Und sie können ein Theoriedesign präsentieren, das an ihre Stelle treten kann, nämlich die Theorie selbstreferentieller, operativ geschlossener Sozialsysteme [13].

Vor allem verlangt die Theorie autopoietischer Systeme, daß man die Operation genau bestimmt, die die Autopoiesis des Systems vollzieht und dadurch sowohl die Elemente (eben diese Operationen) als auch die Differenz von System und Umwelt, also die "Form" (G. S. Brown) des Systems reproduziert. Als soziale Operation kommt nur Kommunikation in Betracht, und zwar Kommunikation nicht im Sinne eines Mitteilungshandelns begriffen, sondern im Sinne einer operativen Einheit von Mitteilung, Information und Verstehen; also im Sinne einer Einheit, die mehr als nur ein Bewußtseinssystem über strukturelle Kopplungen aktiviert [14].

Die Operation Kommunikation stellt, ungeachtet gewisser Randunschärfen, völlig klar, was zur Gesellschaft gehört und was nicht. Zur Gesellschaft gehört nur das, was im Verlauf der Kommunikation als Kommunikation behandelt, also in rekursiver Referenz auf andere Kommunikationen als Operation des Systems produziert wird. Wenn Unklarheit darüber besteht, ob eine Operation Kommunikation ist oder nicht, kann nur im Kommunikationssystem darüber entschieden werden. Alles andere, insbesondere die körperliche und psychische Existenz der Menschen und auch ihr wahrnehmbares Verhalten in seinen nicht als Kommunikation behandelten Aspekten, bleibt Umwelt des Systems. Insofern hat die

Gesellschaft eindeutige, und zwar nichtterritoriale Grenzen. Und insofern gibt es unter heutigen Bedingungen universaler kommunikativer Anschlußfähigkeit nur ein Gesellschafftssystem: die Weltgesellschaft. Alle Unterschiede – und seien es die von Industrieländern und Entwicklungsländern oder die des sozialistischen und des kapitalistischen Lagers – werden in der Gesellschaft und *durch ihre Strukturen* erzeugt. Und sie wirken eben deshalb "anstößig" im Hinblick auf weitere Kommunikation.

Immer noch verführt ein traditionsgeleitetes Denken uns zu der Annahme, "die Menschen" seien für "die Gesellschaft" verantwortlich. Allerdings hatte die alteuropäische Tradition, die von Schöpfergott, von der Natur, von einem feststehenden Essenzenkosmos ausging, so nie gedacht. Der Mensch ist also gewissermaßen nur übriggeblieben, nachdem man nicht mehr an Gott glaubt und die Natur als kontingent begreift. Aber je problematischer die Anblicke werden, die die moderne Gesellschaft bietet, desto unsinniger wird es, Hoffnungen und Verantwortungen auf "die Menschen" abzuladen, ohne genau anzugeben, was damit gemeint ist. Eine geschärfte Begrifflichkeit fordert statt dessen, das Gesellschaftssystem zu beobachten im Hinblick auf das, was unter welchen Bedingungen möglich ist.

Das ist ein gegen die Tradition gerichtetes, deutlich "antihumanistisches" Konzept, was mir böse Kommentare und ständige Hinweise auf die Unentbehrlichkeit und den "Wert" des menschlichen Lebens eingetragen hat. Kritiker (unter ihnen: Jürgen Habermas) sollten jedoch mindestens dies berücksichtigen: daß die Systemtheorie nicht in ein besonderes Objekt vernarrt ist, sondern in eine Differenz, die zwei Seiten hat – System und Umwelt. Es geht im vorliegenden Fall um zwei Seiten einer Form, der Form des Sozialen. Für die Gesellschaftstheorie ist der Mensch die andere Seite, und in der Gesellschaft und ihren Kommunikationen ist er ständig durch die Unterscheidung von Selbstreferenz und Fremdreferenz präsent. Man muß nur das System deutlich im Auge behalten, das diese Unterscheidung praktiziert, eben die sich selbstreferentiell bezeichnende Gesellschaft, für die jede Referenz auf den Menschen Fremdreferenz ist [15]. Außerdem schließt die Systemtheorie ja keineswegs aus, auch psychische Systeme, Gehirne, Zellen, Immunsysteme etc. von Menschen als Systeme-in-ihrer-Umwelt zu beobachten. Sie verlangt nur, daß jeweils bestimmt angegeben wird, auf welches System die Analyse sich bezieht und was, von da her gesehen, Umwelt ist. Wer sich für Individuen interessiert, muß dann eben sagen für welches.

Schließlich ermöglicht die Systemtheorie eine Klarstellung ihrer eigenen erkenntnistheoretischen Voraussetzungen. Sie akzeptiert hier den Zirkelschluß als ohnehin konstitutiv für jedes selbstreferentielle System. Sie behandelt die Gesellschaft als ein sich selbst beobachtendes und beschreibendes System. Sie radikalisiert damit in gewisser Weise die Kritik der politischen Ökonomie und ihre "wissenssoziologischen" Folgetheorien. Es gibt kein gesellschaftsexternes System, kein Bewußtsein (etwa meines), das die Gesellschaft angemessen beobachten und beschreiben könnte. Es fehlte dazu die zureichende Eigenkomplexität oder, im Sinne von Ashby, "requisite variety". Jede Gesellschaftstheorie – auch die hier empfohlene – enthält daher eine "autologische" Komponente [16]. Sie spricht, indem sie von der Gesellschaft spricht, auch von sich selbst, da ihr operativer Vollzug nur als Autopoiesis von Gesellschaft, nur in der rekursiven Vernetzung der gesellschaftlichen Kommunikation möglich ist.

Logiker folgern aus solchen Sachverhalten, daß es notwendig ist, "Ebenen" sprachlicher bzw. erkenntnismäßiger Art zu unterscheiden. Ein solcher Trick ist jedoch nur ein Notbehelf, nur eine schlecht getarnte Paradoxie, weil jede Ebene nur deshalb eine Ebene ist, weil sie auf andere Ebenen verweisen kann. Für soziologische Zwecke ist es daher sinnvoller, zur Kybernetik zweiter Ordnung überzugehen. Das heißt: Es werden nur noch beobachtende Systeme beobachtet. An die Stelle klassischer Objektivitäts- und Konsensansprüche und ihrer Methodologien tritt dann die Unterscheidung des Beobachters, zu sehen, was andere sehen und was sie nicht sehen können. Dem liegt aber keine Entlarvungsabsicht zugrunde, sondern eher eine Generalisierung und Selbstanwendung von "Ideologiekritik". Jede Beobachtung setzt als Operation Instrumente, nämlich für sie spezifische Unterscheidungen voraus, die sie im Moment ihrer Verwendung *nicht* beobachten kann (So im hier vorgeführten Falle: die Unterscheidung von System und Umwelt.). Jede weitere Beobachtung desselben oder eines anderen Beobachters kann diesen "blinden Fleck" erfassen, aber nur dadurch, daß er sich auf einen anderen einläßt. Es gibt deshalb keine "Übereinstimmung von Sein und Denken" im Sinne des einen, von den Musen empfohlenen Weges. Aber es gibt die Möglichkeit, daß sich im ständigen Prozessieren des Beobachtens von Beobachtungen und des Beschreibens von Beschreibungen "Eigenzustände" des Systems stabilisieren, die auch unter diesen Bedingungen einen Orientierungswert behalten.

Von da her begründen sich spezifische Ansprüche an Gesellschaftstheorie, die denen ähneln, die man seit der Romantik an Kunstwerke richtet: klare Exposition ihrer Instrumente, ihrer Beschreibungsmittel, ihrer Leitunterscheidungen, ihrer Machart. Die damit verbundenen Ansprüche an Genauigkeit und an differenztheoretische wie an selbstreferentielle Begrifflichkeit machen die klassische Literatur der Soziologie unlesbar. Aber ohnehin würde eine schlichte historische Reflexion zum selben Ergebnis führen. Wir haben am Ende dieses Jahrhunderts andere Erfahrungen mit dem Gesellschaftssystem und seiner Umwelt. Wir leben in einer anderen Welt. Wir müssen mit ganz anderen Zukunftsperspektiven zurechtkommen. Und um so dringlicher wird die Frage, ob man sich länger dem Taumel postmoderner Beliebigkeiten, den gepflegten Ungenauigkeiten der französischen Philosophen oder der Angstrhetorik sozialer Bewegungen überlassen darf, wenn es bessere Möglichkeiten des Theoriedesigns gibt!

References

1 From Molecular Systems to More Complex Ones
Achim Müller and Klaus Mainzer

[1] Cited from Siegel, R. W. *Physics Today*, **1993**, *46*, 64.

[2] Müller, A. *J. Mol. Struct.* **1994**, *325*, 13.

[3] Ball, P. *Designing the Molecular World*, Princeton University Press, Princeton, USA, **1994**.

[4] Mainzer, K. *Thinking in Complexity: The Complex Dynamics of Matter, Mind, and Mankind*, Springer, Berlin, Germany, **1994**; Nicolis, G.; Prigogine, I. *Exploring Complexity: An Introduction*, Freeman, New York, USA, **1989**; Nicolis, G.; Prigogine, I. *Self-Organization in Nonequilibrium Systems: From Dissipative Structures to Order through Fluctuations*, Wiley, New York, USA, **1977**; Ebeling, W.; Feistel, R. *Chaos und Kosmos: Prinzipien der Evolution*, Spektrum, Heidelberg, Germany, **1994**; Haken, H.; Wunderlin, A. *Die Selbststrukturierung der Materie: Synergetik in der unbelebten Welt*, Vieweg, Braunschweig, Germany, **1991**; Cohen, J.; Stewart, I. *The Collapse of Chaos: Discovering Simplicity in a Complex World*, Penguin, New York, USA, **1994**.

[5] Müller, A. *Nature* **1991**, *352*, 115; Müller, A.; Krickemeyer, E.; Penk, M.; Rohlfing, R.; Armatage, A.; Bögge, H. *Angew. Chem.* **1991**, *103*, 1720; *Angew. Chem. Int. Ed. Engl.* **1991**, *30*, 1674; Müller, A.; Rohlfing, R.; Krickemeyer, E.; Bögge, H. *Angew. Chem.* **1993**, *105*, 916; *Angew. Chem. Int. Ed. Engl.* **1993**, *32*, 909; Müller, A.; Reuter, H.; Dillinger, S. *Angew. Chem.*, **1995**, *107*, in press; *Angew. Chem. Int. Ed. Engl.*, **1995**, *34*, in press.

[6] Müller, A.; Rohlfing, R.; Döring, J.; Penk, M. *Angew. Chem.* **1991**, *103*, 575; *Angew. Chem. Int. Ed. Engl.* **1991**, *30*, 588.

[7] Fischer, K. H.; Hertz, J. A. *Spin Glasses*, Cambridge University Press, Cambridge, United Kingdom, **1991**.

[8] Sessoli, R.; Gatteschi, D.; Caneschi, A.; Novak, A. M. *Nature* **1993**, *365*, 141; Gatteschi, D.; Pardi, L.; Barra, A. L.; Müller, A.; Döring, J. *Nature* **1991**, *354*, 463, and literature cited therein; Kahn, O. *Molecular Magnetism*, VCH, Weinheim, Germany, **1993**.

[9] Vainshtein, B. K. *Fundamentals of Crystals: Symmetry, and Methods of Structural Crystallography*, 2nd ed., Springer, Berlin, Germany, **1994**; Zachariasen, W. H. *Theory of X-Ray Diffraction in Crystals*, Dover Publ., New York, USA, **1967**.

[10] Vögtle, F. *Supramolekulare Chemie*, 2nd ed., Teubner, Stuttgart, Germany, **1992**; *Supramolecular Chemistry*, Wiley, Chichester, United Kingdom, **1993** (reprint); Lehn, J.-M. *Supramolecular Chemistry: Concepts and Perspectives*, VCH, Weinheim, Germany, **1995**.

[11] Pope, M. T.; Müller, A. (eds.), *Polyoxometalates: From Platonic Solids to Anti-Retroviral Activity*, Kluwer, Dordrecht, The Netherlands, **1994**.

[12] Müller, A. et al. *Nature* **1995**, to be published.

[13] Müller, A.; Krahn, E. *Angew. Chem.* **1995**, *107*, 1172; *Angew. Chem. Int. Ed. Engl.* **1995**, *34*, 1071.

[14] Müller, A.; Plass, W.; Krickemeyer, E.; Dillinger, S.; Bögge, H.; Armatage, A.; Proust, A.; Beugholt, C.; Bergmann, U. *Angew. Chem.* **1994**, *106*, 897; *Angew. Chem. Int. Ed. Engl.* **1994**, *33*, 849.

[15] Müller, A.; Krickemeyer, E.; Meyer, J.; Bögge, H.; Peters, F.; Plass, W.; Diemann, E.; Dillinger, S.; Nonnenbruch, F.; Randerath, M.; Menke, C. *Angew. Chem.* **1995**, *107*, in press; *Angew. Chem. Int. Ed. Engl.* **1995**, *34*, in press.

[16] Müller, A.; Krickemeyer, E.; Dillinger, S.; Bögge, H.; Stammler. A, *J. Chem. Soc., Chem. Commun.* **1994**, 2539.

[17] Oppenheim, I.; Shuler, K. E.; Weiss, G. H. *Stochastic Processes*, in: Lerner, R. G.; Trigg, G. L. (eds.), *Encyclopedia of Physics*, 2nd ed., VCH, New York, USA, **1991**, 1177.

[18] Feynman, R. P.; Leighton, R. B.; Sands, M. *The Feynman Lectures on Physics, Vol. 1, (Chapter 6: Probability)*, Addison-Wesley, Reading, Massachusetts, USA, **1966**.

[19] Landau, L. D.; Lifschitz, E. M. *Lehrbuch der Theoretischen Physik, Bd. 5: Statistische Physik, Teil 1*, 8. Aufl., Akademie-Verlag, Berlin, Germany, **1987**.

[20] See also book by Haken et al. in ref. [4].

[21] Mainzer, K. [4], Chapter 6.4; Weidlich, W. *Das Modellierungskonzept der Synergetik für dynamisch sozio-ökonomische Prozesse*, in: Mainzer, K.; Schirmacher, W. (eds.), *Quanten, Chaos und Dämonen*, B.I. Wissenschaftsverlag, Mannheim, Germany, **1994**, 255.

[22] Goodwin, R. M. *Chaotic Economic Dynamics*, Clarendon Press, Oxford, United Kingdom, **1990**; Lorenz, H.-W. *Nonlinear Dynamical Economics and Chaotic Motion*, Springer, Berlin, Germany **1989**; Mainzer, K. [4], Chapter 6.3.

2 Entropies and Lexicographic Analysis of Biosequences

Hanspeter Herzel, Werner Ebeling, Armin O. Schmitt, and Miguel Angel Jiménez-Montaño

[1] Schrödinger, E. *What is Life?*, Cambridge University Press, Cambridge, United Kingdom, **1944**.

[2] Ebeling, W.; Feistel, R. *Physik der Selbstorganisation und Evolution*, Akademie Verlag, Berlin, Germany, **1982**.

[3] Gatlin, L. L. *Information Theory and the Living System*, Columbia University Press, New York, USA, **1972**.

[4] Shannon, C. E. *Bell Syst. Techn. J.* **1948**, *27*, 379.

[5] Wolpert, D.; Wolf, D., *Estimating functions of probability distributions from a finite set of samples*, preprint, Santa Fe Institute, TR-93-07-046, **1993**.

[6] Feistel, R.; Ebeling, W. *Evolution of Complex Systems*, Kluwer, Dordrecht, The Netherlands, **1989**.

[7] Grassberger, P. *Phys. Lett.* **1988**, *128A*, 36.

[8] Herzel, H. *Sys. Anal. Mod. Sim.* **1988**, *5*, 435.

[9] Herzel, H.; Schmitt, A. O.; Ebeling, W. *Chaos Solit. Fract.* **1994**, *4*, 97.

[10] Schmitt, A. O.; Herzel, H.; Ebeling, W. *Europhys. Lett.* **1993**, *23*, 303.

[11] Schmitt, A. O.; Herzel, H.; Ebeling, W. *The Maximum Entropy Principle and Blockentropies*, in preparation.

[12] Oliver, S. G. et al., *Nature* **1992**, *357*, 38.

[13] Herzel, H.; Ebeling, W.; Schmitt, A. O. *Phys. Rev. E.* **1994**, *50*, 5061.

[14] Ebeling, W.; Feistel, R.; Herzel, H. *Physica Scripta* **1987**, *35*, 761.

[15] Rényi, A. *Wahrscheinlichkeitsrechnung,* Verlag der Wissenschaften, Berlin, Germany, **1975**.

[16] Csordás, A.; Györgyi, G.; Szépfalusy, P.; Tél, T. *Chaos* **1993**, *3*, 31.

[17] Eckmann, J.-P.; Ruelle, D. *Rev. Mod. Phys.* **1985**, *57*, 617.

[18] Ebeling, W.; Pöschel, T.; Albrecht, K. *J. Bif. & Chaos*, submitted.

[19] Staden, R. *Nucl. Acid. Res.* **1984**, *12*, 551.

[20] Ficket, J. W.; Chang-Shung Tung, *Nucl. Acid Res.* **1992**, *20*, 6441.

[21] Ebeling, W.; Nicolis, G. *Europhys. Lett.* **1991**, *14*, 191.

[22] Goldberger, A. L. *Phys. Rev. Lett.* **1993**, *70*, 1343.

[23] Mandelbrot, B. B. *The Fractal Geometry of Nature,* Freeman, New York, USA, **1982**.

[24] Weissmann, M. B. *Rev. Mod. Phys.* **1988**, *60*, 537.

[25] Li, W. *Europhys. Lett.* **1992**, *17*, 655.

[26] Peng, C. K.; Buldyrev, S. V.; Goldberger, A. L.; Havlin, S.; Sciortino, F.; Simons, M.; Stanley, H. E. *Nature* **1992**, *356*, 186.

[27] Voss, R. F. *Phys. Rev. Lett.* **1992**, *68*, 3805.

[28] Ebeling, W.; Neiman, A. *Phys. Rev. Lett.*, submitted.

[29] Li, W. *J. Stat. Phys.* **1990**, *60*, 823.

[30] Munson, P. J.; Taylor, R. C.; Michaels, G. S. *Nature* **1992**, *360*, 636.

[31] Berg, P.; Singer, M. *Dealing with Genes,* University Science Books, **1992**.

[32] Knippers, R.; Philippsen, P.; Schäfer, K. P.; Fanning, E. *Molekulare Genetik,* Thieme, Stuttgart, Germany, **1990**.

[33] Trifonov, E. N. *Bull Math. Biol.* **1989**, *51*, 417.

[34] Watson, J. D.; Gilman, M.; Withowski, J.; Zoller, H. *Recombinant DNA,* Freeman, New York, USA, **1992**.

[35] Ioshikhes, I.; Bolshoy, A.; Trifonov, E. N. *J. Biomol. Struct. Dyn.* **1992**, *6*, 1111.

[36] Beckmann, J. S.; Trifonov, E. N. *Proc. Natl. Acad. Sci. USA* **1991**, *88*, 2380.

[37] Borštnik, B.; Pumpernik, D.; Lukman, D. *Europhys. Lett.* **1993**, *23*, 289.

[38] Lewin, B. *Genes IV*, Oxford University Press, United Kingdom, **1990**.

[39] Brendel, V.; Beckmann, J. S.; Trifonov, E. N. *J. Biomol. Struct. Dyn.* **1986**, *4*, 11.

[40] Pietrokovski, S.; Hirshon, J.; Trifonov, E. N. *J. Biomol. Struct. Dyn.* **1990**, *7*, 1251.

[41] Ratner, V. A. *The genetic language* in: Rosen, E.; Snell, F. M. (eds.), *Progress of Theor. Biol.*, Academic Press., New York, USA, **1974**.

[42] Ebeling, W.; Jiménez-Montaño, M. A. *Math. Biosci.* **1980**, *52*, 5371.

[43] Jiménez-Montaño, M. A. *BioSystems* **1993**, *28*, 11.

[44] Chomsky, N. *Topics in the Theory of Generative Grammar.* Mouton, Den Haag, The Netherlands, **1972**.

[45] Chavoya, O.; Garcia, F.; Jiménez-Montaño, M. A. *Memorias IX Rennion Int. Artif. Mexico* **1992**, 243.

[46] Held, G. *Data Compression*, Wiley, Chichester, United Kingdom, **1988**.

3 Coding Coenzyme Handles and the Origin of the Genetic Code
Eörs Szathmáry

[1] Di Giulio, M. *Trends Ecol. Evol.* **1992**, *7*, 176.

[2] Osawa, S.; Jukes, T. H.; Watanabe, K.; Muto, A. *Microbiol. Rev.* **1992**, *56*, 229.

[3] Szathmáry, E. *Proc. Natl. Acad. Sci. USA* **1993**, *90*, 9916.

[4] Wong, J. F.-T. *Microbiol. Sci.* **1988**, *5*, 174.

[5] Szathmáry, E. in: Lukács, B.; Bérczi, Sz.; Molnár, I. (eds.), *Evolution: From cosmogensis to biogenesis*, KFKI-1990-50/C preprint, Budapest, Hungary, **1990**, 77.

[6] Maynard Smith, J.; Szathmáry, E. in: Murphy, M.; O'Neill, L. (eds.), *What is life? The next fifty years*, Cambridge University Press, Cambridge, United Kingdom, **1995**.

[7] Darnell, J. E.; Doolottle, W. F. *Proc. Natl. Acad. Sci. USA* **1986**, *83*, 1271.

[8] Gilbert, W. *Nature* **1986**, *319*, 818.

[9] Benner, S. A.; Allemann, P. K.; Ellington, A. D.; Ge, L.; Glasfeld, A.; Leanz, G. F.; Krauch, T.; MacPherson, L. J.; Moroney, S.; Piccirilli, J. A.; Weinhold, E. *Cold Spring Harbor Symp. Quant. Biol.* **1987**, *52*, 53.

[10] Benner, S. A.; Ellington, A. D.; Tauer, A. *Proc. Natl. Acad. Sci. USA* **1989**, *86*, 7054.

[11] Kacser, H.; Beeby, R. *J. Mol. Evol.* **1984**, *20*, 38.

[12] Szathmáry, E. *Proc. R. Soc. Lond. B* **1991**, *245*, 91.

[13] Szathmáry, E. *Proc. Natl. Acad. Sci. USA* **1992**, *89*, 2614.

[14] Wächtershäuser, G. *Prog. Biophys. molec. Biol.* **1992**, *58*, 85.

[15] Danchin, A. *Prog. Biophys. molec. Biol.* **1989**, *54*, 81.

[16] Yarus, M. *Science* **1988**, *240*, 1751.

[17] Famulok, M. *J. Am. Chem. Soc.* **1994**, *116*, 1698.

[18] White, H. B. *J. Mol. Evol.* **1976**, *7*, 101.

[19] Orgel, L. E. *J. Mol. Evol.* **1989**, *29*, 465.

[20] Kazakov, S.; Altman, S. *Proc. Nat. Acad. Sci. USA* **1992**, *89*, 7939.

[21] Högenauer, G. *Eur. J. Biochem.* **1970**, *12*, 527.

[22] Uhlenbeck, U.C.; Baller, J.; Doty, P. *Nature* **1970**, *225*, 508.

[23] Pongs, O.; Bald, R.; Reinwald, E. *Eur. J. Biochem.* **1973**, *32*, 117.

[24] Musier-Forsyth, K.; Scaringe, S.; Usman, N.; Schimmel, P. *Proc. Natl. Acad. Sci. USA* **1991**, *88*, 209.

[25] Yarus, M. *New Biol.* **1991**, *3*, 183.

[26] Piccirilli, J. A.; McConnell, T. S.; Zaug, A. J.; Noller, H. F.; Cech, T. R. *Science* **1992**, *256*, 1420.

[27] Shimizu, M. *Biophys. Chem.* **1987**, *28*, 169; *J. Phys. Soc. Jpn.* **1987**, *56*, 43; *J. Phys. Soc. Jpn.* **1988**, *57*, 54.

[28] Shimizu, M. *J. Mol. Evol.* **1982**, *18*, 297.

[29] Lacey Jr., J. C.; Mullins Jr., D. W. *Origins of Life* **1983**, *13*, 3.

[30] Lehman, N.; Jukes, T. H. *J. theoret. Biol.* **1988**, *135*, 203.

[32] Wong, J. F.-T. *Trends Biochem. Sci.* **1981**, *6*, 33.

[33] Fitch, W. M. *J. Mol. Biol.* **1966**, *16*, 1.

[34] Woese, C. R. *Proc. Natl. Acad. Sci. USA* **1965**, *52*, 1160; *The genetic code*, Harper and Row, New York, USA, **1967**.

[35] Fitch, W. M.; Upper, K. *Cold Spring Harbor Symp. Quant. Biol.* **1987**, *52*, 759.

[36] Di Giulio, M. *J. Mol. Evol.* **1989**, *29*, 191; *J. Mol. Evol.* **1989**, *29*, 288.

[37] Woese, C. R.; Dugre, S. A.; Kando, M.; Saxinger, W. C. *Cold Spring Harbor Symp. Quant. Biol.* **1966**, *31*, 720.

[38] Szathmáry, E.; Zintzaras, E. *J. Mol. Evol.* **1992**, *35*, 185.

[39] Taylor, F. J. R.; Coates, D. *BioSystems* **1989**, *22*, 177.

[40] Weiner, A.; Maizels, N. *Proc. Natl. Acad. Sci. USA* **1987**, *84*, 7383.

[41] White, H. B. in: Everse, J.; Anderson, B.; You, K. (eds.), *The pyridine nucleotide coenzymes*, Academic Press, New York, USA, **1982**, 1.

[42] Tsuchihashi, Z.; Khosla, M.; Herschlag, D. *Science* **1993**, *262*, 99.

[43] Wilson, E. O. *Sociobiology – The new synthesis*, The Belknap Press of Harvard University Press, Harvard, USA, **1975**.

[44] Wood, P. N. *J. Mol. Evol.* **1991**, *33*, 464.

[45] Grosjean, H.; Söll, D. G.; Crothers, M. *J. Mol. Biol.* **1976**, *103*, 499.

[46] Eisinger, J.; Gross, N. *J. Mol. Biol.* **1974**, *88*, 165.

[47] Grosjean, H. J.; De Henau, S.; Crothers, D. M. *Proc. Natl. Acad. Sci. USA* **1978**, *75*, 610.

[48] Jukes, T. H.; Holmquist, R.; Moise, H. *Science* **1975**, *189*, 50.

[49] Rodin, S.; Ohno, S.; Rodin, A. *Proc. Natl. Acad. Sci. USA* **1993**, *90*, 4723.

[50] Konecny, J.; Eckert, M.; Schöniger, M.; Hofacker, G. L. *J. Mol. Evol.* **1993**, *36*, 407.

[51] Zull, J. E.; Smith, S. K. *Trends Biochem. Sci.* **1990**, *15*, 257.

[52] Fukuchi, S.; Otsuka, J. *J. theor. Biol.* **1992**, *158*, 271.

[53] Hoffmann, G. W. *J. Mol. Biol.* **1974**, *86*, 349.

[54] Bedian, V. *Origins of Life* **1982**, *12*, 181.

[55] Mellersh, A. R. *Orig. Life Evol. Biosph.* **1993**, *23*, 261.

[56] Lacey, Jr., J. C.; Wickramasinghe, N. S. M. D.; Cook, G. W. *Orig. Life Evol. Biosph.* **1992**, *22*, 243.

[57] Weber, A. L.; Lacey, J. C. *J. Mol. Evol.* **1978**, *11*, 199.

[58] Jungck, J. R. *J. Mol. Evol.* **1978**, *11*, 211.

[59] Saks, M. E.; Sampson, J. R.; Abelson, J. N. *Science* **1993**, *263*, 191.

[60] Gibson T. J.; Lamond, A. I. *J. Mol. Evol.* **1990**, *31*, 7.

[61] Szathmáry, E. *Oxf. Surv. Evol. Biol.* **1989**, *6*, 169.

4 RNA Replication and Evolution

Christof K. Biebricher

[1] Eigen, M.; Schuster, P. *Naturwissenschaften* **1977**, *64*, 541.

[2] Orgel, L. E. *Nature* **1992**, *358*, 203.

[3] Spiegelman, S.; Haruna, I.; Holland, I. B.; Beaudreau, G.; Mills, D. R. *Proc. Natl. Acad. Sci. USA* **1965**, *54*, 919.

[4] Mills, D. R.; Peterson R. L.; Spiegelman, S. *Proc. Natl. Acad. Sci. USA* **1967**, *58*, 217.

[5] Biebricher, C. K. *Evol. Biol.* **1983**, *16*, 1.

[6] Weissmann, C.; Feix, G.; Slor, H.; Pollet, R. *Proc. Natl. Acad. Sci. USA* **1967**, *57*, 1870.

[7] Dobkin, C.; Mills, D. R.; Kramer, F. R.; Spiegelman, S. *Biochemistry* **1979**, *18*, 2038.

[8] Biebricher, C. K.; Diekmann, S.; Luce, R. *J. Mol. Biol.* **1982**, *154*, 629.

[9] Haruna, I.; Nozu, K.; Ohtaka, Y.; Spiegelman, S. *Proc. Natl. Acad. Sci. USA* **1963**, *50*, 905.

[10] Biebricher, C. K. *Chemica scripta* **1986**, *26B*, 51.

[11] Biebricher, C. K. *Cold Spring Harb. Symp. Quant. Biol.* **1987**, *52*, 299.

[12] Biebricher, C. K.; Eigen, M.; Luce, R. *J. Mol. Biol.* **1981**, *148*, 391.

[13] Biebricher, C. K.; Eigen, M.; Gardiner, W. C. *Biochemistry* **1983**, *22*, 2544.

[14] Biebricher, C. K.; Eigen, M.; Gardiner, W. C. *Biochemistry* **1984**, *23*, 3186.

[15] Biebricher, C. K.; Eigen, M.; Gardiner, W. C. *Biochemistry* **1985**, *24*, 6550.

[16] Biebricher, C. K.; Eigen, M.; Gardiner, W. C. in: Peliti, L. (ed.), *Biologically Inspired Physics*, NATO ASI Series B, Vol. 263, Plenum Press, New York, USA, **1991**, 317.

[17] August, J. T.; Banerjee, A. K.; Eoyang, L.; Franze de Fernandez, M. T.; Hori, K.; Kuo, C. H.; Rensing, U.; Shapiro, L. *Cold Spring Harbor Symp. Quant. Biol.* **1968**, *33*, 73.

[18] Silverman, P. M. *Arch. Biochem. Biophys.* **1973**, *157*, 234.

[19] Prives, L. C.; Silverman, P. M. *J. Mol. Biol.* **1972**, *71*, 657.

[20] Biebricher, C. K.; Luce, R. *Biochemistry* **1993**, *32*, 4848.

[21] Domingo, E.; Holland, J. J. in: Domingo, E.; Ahlquist, P.; Holland, J. J. (eds.), *RNA Genetics Vol. III: Variability of RNA Genomes*, CRC Press, Boca Raton, USA, **1988**, 3.

[22] Batschelet, E.; Domingo, E.; Weissmann, C. *Gene* **1976**, *1*, 27.

[23] Drake, J. W. *Proc. Natl. Acad. Sci. USA* **1993**, *90*, 4171.

[24] Eigen, M.; Biebricher, C. K. in: Domingo, E.; Ahlquist, P.; Holland, J. J. (eds.), *RNA Genetics* Vol. III: *Variability of RNA Genomes*, CRC Press, Boca Raton, USA, **1988**, 211.

[25] Hilliger, N. *Bestimmung der Selektionswerte der Mutanten einer Quasispeziesverteilung*, Master thesis, Göttingen University, Germany, **1990**.

[26] Biebricher, C. K.; Eigen, M.; Luce, R. *J. Mol. Biol.* **1981**, *148*, 369.

[27] Biebricher, C. K.; Eigen, M.; Luce, R. *Nature* **1986**, *321*, 89.

[28] Axelrod, V. D.; Brown, E.; Priano, C.; Mills, D. R. *Virology* **1991**, *184*, 595.

[29] Giegé, R.; Puglisi, J. P.; Florentz, C. *Progr. Nucl. Acid Res. Mol. Biol.* **1993**, 45, 129.

[30] Zuker, M.; Stiegler, P. *Nucleic Acid Res.* **1981**, *9*, 133.

[31] McCaskill, J. S. *Biopolymers* **1990**, *29*, 1105.

[32] Biebricher, C. K.; Luce, R. *EMBO J.* **1992**, *11*, 5129.

[33] Tuerk, C.; Gauss, P.; Thermes, C.; Groebe, D. R.; Gayle, M.; Guild, N.; Stormo, G.; d'Aubenton-Carafa, Y.; Uhlenbeck, O. C.; Tinoco, I.; Brody, E. N.; Gold, L. *Proc. Natl. Acad. Sci. USA* **1988**, *85*, 1364.

[34] Biebricher, C. K.; Orgel, L. E. *Proc. Natl. Acad. Sci. USA* **1973**, *70*, 934.

[35] Konarska, M. M.; Sharp, P. A. *Cell* **1989**, *57*, 423.

[36] Biebricher, C. K.; Eigen, M.; McCaskill, J. S. *J. Mol. Biol.* **1994**, *231*, 175.

[37] Doudna, J. A.; Szostak, J. W. *Nature* **1989**, *339*, 519.

[38] Doudna, J. A.; Usman , N.; Szostak, J. W. *Biochemistry* **1993**, *32*, 2111.

[39] Altman, S.; Kirsebom, L.; Talbot, S. *FASEB J.* **1993**, *7*, 7.

[40] Cech, T. R.; Herschlag, D.; McConnell, T. S.; Piccirilli, J. A. *J. Cell. Biochem.* **1993**, *1993*, 159.

5   DNA Binding and Bending by two different Proteins: Factor for Inversion Stimulation (FIS) and Tetracycline Repressor (TetR)

Wolfram Saenger, Dirk Kostrewa, Joachim Granzin, Frank Cordes, Claus Sandmann, Caroline Kisker, and Winfried Hinrichs

[1] Schulz, G. E.; Schirmer, R. H. *Principles of Protein Structure.* Springer, New York, USA, **1978**.

[2] Saenger, W. *Principles of Nucleic Acid Structure.* Springer, New York, USA, **1983**.

[3] Brennan, R. G.; Matthews, B. W. *J. Biol. Chem.* **1989**, *264*, 1903.

[4]   Blundell, T. L.; Johnson, L. N. *Protein Crystallography,* Academic Press, London, United Kingdom, **1976.**

[5]   Kostrewa, D.; Granzin, J.; Koch, C.; Choe, H.-W.; Raghunathan, S.; Wolf, W.; Labahn, J.; Kahmann, R.; Saenger, W. *Nature* **1991,** *349,* 178.

[6]   Saenger, W.; Sandmann, C.; Theis, K.; Starikow, E. B.; Kostrewa, D.; Labahn, J.; Granzin, J. *Nucl. Acids Mol. Biol.* **1993,** *7,* 158.

[7]   Schultz, S. C.; Shields, G. C.; Steitz, T. A. *Science* **1991,** *253,* 1001.

[8]   Zhurkin, V. B.; Ulyanov, N. B.; Gorin, A. A.; Jernigan, R. L. *Proc. Natl. Acad. Sci. USA* **1991,** *88,* 7046.

[9]   Osuna, R.; Finkel, S. E.; Johnson, R. C. *EMBO J.* **1991,** *10,* 1593.

[10]   Koch, C.; Ninnemann, O.; Fuss, H.; Kahmann, R. *Nucl. Acids Res.* **1991,** *19,* 5915.

[11]   Kostrewa, D.; Granzin, J.; Stock, D.; Choe, H.-W.; Labahn, J.; Saenger, W. *J. Mol. Biol.* **1992,** *226,* 209.

[12]   Hinrichs, W.; Kisker, C.; Düvel, M.; Müller, A.; Tovar, K.; Hillen, W.; Saenger, W. *Science* **1994,** *264,* 418.

6   Function Based on Organization and Recognition: From "Complicated" Biological Systems to "Simple" Synthetic Systems

Steffen Denzinger, Achim Dittrich, Wolfgang Paulus, and Helmut Ringsdorf

[1]   a) Lehn, J.-M. *Science* **1985,** *227,* 849; b) *Angew. Chem.* **1988,** *100,* 91; *Angew. Chem. Int. Ed. Engl.* **1988,** *27,* 89; c) Ringsdorf, H.; Schlarb, B.; Venzmer, J. *Angew. Chem.* **1988,** *100,* 117; *Angew. Chem. Int. Ed. Engl.* **1988,** *27,* 113.

[2]   Alberts, B.; Bray, D.; Lewis, J.; Raff, M.; Roberts, K.; Watson: J. D. *Molecular Biology of the Cell,* Garland, New York, USA, **1983;** *Molekularbiologie der Zelle,* VCH, Weinheim, Germany, **1986.**

[3]   Fischer, E. *Ber. Dtsch. Chem. Ges.* **1894,** *27,* 2985.

[4]   a) Grainger, D. W.; Reichert, A.; Ringsdorf, H.; Salesse, C.; Davies, D.; Lloyd, J. B. *Biochim. Biophys. Acta* **1990,** *1022,* 146; b) Kitano, H.; Ringsdorf, H. *Bull. Chem. Soc. Jpn.* **1985,** *58,* 2826; c) Ahlers, M.; Ringsdorf, H.; Rosemeyer, H.; Seela, F. *Colloid Polym. Sci.* **1990,** *268,* 132; d) Krauch, T.; Zaitsev, S. Y.; Zubov, V. B. Colloid Surf. **1991,** *57,* 383; Honda, Y.; Kurihara, K.; Kunitake, T. *Chem. Lett.* **1991,** 681; e) Rotello, V. M.; Viani, E. A.; Deslongchamps, G.; Murray, B. A.; Rebek, Jr., J. *J. Am. Chem. Soc.* **1993,** *115,* 797; f) Ikeura, Y.; Kurihara, K.; Kunitake, T. *J. Am. Chem. Soc.* **1991,** *113,* 7342; g) Sasaki, D. Y.; Kurihara, K.; Kunitake, T. *J. Am. Chem. Soc.* **1991,** *113,* 9685; h) Mertesdorf, C.; Plesnivy, T.; Ringsdorf, H.; Suci, P. A. *Langmuir* **1992,** *8,* 2531.

[5]   McConnell, N. M.; Watts, T. H.; Weis, R. M.; Brian, A. A. *Biochim. Biophys. Acta* **1986,** *864,* 95

[6]   Waite, M. *The Phospholipases,* Plenum, New York, USA, **1987.**

[7]   a) Dennis, E. A. in: Boyer, P. D. (ed.), *The Enzymes, Vol. 16,* 3rd ed., Academic Press, New York, USA, **1983,** 307; b) Van den Bosch, H. *Biochim. Biophys. Acta* **1980,** *604,* 191.

[8]   a) Verheij, H. M.; Slotboom, A. J.; de Haas, G. H. *Rev. Physiol. Biochem. Pharmacol.* **1981,** *91,* 91; b) Achari, A.; Scott, D.; Barlow, P.; Vidal, J. C.; Otwinowski, Z.; Brunie, S.; Sigler, P. B. *Cold Spring Harbor Symp. Quant. Biol.* **1987,** *52,* 441; c) Drenth, J.; Dijkstra, B. W.; Renetseder, R. in: Jurnak, F. A.; McPherson, A. (eds.), *Biological Macromolecules and Assemblies, Vol. 3 - Active*

Sites of Enzymes, Wiley, New York, USA, **1987**, 287; d) Jain, M. K.; Berg, O. G. *Biochim. Biophys. Acta* **1989**, *1002*, 127.

[9]   a) Gheriani-Gruszka, N.; Almog, S.; Biltonen, R. L.; Lichtenberg, D. *J. Biol. Chem.* **1987**, *263*, 11808; b) Thuren, T.; Vainio, P.; Virtanen, J. A.; Somerharju, P.; Blomquist, K.; Kinnunen, P. K. J. *Biochemistry* **1984**, *23*, 5129; c) Israelachvili, J. N.; Marcelja, S.; Horn, G. *Q. Rev. Biophys.* **1980**, *13*, 121; d) Barlow, P. N.; Vidal, J.-C.; Lister, M. D.; Dennis, E. A.; Sigler, P. B. *J. Biol. Chem.* **1988**, *263*, 12954; e) Romero, G.; Thompson, K.; Biltonen, R. L. *J. Biol. Chem.* **1987**, *262*, 13476; f) Sundler, R. *Chem. Phys. Lipids* **1984**, *34*, 153; g) Lichtenberg, D.; Romero, G.; Menashe, M.; Biltonen, R. L. *J. Biol. Chem.* **1986**, *261*, 5328; h) Lichtenberg, D.; Romero, G.; Menashe, M.; Biltonen, R. L. *J. Biol. Chem.* **1987**, *261*, 5334; i) Alsina, A.; Valls, O.; Rieroni, G.; Verger, R.; Garcia, S. *Colloid Polymer Sci.* **1983**, *261*, 923; k) Alsina, A.; Garcia, M. L.; Valls, O.; Garcia, S. *Colloids Surf.* **1985**, *14*, 277; l) Alsina, A.; Garcia, M. L.; Espina, M.; Valls, O. *Colloid Polymer Sci.* **1989**, *267*, 923, m) Grainger, D. W.; Reichert, A.; Ringsdorf, H.; Salesse, C.; Davies, D. E.; Lloyd, J. B. *Biochim. Biophys. Acta* **1990**, *1022*, 146.

[10] a) von Tscharner, V.; McConnel, H. M. *Biophys. J.* **1981**, *36*, 409; b) Möhwald, H.; Lösche, M. **1984**, *55*, 1968; c) Meller, P. *Rev. Sci. Instrum.* **1988**, *59*, 2225.

[11] a) Wilschut, J. C.; Regts, J.; Westenberg, H.; Scherphof, G. *Biophys. Acta* **1978**, *508*, 185; b) Upreti, G. C.; Jain, M. K. *J. Membr. Biol.* **1980**, *55*, 113.

[12] a) Grainger, D. W.; Reichert, A.; Ringsdorf, H.; Salesse, C. *FEBS Lett.* **1989**, *252*, 74; b) Grainger, D. W.; Reichert, A.; Ringsdorf, H.; Salesse, C. *Biochim. Biophys. Acta* **1990**, *1023*, 365.

[13] McConnell, H. M.; Tamm, L. K.; Weis, R. M. *Proc. Natl. Acad. Sci. USA* **1984**, *81*, 3249.

[14] a) Plückthun, A.; Dennis, E. A. *J. Biol. Chem.* **1985**, *260*, 11099; b) Jain, M. K.; Jahagirdar, D. V. *Biochim. Biophys. Acta* **1985**, *814*, 313; c) Jain, M. K.; Egmont, M. R.; Verheij, H. M.; Apitz-Castro, R.; Dijkman, R.; De Haas, G. H. *Biochim. Biophys. Acta* **1982**, *688*, 341.

[15] a) Hazlett, T. L.; Dennis, E. A. *Biochemistry* **1985**, *24*, 6152; b) Hazlett, T. L.; Dennis, E. A. *Biochim. Biophys. Acta* **1988**, *961*, 22; c) Kensil, C. R.; Dennis, E. A. *J. Biol. Chem.* **1979**, *254*, 5843; d) Romero, G.; Thompson, K.; Biltonen, R. L. *J. Biol. Chem.* **1987**, *262*, 13476; e) Smith, C. M.; Wells, M. A. *Biochim. Biophys. Acta* **1981**, *663*, 687; f) Deems, R. A.; Dennis, E. A. *Methods Enzymol.* **1981**, *71*, 703.

[16] Cho, W.; Tomaselli, A. G.; Henrickson, R. L.; Kezdy, F. J. *J. Biol. Chem.* **1988**, *263*, 11237.

[17] e.g., for a phase separation of the hydrolyzed products in vesicles see: Yu, B.-Z.; Kozubek, A.; Jain, M. K. *Biochim. Biophys. Acta* **1989**, *980*, 23.

[18] a) Lehn, J.-M.; Mascal, M.; Decian, A.; Fischer, J. *J. Chem. Soc., Chem. Commun.* **1990**, 479; b) Zerkowsky, J. A.; Seto, C. T.; Wierda, D. A.; Whitesides, G. M. *J. Am. Chem. Soc.* **1990**, *112*, 9025.

[19] Ahuja, R.; Caruso, P. L.; Moebius, D.; Paulus, W.; Ringsdorf, H.; Wildburg, G. *Angew. Chem.* **1993**, *105*, 1082; *Angew. Chem. Int. Ed. Engl.* **1993**, *32*, 1033.

[20] Bohanon, T. M.; Denzinger, S.; Fink, R.; Ringsdorf, H.; Weck, M. *Angew. Chem.* **1995**, *107*, 102; *Angew. Chem. Int. Ed. Engl.* **1995**, *34*, 48.

[21] Lehn, J.-M. *Angew. Chem.* **1990**, *102*, 1347; *Angew. Chem. Int. Ed. Engl.* **1990**, *29*, 1304.

7   Self-Organization of Molecules en Route to Crystal Formation

Ronit Popovitz-Biro, Isabelle Weissbuch, Jarek Majewski, Leslie Leiserowitz, and Meir Lahav

[1]   Mullin, J. W. *Crystallization*; Butterworth-Heinemann, Oxford, United Kingdom, **1993**.

[2]   Staab, E.; Addadi, L.; Leiserowitz, L.; Lahav, M. *Adv. Mater.* **1990**, *2*, 40.

[3]   Weissbuch, I.; Leiserowitz, L.; Lahav, M. *Adv. Mater.* **1994**, *6*, 952.

[4]   Zbaida, D.; Weissbuch, I.; Shavit-Gati, E.; Addadi, L.; Leiserowitz, L.; Lahav, M. *Reactive Polymers* **1987**, *6*, 241.

[5]   Weissbuch, I.; Zbaida, D.; Addadi, L.; Lahav, M.; Leiserowitz, L. *J. Am. Chem. Soc.* **1987**, *109*, 1869.

[6]   Weissbuch, I.; Addadi, L.; Leiserowitz, L.; Lahav, M. *J. Am. Chem. Soc.* **1990**, *112*, 7718.

[7]   Landau, E. M.; Grayer Wolf, S.; Levanon, M.; Leiserowitz, L.; Lahav, M.; Sagiv, J. *J. Am. Chem. Soc.* **1989**, *111*, 1436.

[8]   Weissbuch, I.; Berkovic, G.; Leiserowitz, L.; Lahav, M. *J. Am. Chem. Soc.* **1990**, *111*, 5874.

[9]   Weissbuch, I.; Majewski, J.; Kjaer, K.; Als-Nielsen, J.; Lahav, M.; Leiserowitz, L. *J. Phys. Chem.* **1993**, *97*, 12848.

[10]   Gavish, M.; Popovitz-Biro, R.; Lahav, M.; Leiserowitz, L. *Science* **1990**, *250*, 973.

[11]   Jacquemain, D.; Grayer Wolf, S.; Leveiller, F.; Deutsch, M.; Kjaer, K.; Als-Nielsen, J.; Lahav, M.; Leiserowitz, L. *Angew. Chem.* **1992**, *104*, 134; *Angew. Chem. Int. Ed. Engl.* **1992**, *31*, 130.

[12]   Grayer Wolf, S.; Leiserowitz, L.; Lahav, M.; Deutsch, M.; Kjaer, K.; Als-Nielsen, J. *Nature* **1987**, *328*, 63.

[13]   Grayer Wolf, S.; Deutsch, M.; Landau, E. M.; Lahav, M.; Leiserowitz, L.; Kjaer, K.; Als-Nielsen, J. *Science* **1988**, *242*, 1286.

[14]   Berfeld, M.; Weissbuch, I. **1994**, work in progress.

[15]   Popovitz-Biro, R.; Wang, J. L.; Majewski, J.; Shavit-Gati, E.; Leiserowitz, L.; Lahav, M. *J. Am. Chem. Soc.* **1994**, *116*, 1179.

[16]   Wang, J. L.; Leveiller, F.; Jacquemain, D.; Kjaer, K.; Als-Nielsen, J.; Lahav, M.; Leiserowitz, L. *J. Am. Chem. Soc.* **1994**, *116*, 1192.

[17]   Majewski, J.; Lahav, M.; Leiserowitz, L.; Kjaer, K.; Als-Nielsen, J. *J. Phys. Chem.* **1994**, *98*, 4087.

[18]   Majewski, J.; Margulis, L.; Jacquemain, D.; Leveiller, F.; Böhm, C.; Arad, T.; Talmon, Y.; Lahav, M.; Leiserowitz, L. *Science* **1993**, *261*, 899.

[19]   Jacquemain, D.; Grayer Wolf, S.; Leveiller, F.; Lahav, M.; Leiserowitz, L.; Deutsch, M.; Kjaer, K.; Als-Nielsen, J. *J. Am. Chem. Soc.* **1990**, *112*, 7724.

8   Towards Molecular Devices
J. Fraser Stoddart and Natalie M. Rowley

[1]   Lehn, J.-M. *Angew. Chem.* **1988**, *100*, 91; *Angew. Chem. Int. Ed. Engl.* **1988**, *27*, 89;
      J. Inclusion Phenom. **1988**, *6*, 351; *Angew. Chem.* **1990**, *102*, 1347; *Angew. Chem.
      Int. Ed. Engl.* **1990**, *29*, 1304; in: Schneider, H.-J.; Dürr, H. (eds.), *Frontiers in
      Supramolecular Organic Chemistry and Photochemistry*, VCH, Weinheim,
      Germany, **1991**, 1.

[2]   Stoddart, J. F.; Mathias, P. J.; Kohnke, F. H. *Adv. Mater.* **1989**, *101*, 1129.

[3]   Cybernetics (Gk. kybernetes; steersman, governor) can be defined as the study of the
      self-organising machine or mechanical brain. See: *Webster's Third New Inter-
      national Dictionary*, G & C Merriam, Springfield, MA, USA, **1976**, 563.

[4]   (a) Carter, F. L.; Siatowski, R. E. (eds.), *Molecular Electronic Devices*, North
      Holland, Amsterdam, The Netherlands, **1988**; (b). Balzani, V.; Scandola, F.
      Supramolecular Photochemistry, Horwood, Chichester, United Kingdom, **1991**.

[5]   Carter, F. L. *Physica*, **1984**, *10D*, 175.

[6]   a) Lehn, J.-M. *Struct. Bonding (Berlin)* **1973**, *16*, 1; *Pure Appl. Chem.* **1977**, *49*, 857;
      1979, *51*, 979; **1980**, *52*, 2303 and 2441; *Acc. Chem. Res.* **1978**, *11*, 49; Leçon
      Inaugurale, Collège de France, Paris, France, **1980**; *Science* **1985**, *227*, 849; in:
      Balzani, V. (ed.), *Supramolecular Photochemistry*, Reidel, Dordrecht, The
      Netherlands, **1987**, 29; (b) Popov, A. I.; Lehn, J.-M. in: Melson, G. A. (ed.),
      Coordination Chemistry of Macrocyclic Compounds, Plenum, New York, USA,
      1979, 537; (c) Potvin, P. G.; Lehn, J.-M. *Prog. Macrocyclic. Chem.* **1987**, *3*, 167; (d)
      Vögtle, F. *Supramolekulare Chemie*, 2nd ed., Teubner, Stuttgart, Germany, **1992**;
      Supramolecular Chemistry, Wiley, Chichester, United Kingdom, **1993** (reprint).

[7]   (a) Cram, D. J.; Cram, J. M. *Science* **1974**, *183*, 803; *Acc. Chem. Res.* **1978**, *11*, 8; in:
      Bartman, W.; Trost, B. M. (eds.), *Selectivity, A Goal for Synthetic Efficiency*, VCH,
      Weinheim, Germany, **1983**, 42; (b) Cram, D. J.; Helgeson, R. C.; Sousa, L. R.;
      Timko, J. M.; Newcombe, M.; de Jong, F.; Gokel, G. W.; Hoffman, D. H.; Domeier,
      L. A.; Peacock, S. C.; Kaplan, L. *Pure Appl. Chem.* **1975**, *43*, 327; (c) Cram, D. J.;
      Trueblood, K. N. *Top. Curr. Chem.* **1981**, *98*, 43; in: Vögtle, F.; Weber, E. (eds.),
      Host Guest Chemistry / Macrocycles, Springer, Berlin, Germany, **1985**, 125; (d)
      Cram, D. J. in: Jones, J. B.; Sih, C. J.; Perlman, D. (eds.), *Applications of Biomedical
      Systems in Chemistry*, Wiley Interscience, New York, USA, **1976**, 815; *Science*
      1983, *219*, 1177; *Science* **1988**, *240*, 760; *Angew. Chem.* **1986**, *98*, 1041; *Angew.
      Chem. Int. Ed. Engl.* **1986**, *25*, 1039; *Angew. Chem.* **1988**, *100*, 1041; *Angew. Chem.
      Int. Ed. Engl.* **1988**, *27*, 1009; *CHEMTECH* **1987**, 120; *J. Inclusion Phenom.* **1988**,
      6, 397.

[8]   Pedersen, C. J. *J. Am. Chem. Soc.* **1967**, *89*, 2495, 7017; *Angew. Chem.* **1988**, *100*,
      1053; *Angew. Chem. Int. Ed. Engl.* **1988**, *27*, 1021; *J. Inclusion Phenom.* **1988**, *6*,
      337.

[9]   (a) Pedersen, C. J. *Aldrichimica Acta* **1971**, *4*, 1 in: *Selections from Aldrichimica
      Acta : Fifteen Years* **1984**, 15; (b) Pedersen, C. J.; Truter, M. R. *Endeavour* **1971**, *30*,
      142; (c) Pedersen, C. J.; Frensdorff, H. K. *Angew. Chem.* **1972**, *84*, 16; *Angew.
      Chem. Int. Ed. Engl.*, **1972**, *11*, 16; (d) Pedersen, C. J. in: Izatt, R. M.; Christenson, J.
      J. (eds.), *Synthetic Multidentate Macrocyclic Compounds*, Academic Press, New
      York, USA, **1978**, 1; in: Kimura, E. (ed.), *Current Topics in Macrocyclic Chemistry*

in Japan, Hiroshima University, Hiroshima, Japan, **1987**, 1; Schroder, H. E. *Pure Appl. Chem.* **1988**, *60*, 445.

[10] Gokel, G. W.; Korzeniowski, S. H. in: *Macrocyclic Polyether Synthesis*, Springer, Berlin, Germany, **1982**; Lindoy, L. F. in: *The Chemistry of Macrocyclic Ligand Complexes*, Cambridge University Press, Cambridge, United Kingdom, **1989**; Laidler, D. A.; Stoddart, J. F. in: Patai, S. (ed.), *The Chemistry of the Functional Groups. Supplement E. The Chemistry of Ethers, Hydroxyl Groups, and their Sulphur Analogues*, Wiley, Chichester, United Kingdom **1980**, Part 1, 1; Patai, S.; Rappoport, Z. (eds.), *Crown Ethers and Analogs*, Wiley, Chichester, United Kingdom, **1989**, 1; Krakowiak, K. E.; Bradshaw, J. S.; Zamecka-Krakowiak, D. J. *Chem. Rev.* **1989**, *89*, 929.

[11] Ashton, P. R.; Isaacs, N. S.; Kohnke, F. H.; Mathias, J. P.; Stoddart, J. F. *Angew. Chem.* **1989**, *101*, 1266; *Angew. Chem. Int. Ed. Engl.* **1989**, *28*, 1258; Ellwood, P.; Mathias, J. P.; Stoddart, J. F.; Kohnke, F. H. *Bull. Soc. Chim. Belg.* **1988**, *97*, 669; Kohnke, F. H.; Mathias, J. P.; Stoddart, J. F. in: Roberts, S. M. (ed.), *Molecular Recognition: Chemical and Biochemical Problems*, RSC Special Publication No. 78, Cambridge, United Kingdom, **1989**, 241; Eschenmoser, A. *Angew. Chem.* **1988**, *100*, 5; *Angew. Chem. Int. Ed. Engl.* **1988**, *27*, 5.

[12] (a) Schill, G. *Catenanes, Rotaxanes and Knots*, Academic Press, New York, USA, **1971**; in: Chiurdoglu, G. (ed.), *Conformational Analysis*, Academic Press, New York, USA, **1971**, 229. (b) Boeckmann, J.; Schill, G. *Tetrahedron* **1974**, *30*, 1945; (c) Rouvray, D. H. *Educ. Chem.* **1971**, *8*, 134; (d) Sokolov, V. I. *Russ. Chem. Rev. (Engl. Transl).* **1973**, *42*, 452.

[13] Harrison, I. T.; Harrison, S. *J. Am. Chem. Soc.* **1967**, *89*, 5723; Harrison, I. T. *J. Chem. Soc., Perkin Trans 1* **1974**, 301; Schill, G.; Beckmann, W.; Schweickert, N.; Fritz, H. *Chem. Ber.* **1986**, *119*, 2647; Agam, G.; Graiver, D.; Zilkha, A. *J. Am. Chem. Soc.* **1976**, *98*, 5206; Agam, G.; Zilkha, A. *J. Am. Chem. Soc.* **1976**, *98*, 5214; Schill, G.; Schweickert, N.; Fritz, H.; Vetter, W. *Angew. Chem.* **1983**, *95*, 909; *Angew. Chem. Int. Ed. Engl.* **1983**, *22*, 889; Schill, G.; Zollenkopf, G. *Liebigs Ann. Chem.* **1969**, *721*, 53; Schill, G.; Zürcher, C. *Chem. Ber.* **1977**, *110*, 2046; Rissler, K.; Schill, G.; Fritz, H.; Vetter, W. *Chem. Ber.* **1986**, *119*, 1374; Schill, G.; Schweickert, N.; Fritz, H.; Vetter, W. *Chem. Ber.* **1988**, *121*, 961; Ogino, H. *J. Am. Chem. Soc.* **1981**, *103*, 1303; Ogino, H.; Ohata, K. *Inorg. Chem.* **1984**, *23*, 3312; Yamanari, K.; Shimura, Y. *Bull. Chem. Soc. Jpn.* **1983**, *56*, 2283; Pajerski, A. D.; Bergstresser, G. L.; Parvez, M.; Richey, H. G. Jr. *J. Am. Chem. Soc.* **1988**, *110*, 4844; Markies, P. R.; Nomoto, T.; Akkerman, O. S.; Bickelhaupt, F.; Smeets, W. J. J.; Spek, A. L. *J. Am. Chem. Soc.* **1988**, *110*, 4845.

[14] Stoddart J. F. in: *Host-Guest Molecular Interactions: From Chemistry to Biology*, Ciba Foundation Symposium 158, Wiley, Chichester, United Kingdom, **1991**, 5; Philp, D.; Stoddart, J. F. *Synlett* **1991**, 445; Stoddart, J. F. *Chem. Aust.* **1992**, *59*, 476; *An. Quim.* **1993**, *89*, 51; *Angew. Chem.* **1992**, *104*, 860; *Angew. Chem. Int. Ed. Engl.*, **1992**, *31*, 846; Gibson, H. W.; Marand, H. *Adv. Mater.* **1993**, *5*, 11.

[15] Anelli, P.-L.; Ashton. P. R.; Ballardini, R.; Balzani, V.; Delgado, M.; Gandolfi, M. T.; Goodnow, T. T.; Kaifer, A. E.; Philp, D.; Pietraskiewicz, M.; Prodi, L.; Reddington, M. V.; Slawin, A. M. Z.; Spencer, N.; Stoddart, J. F.; Vicent, C.; Williams, J. D. *J. Am. Chem. Soc.* **1992**, *114*, 193.

[16] Córdova, E.; Bissell, R. A.; Spencer, N.; Ashton, P. R.; Stoddart, J. F.; Kaifer, A. E. *J. Org. Chem.* **1993**, *58*, 6550.

[17] Bissell, R. A.; Córdova, E.; Kaifer, A. E.; Stoddart, J. F. *Nature* **1994**, *369*,133.

[18] Anelli, P.-L.; Spencer, N.; Stoddart, J. F. *J. Am. Chem. Soc.* **1991**, *113*, 5131.

[19] Ashton, P. R.; Grognuz, M.; Slawin, A. M. Z.; Stoddart, J. F.; Williams, D. J. *Tetrahedron Lett.* **1991**, *32*, 6235;

[20] Ashton, P. R.; Johnston, M. R.; Stoddart, J. F.; Tolley, M. S.; Wheeler, J. W. *J. Chem. Soc., Chem. Commun.* **1992**, 1128.

[21] Ashton, P. R.; Bissell, R. A.; Spencer, N.; Stoddart, J. F.; Tolley, M. S. *Synlett* **1992**, 914.

[22] Ashton, P. R.; Bissell, R. A.; Górski, R.; Philp, D.; Spencer, N.; Stoddart, J. F.; Tolley, M. S. *Synlett* **1992**, 919.

[23] Ashton, P. R.; Bissell, R. A.; Spencer, N.; Stoddart, J. F.; Tolley, M. S. *Synlett* **1992**, 923.

[24] Borman, S. *C & E News*, **1991**, July 1, 4.

[25] Stoddart, J. F. *Chem. Ber.* **1991**, *27*, 714.

[26] Ashton, P. R.; Philp, D.; Spencer, N.; Stoddart, J. F. *J. Chem. Soc., Chem. Commun.* **1992**, 1124.

[27] Ashton, P. R.; Belohradsky, M.; Philp, D.; Stoddart, J. F. *J. Chem. Soc., Chem. Commun.* **1993**, 1269.

[28] Ashton, P. R.; Belohradsky, M.; Philp, D.; Spencer, N.; Stoddart, J. F. *J. Chem. Soc., Chem. Commun.* **1993**, 1274.

[29] Harrison, I. T. *J. Chem. Soc., Chem. Commun.* **1972**, 231; Harrison, I. T. *J. Chem. Soc., Perkin Trans. 1* **1974**, 301.

[30] Ballardini, R.; Balzani, V.; Gandolfi, M. T.; Prodi, L.; Ventura, M.; Philp, D.; Ricketts, H. G.; Stoddart, J. F. *Angew. Chem.* **1993**, *105*, 1362; *Angew. Chem. Int. Ed. Engl.* **1993**, *32*, 1301.

[31] (a) Langton, C. G. (ed.), *Artificial Life*, Addison-Wesley, Redwood City, California, USA, **1989**; (b) Drexler, K. E. *Engines of Creation*, Fourth Estate, London, United Kingdom, **1990**.

[32] Hopfield, J. J.; Onuchic, J. N.; Beratan, D. N. *Science* **1988**, *241*, 817; *J. Phys. Chem.* **1989**, *93*, 6350.

[33] Pimentel, G. C. *Opportunities in Chemistry*, National Academy Press, Washington DC, USA, **1985**.

9 Non-Covalent Interactions Between Aromatic Molecules

Christopher Hunter

[1] This is an extended abstract of a talk presented in Bielefeld in July **1993**.

[2] We will use the term π-π *interaction* as a shorthand notation to describe non-covalent interactions between delocalized π-systems, including interactions between aromatic molecules. It is not intended to imply anything about the nature of the interaction.

[3] Desiraju, G. R.; Gavezzotti, A. *J. Chem. Soc., Chem. Commun.* **1989**, 621.

[4] Hunter, C. A. *J. Mol. Biol.* **1993**, *230*, 1025.

[5] Burley, S. K.; Petsko, G. A. *Science* **1985**, *229*, 23.

[6] Diederich, F. *Angew. Chem.* **1988**, *100*, 372; *Angew. Chem. Int. Ed. Engl.* **1988**, *27*, 362.

[7] Leighton, P.; Cowan, J. A.; Abraham, R. J.; Sanders, J. K. M. *J. Org. Chem.* **1988**, *53*, 733.

[8] Hunter, C. A.; Meah, M. N.; Sanders, J. K. M. *J. Am. Chem. Soc.* **1990**, *112*, 5773.

[9]   Smithrud, D. B.; Diederich, F. *J. Am. Chem. Soc.* **1990**, *112*, 339.

[10]   Hunter, C. A. *Angew. Chem.* **1993**, *105*, 1653; *Angew. Chem. Int. Ed. Engl.* **1993**, *32*, 1584.

[11]   Hunter, C. A.; Sanders, J. K. M. *J. Am. Chem. Soc.* **1990**, *112*, 5525.

[12]   Hunter, C. A.; Singh, J.; Thornton, J. M. *J. Mol. Biol.* **1991**, *218*, 837.

[13]   Hunter, C. A. *J. Chem. Soc., Chem. Commun.* **1991**, 749.

[14]   Hunter, C. A. *J. Am. Chem. Soc.* **1992**, *114*, 5303.

[15]   Weber, P.; Guillon, D.; Skoulios, A. *Liq. Cryst.* **1991**, *9*, 369.

[16]   Drew, H. R.; McCall, M. J.; Calladine, C. R. in: *DNA Topology and Its Biological Effects*, Cold Spring Harbour, USA, **1990**, 1.

[17]   Travers, A. A.; Klug, A. in: *DNA Topology and Its Biological Effects*, Cold Spring Harbour, USA, **1990**, 57.

[18]   Calladine, C. R.; Drew, H. R. *J. Mol. Biol.* **1984**, *178*, 773.

[19]   Diekmann, S. *J. Mol. Biol.* **1989**, *205*, 787.

[20]   Calladine, C. R.; Elhasssan, M. A. unpublished results.

[21]   Nelson, H. C. M.; Finch, J. T.; Luisi, B. F.; Klug, A. *Nature* **1987**, *330*, 221.

[22]   McCall, M.; Brown, T.; Kennard, O. *J. Mol. Biol.* **1985**, *183*, 385.

[23]   Arnott, S.; Hukins, D. W. L.; Dover, S. D.; Fuller, W.; Hodgson, A. R. *J. Mol. Biol.* **1973**, *81*, 107.

[24]   Dock-Bregon, A. C.; Chevrier, B.; Podjarny, A.; Moras, D.; de Bear, J. S.; Gough, G. R.; Gilham, P. T. *Nature* **1988**, *335*, 375.

[25]   Yanagi, K.; Privé, G. G.; Dickerson, R. E. *J. Mol. Biol.* **1991**, *217*, 201.

10   Design, Syntheses, and Supramolecular Chemistry of Smart Cascade Polymers

George R. Newkome and Charles N. Moorefield

[1]   Moorefield, C. N.; Newkome, G. R. in: Newkome, G. R. (ed.), *Advances in Dendritic Macromolecules*, Vol. 1, Chapter 1, JAI Press, Greenwich, Conn., USA, **1994.**

[2]   Hallé, F.; Oldeman, R. A. A.; Tomlinson, P. B. in: *Tropical Trees and Forests: An Architectural Analysis*, Springer, Berlin, Germany, **1982**.

[3]   Newkome, G. R.; Gupta, V. K.; Baker, G. R.; Yao, Z-q. *J. Org. Chem.* **1985**, *50*, 2003.

[4]   Newkome, G. R.; Lin, X. *Macromolecules* **1991**, *24*, 1443.

[5]   Newkome, G. R.; Behera, R. K.; Moorefield, C. N.; Baker, G. R. *J. Org. Chem.* **1991**, *56*, 7162.

[6]   Klausner, Y. S.; Bodansky, M. *Synthesis* **1972**, 453; deTar, D. F.; Silverstein, R.; Rogers, Jr., F. F. *J. Am. Chem. Soc.* **1966**, *88*, 1024; König, W.; Geiger, R. *Chem. Ber.* **1970**, *103*, 788.

[7]   Cooper, D. S.; Potter, R. L. unpublished results.

[8]   Nomenclature: Newkome, G. R.; Baker, G. R.; Young, J. K.; Traynham, J. G. *J. Polym. Sci. Part A: Polym. Chem.* **1992**, *31*, 641.

[9]   Newkome, G. R.; Lin, X. *Tetrahedron Asymmetry* **1991**, *2*, 957.

[10]   Young, J. K. PhD Dissertation (USF) **1993**.

[11]   Young, J. K.; Baker, G. R.; Newkome, G. R.; Morris, K. F.; Johnson, Jr., C. S. *Macromolecules* **1994**, *27*, in press.

[12]   Newkome, G. R.; Moorefield, C. N.; Theriot, K. J. *J. Org. Chem.* **1988**, *53*, 5552.

[13]   Morris, K. F.; Johnson, Jr., C. S. *J. Am. Chem. Soc.* **1993**, *115*, 4291.

[14] Computations performed on a Silicon Graphics Indigo workstation.

[15] Baker, G. R. unpublished results.

[16] Newkome, G. R.; Moorefield, C. N.; Baker, G. R.; Johnson, A. L.; Behera, R. J. *Angew. Chem.* **1991**, *103*, 1205; *Angew. Chem. Int. Ed. Engl.* **1991**, *30*, 1176; Newkome, G. R.; Moorefield, C. N. U.S. Pat. 5,154,853 (oct. 13, **1992**).

[17] Newkome, G. R.; Moorefield, C. N.; Baker, G. R.; Saunders, M. J.; Grossman, S. H. *Angew. Chem.* **1991**, *103*, 1207; *Angew. Chem. Int. Ed. Engl.* **1991**, *30*, 1178.

[18] Newkome, G. R.; Moorefield, C. N.; Keith, J. M.; Baker, G. R.; Escamilla, G. H. *Angew. Chem.* **1994**, *106*, 701; *Angew. Chem. Int. Ed. Engl.* **1994**, *33*, 666.

[19] Newkome, G. R.; Moorefield, C. N. *Polym. Preprints* **1993**, *34*, 75.

11 Polyoxometallate Cluster Anions
Michael T. Pope

[1]   (a) Pope, M. T. *Heteropoly and Isopoly Oxometalates*, Springer, New York, USA, **1983**; (b) Day, V. W.; Klemperer, W. G. *Science* **1985**, *228*, 533; (c) Pope, M. T.; Müller, A. *Angew. Chem.* **1991**, *103*, 56; *Angew. Chem. Int. Ed. Engl.* **1991**, *30*, 34; (d) Pope, M. T.; Müller, A. (eds.), *Polyoxometalates. From Platonic Solids to Anti-Retroviral Activity*, Kluwer, Dordrecht, The Netherlands, **1994**.

[2]   (a) Pettersson, L.;Hedman, B.; Andersson, I.;Ingri, N. *Chem. Scr.* **1983**, *22*, 254; (b) Pettersson, L.; Hedman, B.; Nenner, A.-M.; Andersson, I. *Acta Chem. Scand.* **1985**, *A39*, 499.

[3]   Berkowitz, J.; Ingraham, M. G.; Chupta, W. A. *J. Chem. Phys.* **1957**, *26*, 842

[4]   Day, V. W.; Fredrich, M. F.; Klemperer, W. G.; Shum, W. *J. Am. Chem. Soc.* **1977**, *99*, 952.

[5]   Keggin, J. F. *Nature* **1933**, *131*, 908. This is now known as the α-isomer (T_d symmetry). Four other skeletal isomers (β-ε) are hypothetically possible; two (β (C_{3v}), γ (C_{2v})) have been identified [1(a)].

[6]   The terms "monovacant", "trivacant", etc. were introduced by Finke to denote lacunary structures derived by formal removal of one, three, etc. MO_6 octahedra from parent ("plenary") structures.

[7]   Lipscomb, W. N. *Inorg. Chem.* **1965**, *4*, 132.

[8]   Pope, M. T. *Inorg. Chem.* **1972**, *11*, 1973.

[9]   (a) Ma, L.; Liu, S.; Zubieta, J. *Inorg. Chem.* **1989**, *28*, 175; (b) Müller, A.; Krickemeyer, E.; Penk, M.; Wittneben, V.; Döring, J. *Angew. Chem.* **1990**, *102*, 85; *Angew. Chem. Int. Ed. Engl.* **1990**, *29*, 88; (c) Krebs, B.; Klein, R. in: Pope, M. T.; Müller, A. (eds.), *Polyoxometalates. From Platonic Solids to Anti-Retroviral Activity*, Kluwer, Dordrecht, The Netherlands, **1994**, 41-57.

[10] As for the Keggin anion, various skeletal isomers of the Dawson structure (α, D_{3h} symmetry) are possible, and some of these have been confirmed.

[11] Contant, R.; Tézé, A. *Inorg. Chem.* **1985**, *24*, 4610.

[12] Michelon, M.; Hervé, G.; Leyrie, M. *J. Inorg. Nucl. Chem.* **1980**, *42*, 1583; **1985**, *74*, 4610.

[13] Fischer, J.; Ricard, L.; Weiss, R. *J. Am. Chem. Soc.* **1976**, *98*, 3050.

[14] Yamase, T.; Naruke, H.; Sasaki, Y. *J. Chem. Soc., Dalton Trans.* **1990**, 1687.

[15] Leyrie, M.; Hervé, G. *Nouv. J. Chim.* **1978**, *2*, 233.

[16]   The unusual five-fold symmetry arises from geometrical constraints: each PW_6 component subtends an angle corresponding to that between two faces of a tetrahedron, i.e. 70.53°, which is close to 360/5.

[17]   Creaser, I.; Heckel, M.; Neitz, R. J.; Pope, M. T. *Inorg. Chem.* **1993**, *32*, 1573.

[18]   Although the structurally analogous heptatungstate is known, there have been no large complexes incorporating this unit yet reported.

[19]   Fedoseev, A. M.; Grigor'ev, M. S.; Yanovskii, A. I.; Struchkov, Yu. T.; Spitsyn, V. I. *Dokl. Chem. Engl. Trans.* **1987**, *297*, 477.

[20]   Samokhvalova, E. P.; Molchanov, V. N.; Tat'yanina, I. V.; Torchenkova, E. A. *Vestn. Mosk. Univ. Ser. 2 Khim.* **1989**, *30*, 108; [*Chem. Abs.* **1989**, *111*, 89303g].

[21]   Naruke, H.; Ozeki, T.; Yamase, T. *Acta Crystallogr.* **1991**, *C47*, 489.

[22]   (a) Krebs, B.; Paulat-Böschen, I. *Acta Crystallogr.* **1982**, *B38*, 1710; (b) Krebs, B.; Stiller. S.; Tytko, K.-H.; Mehmke, J. *Eur. J. Solid State Inorg. Chem.* **1991**, *28*, 883.

[23]   Müller, A.; Krickemeyer, E.; Dillinger, S.; Bögge, H.; Plass, W.; Proust, A.; Dloczik, L.; Menke, C.; Meyer, J.; Rohlfing, R. *Z. anorg. Chem.* **1994**, *620*, 599; Müller, A.; Plass, W.; Krickemeyer, E.; Dillinger, S.; Bögge, H.; Armatage, A.; Proust, A.; Beugholt, C.; Bergmann, U. *Angew. Chem.* **1994**, *106*, 897; *Angew. Chem. Int. Ed. Engl.* **1994**, *33*, 849.

12   At the Boundary of the Metallic State
Günter Schmid

[1]   Larpent, C.; Patin, H. *J. Mol. Catal.* **1988**, *44*, 191.

[2]   Turkevich, J.; Stevenson, P. C.; Hillier, J. *Disc. Faraday Soc.* **1951**, *11*, 55.

[3]   Mabuchi, M.; Takenaka, T.; Fujiyoshi, Y.; Uyeda, N. *Surf. Sci.* **1982**, *119*, 150.

[4]   Frens, G. *Kolloid Z. Z. Polym.* **1972**, *250*, 736.

[5]   Duff, D. G.; Curtis, A. C.; Edwards, P. P.; Jefferson, D. A.; Johnson, B. F. G.; Logan, D. E. *Angew. Chem.* **1987**, *99*, 688; *Angew. Chem. Int. Ed. Engl.* **1987**, *26*, 676.

[6]   Duff, D. G.; Curtis, A. C.; Edwards, P. P.; Jefferson, D. A.; Johnson, B. F. G.; Logan, D. E. *J. Chem. Soc., Chem. Comm.* **1987**, 1264.

[7]   Freund, P. L.; Spiro, M. *J. Phys. Chem.* **1985**, *89*, 1074.

[8]   Westerhausen, J.; Henglein, A.; Lilie, J. *Ber. Bunsen-Ges. Phys. Chem.* **1981**, 85, 182.

[9]   Hirau, H.; Nakao, Y.; Toshima, N. *J. Macromol. Sci. Chem.* **1979**, *A13*, 633.

[10]   Hirau, H.; Nakao, Y.; Toshima, N. *J. Macromol. Sci. Chem.* **1979**, *A13*, 727.

[11]   Hirau, H.; Nakao, Y.; Toshima, N. *J. Macromol. Sci. Chem.* **1979**, *A13*, 1117.

[12]   Meguro, K.; Nakamura, Y.; Hayashi, Y.; Torizuka, M.; Esumi, K. *Bull. Chem. Soc. Jpn.* **1988**, *61*, 347.

[13]   Komiyama, M.; Hirai, H. *Bull. Chem. Soc. Jpn.* **1983**, *56*, 2833.

[14]   Yonezawa, Y.; Sato, T.; Ohno, M.; Hada, H. *J. Chem. Soc., Faraday Trans. 1* **1987**, 1559.

[15]   Schmid, G.; Lehnert, A. *Angew. Chem.* **1989**, *101*, 773; *Angew. Chem. Int. Ed. Engl.* **1989**, *28*, 780.

[16]   Schmid, G.; Lehnert, A.; Kreibig, U.; Adamczyk, Z.; Belouschek, P. *Z. Naturforsch.* **1990**, *45b*, 989.

[17]   Schmid, G. *Chem. Rev.* **1992**, *92*, 1709.

[18] Furlong, D. N.; Launikonis, A.; Sasse, W. H. F. *J. Chem. Soc., Faraday Trans. 1* **1984**, *80*, 571.

[19] Michel, J. B.; Schwartz, J. T. *Studies Surf. Sci. Catal.* **1987**, *31*, 669.

[20] Baddely, C. J.; Jefferson, D. A.; Lambert, R. M.; Ormerod, R. M.; Rayment, T.; Schmid, G.; Walker, A. P. *Mat. Res. Symp. Proc.* **1992**, *272*, 85.

[21] Schmid, G.; Lehnert, A.; Malm, J.-O.; Bovin, J.-O. *Angew. Chem.* **1991**, *103*, 852; *Angew. Chem. Int. Ed. Engl.* **1991**, *30*, 874.

[22] Schmid, G.; Harms, M.; Malm, J.-O.; Bovin, J.-O.; van Ruitenbeck, J.; Zandbergen, H. W.; Fu, W. T. *J. Am. Chem. Soc.* **1993**, *115*, 2046.

[23] Schmid, G. *Polyhedron* **1988**, *7*, 2321.

[24] Schmid, G. *Mater. Chem. Phys.* **1991**, *29*, 133.

[25] Vargaftik, M. N.; Zagarodnikov, V. P.; Stolyarov, I. P.; Moiseev, I. I.; Likhobolov, V. I.; Kochubey, D. I.; Chuvelin, A. L.; Zaikowsky, V. I.; Zamaraev, K. I.; Timofeeva, G. I. *J. Chem. Soc., Chem. Commun.* **1985**, 937.

[26] Schmid, G.; Morun, B.; Malm, J.-O. *Angew. Chem.* **1989**, *101*, 772; *Angew. Chem. Int. Ed. Engl.* **1989**, *28*, 778.

[27] Schmid, G.; Boese, R.; Pfeil, R.; Bandermann, F.; Meyer, S.; Calis, G. H. M.; van der Velden, J. W. A. *Chem. Ber.* **1981**, *114*, 3634.

[28] Schmid, G.; Giebel, U.; Huster, W.; Schwenk, A. *Inorg. Chim. Acta* **1984**, *85*, 97.

[29] Schmid, G.: Huster, W. *Z. Naturforsch.* **1986**, *41b*, 1028.

[30] Hartmann, U., unpublished results

[31] Elock, E. W.; Rhodes, P.; Teviodtdale, A. *Proc. Roy. Soc. (London)* **1954**, *A53*, 53.

[32] Liu, K. L.; MacDonald, A. H.; Daams, J. M.; Vosko, S. H. *J. Magnetism and Magn. Mat.* **1979**, *12*, 43.

[33] Sänger, W.; Voitländer, J. *Z. Phys.* **1978**, *B30*, 13.

[34] van Leeuwen, D. A.; van Ruitenbeck, J. M.; Schmid, G.; de Jongh, L. J. *Physics Letters* **1992**, *A170*, 325.

[35] van der Putten, D; Brom, H. B.; Wittereen, J.; de Jongh, L. J.; Schmid, G. *Z. Physik D,* **1993**, *26*, 21.

[36] Mulder, F. M.; Stegink, T. A.; Thiel, R. C.; de Jongh, L. J.; Schmid G. *Nature* **1994**, *367*, 716.

[37] Thiel, R. C.; Benfield, R. E.; Zanoni, R.; Smit, H. H. A.; Dirken, M. W. *Z. Phys. D,* **1993**, *26*, 162.

[38] Thiel, R. C.; Dirken, M. W.; Benfield, R. E.; Zanoni, R.; Smit, H. A. A. *Struct. Bonding (Berlin)* **1993**, *81*, 1.

[39] Smit, H. H.; Thiel, R. C.; de Jongh, L. J. *Z. Phys.* **1989**, *D12*, 193.

[40] Schmid, G.; Pfeil, R.; Boese, R.; Bandermann, F.; Meyer, S.; Calis, G. H. M.; van der Velden, J. W. A. *Chem. Ber.* **1981**, *114*, 5634.

[41] Simon, U.; Schön, G.; Schmid, G. *Angew. Chem.* **1993**, *105*, 264; *Angew. Chem. Int. Ed. Engl.* **1993**, *32*, 250.

[42] van Steveren, M. P. J.; Moonen, J. T.; de Jongh, L. J.; Schmid, G. *Z. Phys.* **1989**, *D12*, 461.

13 Physical Properties of High-nuclearity Metal Cluster Compounds: Model Systems for Uniform-sized Metal Particles

L. Jos de Jongh

[1] Longoni, G.; Ceriotti, A.; Marchionna, M.; Piro, G. in: Basset, J. M. et al. (eds.), *Surface Organometallic Chemistry: Molecular Approaches to Surface Catalysis*, Kluwer, Dordrecht, The Netherlands, **1988**, 157 (and references in this review).

[2] Schmid, G. *Struct. Bonding (Berlin)* **1985**, *62*, 51; *Polyhedron* **1988**, *7*, 2321; *Endeavour, New Series* **1990**, *14*, 172.

[3] Schmid, G. *Nachr. Chem. Tech. Lab.* **1987**, *34*, 249.

[4] Schmid, G.; Morun, B.; Malm, J.-O. *Angew. Chem.* **1989**, *101*, 772; *Angew. Chem. Int. Ed. Engl.* **1989**, *28*, 778.

[5] Schmid, G. (ed.), *Clusters and Colloids: From Theory to Applications*, VCH, Weinheim, Germany, **1994**.

[6] Wigner, E. P. *Ann. Math.* **1951**, *53*, 36.

[7] Dyson, F. J. *J. Math. Phys.* **1962**, *3*, 140.

[8] Gorkov, L. P.; Eliashberg, G. M. *Sov. Phys. JETP* **1965**, *21*, 940.

[9] See e.g. Bjornholm, S. *Europhys. News* **1994**, *25*, 7.

[10] Berry, M. V. *Proc. Roy. Soc. Lond. A* **1987**, *413*, 183.

[11] Efetov, K. B.; Prigodin, V. N. *Phys. Rev. Lett.* **1993**, *70; Mod. Phys. Lett.* **1993**, *7*, 981.

[12] Schönenberger, C.; van Houten, H.; Donkersloot, H. C. *Europhys. Lett.* **1992**, *20*, 249.

[13] Dubois, I. G. A.; Verheijen, E. N. G.; Gerritsen, J. W.; van Kempen, H. *Phys. Rev. B* **1993**, *48*, 11260.

[14] de Jongh, L. J.; Brom, H. B.; Longoni, G.; Nugteren, P. R.; Pronk, B. J.; Schmid, G.; Smit, H. H. A.; van Staveren, M. P. J.; Thiel, R. C. in: Jena, P. et al. (eds.), *Physics and Chemistry of Small Clusters*, Plenum, *NATO-ASI Series B* **1986**, *158*, 807; *J. Chem. Research* (S) **1987**, 150.

[15] de Jongh, L. J.; Brom, H. B.; van Ruitenbeek, J. M.; Thiel, R. C.; Schmid, G.; Longoni, G.; Ceriotti, A.; Benfield, R. E.; Zanoni, R. in: Pacchioni, G. et al. (eds.), *Cluster Models for Surface and Bulk Phenomena*, Plenum, *NATO-ASI Series B* **1992**, *283*, 151.

[16] de Jongh, L. J.; Baak, J.; Brom, H. B.; van der Putten, D.; van Ruitenbeek, J. M.; Thiel, R. C. in: Jena, P. et al. (eds.), *Physics and Chemistry of Finite Systems: From Clusters to Crystals*, Vol. II, Kluwer, Dordrecht, The Netherlands, *NATO-ASI Series C*, **1992**, *374*, 839.

[17] de Jongh, L. J. in: Hadjipanayis, G. C. (ed.), *Nanophase Materials: Synthesis, Properties, Applications*, Corfu, June **1993**, Kluwer, Dordrecht, The Netherlands, *Proceedings NATO-ASI*, to be published.

[18] Rösch, N.; Ackermann, L.; Pacchioni, G.; Dunlap, B. I. *J. Chem. Phys.* **1991**, *95*, 7004; *J. Quant. Chem.* **1992**, *S26*, 605.

[19] Mulder, F. M.; v. d. Zeeuw, E. A.; Thiel, R. C. *State Comm.* **1992**, *85* (2), 93.

[20] Thiel, R. C., Benfield, R. E.; Zanoni, R.; Smit, H. H. A.; Dirken, M. W. *Z. Phys. D* **1993**, *26*, 162.

[21] Thiel, R. C., Benfield, R. E.; Zanoni, R.; Smit, H. H. A.; Dirken, M. W. *Struct. Bonding (Berlin)* **1993**, *81*, 1.

[22]   Mulder, F. M., Stegink, T. A.; Thiel, R. C.; de Jongh, L. J.; Schmid, G. *Nature* **1994**, *367*, 716.

[23]   van der Putten, D.; Brom, H. B.; de Jongh, L. J.; Schmid, G. in: Jena, P. et al. (eds.), *Physics and Chemistry of Finite Systems: From Clusters to Crystals*, Vol. II, Kluwer, Dordrecht, The Netherlands, **1992**, 1007.

[24]   van der Putten, D.; Brom, H. B.; Witteveen, J.; de Jongh, L. J.; Schmid, G. *Suppl. Z. Phys. D* **1993**, *26*, 21.

[25]   Baak, J., Brom, H. B.; de Jongh, L. J.; Schmid: G. in: Jena, P. et al. (eds.), *Physics and Chemistry of Finite Systems: From Clusters to Crystals*, Vol. II, Kluwer, Dordrecht, The Netherlands, **1992**, 339.

[26]   Baak, J., Brom, H. B.; de Jongh, L. J.; Schmid, G. *Z. Phys. D* **1993**, *26*, 30.

[27]   Peerenboom, J. A. A. J.; Wyder, P.; Meider, F. *Phys. Rep.* **1981**, *78*, 173.

[28]   Halperin, W. P. *Rev. Mod. Phys.* **1986**, *58*, 533.

[29]   van Leeuwen, D. A.; van Ruitenbeek, J. M.; Schmid, G.; de Jongh, L. J. *Phys. Lett. A* **1992**, *170*, 325.

[30]   de Jongh, L. J. (ed.), *Physics and chemistry of metal cluster compounds. Model systems for small metal particles*. Series on physics and chemistry of materials with low-dimensional structures, Reidel, Dordrecht, The Netherlands, **1994**, to be published, .

14   Towards Nanoscale Molecular Magnetic Materials

Dante Gatteschi, Andrea Caneschi, Luca Pardi, and Roberta Sessoli

[1]   Carlin, R. L. *Magnetochemistry*, Springer, Berlin, Germany, **1986**.

[2]   Morrish, A. H. *The Physical Principle of Magnetism*, Wiley, New York, USA, **1966**.

[3]   Wieghardt, K.; Pohl, K.; Jibril, I.; Huttner, G. *Angew. Chem.* **1984**, *96*, 66; *Angew. Chem. Int. Ed. Engl.* **1984**, *23*, 77.

[4]   Chien, C. L.; Xiao, J. Q.; Jiang, J. S. *J. Appl. Phys.* **1993**, *73*, 5309.

[5]   Mc Michael, R. D.; Shull, R. D.; Swatzendruber, L. J.; Bennett, L. H.; Watson, R. E. *J. Magn. Magn. Mat.* **1992**, *111*, 29.

[6]   Ziolo, R. F.; Giannellis, E. P.; Weinstein, B. A.; O'Horo, M. P.; Granguly, B. N.; Mehrotra, V.; Russel, M. V.; Huffman, D. R. *Science* **1992**, *257*, 219.

[7]   Ford, G. C.; Harrison, P. M.; Rice, D. W.; Smith, J. M. A.; Treffry, A.; White, J. L.; Yariv, J. *Philos. Trans. R. Soc. London* **1984**, *B304*, 551.

[8]   Harrison, P. M.; Artymiuk, P. J.; Ford, G. C.; Lawson, D. M.; Smith, J. M. A.; Treffry, A.; White J. L. in: Mann, S.; Webb, J.; Williams, R. J. P. (eds.), *Biomineralization: Chemical and Biochemical Perspectives*, VCH, Weinheim, Germany, **1989**, 252.

[9]   St Pierre, T. G.; Webb, J.; Mann S. in: Mann, S.; Webb, J.; Williams, R. J. P. (eds.), *Biomineralization: Chemical and Biochemical Perspectives*, VCH, Weinheim, Germany, **1989**, 295.

[10]   Mann, S.; Frankel R. B. in: Mann, S.; Webb, J.; Williams, R. J. P. (eds.), *Biomineralization: Chemical and Biochemical Perspectives*, VCH, Weinheim, Germany, **1989**, 389.

[11]   Leggett, A. J. *Re. Mod. Phys.* **1987**, *59*, 1.

[12]   Stamp, P. C. E.; Chudnovsky, E. M.; Barbara, B. *Int. J. Mod. Phys. B* **1992**, *6*, 1355.

[13]   vam Staveren, M. P. J.; Brom, H. B.; de Jongh, L. J. *Phys. Reports* **1991**, *208*, 3.

[14]   Schmid, G. Chapter 12, 149–162 in this same book.

[15] Fujita, I.; Teki, Y.; Takui, T.; Kinoshita, T.; Itoh, K.; Miko, F.; Sawaki, Y.; Iwamura, H.; Izuoka, A.; Sugawara, T. *J. Am. Chem. Soc.* **1990**, *112*, 4074.

[16] Nakamura, N.; Inoue, K.; Iwamura, H.; Fujioka, T. *J. Am. Chem. Soc.* **1992**, *114*, 1484.

[17] Nakamura, N.; Inoue, K.; Iwamura, H. *Angew. Chem.* **1993**, *105*, 900; *Angew. Chem. Int. Ed. Engl.* **1993**, *32*, 872.

[18] Tomalia, D. A.; Baker, H.; Dewall, J.; Hall, M.; Kallos, G.; Martin, S.; Roeck, J.; Smith, P. *Polym. J. (Tokyo)* **1985**, *17*, 117.

[19] Veciana, J.; Rovira, C.; Ventosa, N.; Crespo, M. I.; Palacio, F. *J. Am. Chem. Soc.* **1993**, *115*, 57.

[20] Pope, M. T. *Heteropoly and Isopoly Oxometalates*, Springer, New York, USA, **1983**.

[21] Müller, A.; Krickemeyer, E.; Penk, M.; Walberg, M. J.; Bögge, H. *Angew. Chem.* **1987**, *99*, 1060; *Angew. Chem. Int. Ed. Engl.* **1987**, *26*, 1045.

[22] Müller, A.; Döring, J. *Angew. Chem.* **1988**, *100*, 1789; *Angew. Chem. Int. Ed. Engl.* **1988**, *27*, 1721.

[23] Müller, A.; Penk, M.; Krickemeyer, E.; Bögge, H.; Walberg, M. J. *Angew. Chem.* **1988**, *100*, 1787; *Angew. Chem. Int. Ed. Engl.* **1988**, *27*, 1719.

[24] Müller, A. *Nature* **1991**, *352*, 115.

[25] Müller, A.; Döring, J.; Khan, M. I.; Wittneben, V. *Angew. Chem.* **1991**, *103*, 203; *Angew. Chem. Int. Ed. Engl.* **1991**, *30*, 210.

[26] Müller, A.; Penk, M.; Döring, J. *Inorg. Chem.* **1991**, *30*, 4935.

[27] Berg, J. M.; Holm R. H. in: Spiro, T. G. (ed.), *Iron Sulfur Proteins*, Wiley Interscience, New York, USA, **1982**.

[28] Fenske, D.; Krautscheid, H. *Angew. Chem.* **1990**, *102*, 1513; *Angew. Chem. Int. Ed. Engl.* **1990**, *29*, 1452.

[29] Fenske, D.; Ohmer, J.; Hachgenei, J.; Merzweiler, K. *Angew. Chem.* **1988**, *100*, 1300; *Angew. Chem. Int. Ed. Engl.* **1988**, *27*, 1277.

[30] Bencini, A.; Ghilardi, C. A.; Orlandini, A.; Midollini, S.; Zanchini, C. *J. Am. Chem. Soc.* **1992**, *114*, 9898.

[31] You, J. F.; Holm, R. H. *Inorg. Chem.* **1991**, *30*, 1431.

[32] You, J. F.; Papaefthymiou, G. C.; Holm, R. H. *J. Am. Chem. Soc.* **1992**, *114*, 2697.

[33] You, J. F.; Papaefthymiou, G. C.; Holm, R. H. *J. Am. Chem. Soc.* **1990**, *112*, 1076.

[34] Gatteschi, D.; Pardi, L. *Gazz. Chim. It.* **1993**, *123*, 231.

[35] Delfs, C. D.; Gatteschi, D.; Pardi, L.; Sessoli, R.; Wieghardt, K.; Hanke, D. *Inorg. Chem.* **1993**, *32*, 3099.

[36] Vannimenous, J.; Toulouse, G. *J. Phys. C. Solid State Phys.* **1977**, *10*, L537.

[37] Gatteschi, D.; Pardi, L.; Barra, A. L.; Müller A. in: Pope, M. T.; Müller, A. (eds.), *Polyoxometalates. From Platonic Solids to Anti-Retroviral Activity*, Kluwer, Dordrecht, The Netherlands, **1994**.

[38] Müller, A.; Penk, M.; Rohlfing, R.; Krickemeyer, E.; Döring, J. *Angew. Chem.* **1990**, *102*, 927; *Angew. Chem. Int. Ed. Engl.* **1990**, *29*, 926; Müller, A.; Reuter H., Dillinger, S. *Angew. Chem.* **1995**, *107*, in press; *Angew. Chem. Int. Ed. Engl.* **1995**, *34*, in press.

[39] Müller, A.; Döring, J.; Penk, M. *Z. Anorg. Allg. Chemie* **1991**, *595*, 251.

[40] Barra, A. L.; Gatteschi, D.; Pardi, L.; Müller, A.; Döring, J. *J. Am. Chem. Soc.* **1992**, *114*, 8059.

[41] Taft, K. L.; Lippard, S. J. *J. Am. Chem. Soc.* **1990**, *112*, 9629.

[42] Fischer, M. E. *Am. J. Phys.* **1964**, *32*, 343.

[43] Taft, K. L.; Delfs, C. D.; Papaefthymiou, G. C.; Foner, S.; Gatteschi, D.; Lippard, S. J. *J. Am. Chem. Soc.* **1994**, *116*, 823.

[44] Heath, S. L.; Powell, A. K. *Angew. Chem.* **1992**, *104*, 191; *Angew. Chem. Int. Ed. Engl.* **1992**, *31*, 191.

[45] Powell, A. K.; Heath, S. L.; Gatteschi, D.; Pardi, L.; Sessoli, R.; Spina, G.; Del Giallo, F.; Pieralli, F. submitted for publication.

[46] Low, D. W.; Eichhorn, D. M.; Draganescu, A.; Armstrong, W. H. *Inorg. Chem.* **1991**, *30*, 878.

[47] Christmas, C.; Vincent, J. B.; Chang, H. R.; Huffmann, J. C.; Christou, G.; Hendrickson, D. N. *J. Am. Chem. Soc.* **1988**, *110*, 823.

[48] Cavalluzzo, M.; Chen, Q.; Zubieta, J. *J. Chem. Soc., Chem. Commun.* **1993**, 131.

[49] Hagen, K. S.; Armstrong, W. H.; Olmstead, M. M. *J. Am. Chem. Soc.* **1989**, *111*, 774.

[50] Schake, A. R.; Vincent, J. B.; Li, Q.; Boyd, P. D. W.; Folting, K.; Huffmann, J. C.; Hendrickson, D. N.; Christou, G. *Inorg. Chem.* **1989**, *28*, 1915.

[51] Bhula, R.; Collier, S.; Robinson, W. T.; Weatherburn, D. C. *Inorg. Chem.* **1990**, *29*, 4027.

[52] Bhula, R.; Weatherburn, D. C. *Angew. Chem.* **1991**, *103*, 715; *Angew. Chem. Int. Ed. Engl.* **1991**, *30, 688.*

[53] Goldberg, D. P.; Caneschi, A.; Lippard, S. J. *J. Am. Chem. Soc.* **1993**, *115*, 9299.

[54] Boyd, P. D. W.; Li, Q.; Vincent, J. B.; Folting, K.; Chang, H. R.; Streib, W. E.; Huffmann, J. C.; Christou, G.; Hendrickson, D. N. *J. Am. Chem. Soc.* **1988**, *110*, 8537.

[55] Lis, T. *Acta Crystallogr. Sect. B.* **1980**, *36*, 2042.

[56] Caneschi, A.; Gatteschi, D.; Sessoli, R.; Barra, A. L.; Brunel, L. C.; Guillot, M. *J. Am. Chem. Soc.* **1991**, *113*, 5873.

[57] Sessoli, R.; Tsai, H. L.; Schake, A. R.; Wang, S.; Vincent, J. B.; Folting, K.; Gatteschi, D.; Christou, G.; Hendrickson, D. N. *J. Am. Chem. Soc.* **1993**, *115*, 1804.

[58] Schake, A. R.; Tsai, H. L.; De Vries, N.; Webb, R. J.; Folting, K.; Hendrickson, D. N.; Christou, J. *J. Chem. Soc., Chem. Commun.* **1992**, 181.

[59] Sessoli, R.; Gatteschi, D.; Caneschi, A.; Novak, M. A. *Nature* **1993**, *365*, 141.

[60] Villain, J.; Hartman-Boutron, F.; Sessoli, R.; Rettori, A. *Europhys. Lett.* **1994**, *27*, 159.

15 Philosophische Aspekte der Chemie – Ihr Wesen: Universalität und Beständigkeit des Wandels

Achim Müller und Herbert Hörz

[1] Hinshelwood, C. N. *J. Chem. Soc.* **1947**, 1271.

[2] d'Holbach, P. T. *System der Natur oder von den Gesetzen der physischen und der moralischen Welt*, Suhrkamp, Frankfurt am Main, Deutschland, **1978**, 39.

[3] Feynman, R. P.; Leighton, R. B.; Sands, M. *The Feynman Lectures on Physics*, Vol. 1, Chapter: 1-2 (*Matter is made of atoms*), Addison-Wesley, Reading, Massachusetts, USA, **1963**.

[4] Liebig, J. *Über das Studium der Naturwissenschaften und über den Zustand der Chemie in Preußen*, Braunschweig, Deutschland, **1840**, zitiert nach Böhme, G. (Einleitung) in: Böhme, G. (Hrsg.), *Klassiker der Naturphilosophie: Von den Vorsokratikern bis zur Kopenhagener Schule*, Beck, München, Deutschland, **1989**, 7, 9.

[5]   Müller, A. *Gaia* **1992**, *1*, 190.

[6]   Hörz, H.; Wessel, K. F. *Philosophische Entwicklungstheorie*, Deutscher Verlag der Wissenschaften, Berlin, Deutschland, **1983**.

[7]   Schelling, F. W. J. *Allgemeine Deduktion des dynamischen Processes oder der Kategorien der Physik* (1800), in: Schröter, M. (Hrsg.), *Schellings Werke*, Zweiter Hauptband, *Schriften zur Naturphilosophie*, Beck, München, Deutschland, **1958**, 637.

[8]   a) Lindsey, J. S. *New J. Chem.* **1991**, *15*, 153; b) Müller, A. *J. Mol. Struct.* **1994**, *325*, 13.

[9]   Haken, H. *Synergetics: An Introduction*, Springer, Berlin, Deutschland, **1977**, 3. Aufl., **1983**; Haken, H.; Wunderlin, A. *Die Selbststrukturierung der Materie: Synergetik in der unbelebten Welt*, Vieweg, Braunschweig, Deutschland, **1991**; Haken, H.; Haken-Krell, M. *Entstehung von biologischer Information und Ordnung*, Wiss. Buchgesellschaft, Darmstadt, Deutschland, **1989**.

[10] Nicolis, G.; Prigogine, I. *Die Erforschung des Komplexen: Auf dem Weg zu einem neuen Verständnis der Naturwissenschaften*, Piper, München, Deutschland, **1987**; Prigogine, I. *Vom Sein zum Werden: Zeit und Komplexität in den Naturwissenschaften*, Piper, München, Deutschland, **1979**, 5. Aufl. 1988; Prigogine, I.; Stengers, I. *Dialog mit der Natur: Neue Wege naturwissenschaftlichen Denkens*, Piper, München, Deutschland, **1981**, 4. Aufl. 1983.

[11] Schleiden, M. J. *Schelling's und Hegel's Verhältnis zur Naturwissenschaft: Zum Verhältnis der physikalistischen Naturwissenschaft zur spekulativen Naturphilosophie*, (Breidbach, O. (Hrsg.)), VCH, Weinheim, Deutschland, **1988**.

[12] Förster, W. *Schelling als Theoretiker der Dialektik der Natur* und Pälike, D. *Wissenschaftstheoretische Aspekte in Schellings naturphilosophischer Interpretation der Chemie*, in: Sandkühler, H. J. (Hrsg.), *Natur und geschichtlicher Prozeß: Studien zur Naturphilosophie F. W. J. Schellings*, Suhrkamp, Frankfurt am Main, Deutschland, **1984**, 175, 261.

[13] Heisenberg, W. *Ordnung der Wirklichkeit* (1942), in: Blum, W.; Dürr, H.-P.; Rechenberg, H. (Hrsg.), *Werner Heisenberg, Gesammelte Werke*, Abt. C: *Allgemeinverständliche Schriften*, Vol. 1: *Physik und Erkenntnis 1927–1955*, Piper, München, Deutschland, **1984**, 252.

[14] Davies, P. C. W.; Brown, J. R. (Hrsg.), *Der Geist im Atom: Eine Diskussion der Geheimnisse der Quantenphysik*, Birkhäuser, Basel, Schweiz, **1988**.

[15] Störig, H. J. *Weltgeschichte der Wissenschaft* Vol. 2: *Natur- und Geisteswissenschaften des 19. und 20. Jahrhunderts*, Weltbild, Augsburg, Deutschland, **1992**, 155.

[16] Wachtel, S.; Jendrusch, A. *Der Linksdrall in der Natur: Eine Entdeckung und ihr Schicksal*, dtv, München, Deutschland, **1993**; Derflinger, G. in: Janoschek, R. (Hrsg.), *Chirality – From Weak Bosons to the a-Helix*, Springer, Heidelberg, Deutschland, **1991**; Buda, A. B.; Auf der Heyde, T.; Mislow, K. *Angew. Chem.* **1992**, *104*, 1012; *Angew. Chem. Int. Ed. Engl.* **1992**, *31*, 989.

[17] Ludwig, C. *Einige neue Beziehungen zwischen dem Bau und der Funktion der Niere*, in: Sitzungsberichte der Kaiserlichen Akademie der Wissenschaften, Wien, Math.-naturwiss. Klasse, **1863**, *48*, 725.

[18] Hörz, H. *Physiologie und Kultur in der zweiten Hälfte des 19. Jahrhunderts*, Basilisken-Presse, Marburg (Lahn), Deutschland, **1994**, 311.

[19] Hyrtl, J. *Die materialistische Weltanschauung unserer Zeit*, Holzhausen, Wien, Österreich, **1865**, 5.

[20] Ebenda, S. 4.

[21] Black, I. B. *Symbole, Synapsen und Systeme: Die molekulare Biologie des Geistes*, Spektrum, Heidelberg, Deutschland, **1993**; Snyder, S. H. *Chemie der Psyche*, Spektrum, Heidelberg, Deutschland, **1988**; Dittrich, A.; Hofmann, A.; Leuner, H. (Hrsg.), *Welten des Bewußtseins*, Vol. 3: *Experimentelle Psychologie, Neurobiologie und Chemie*, VWB, Berlin, Deutschland, **1994**.

[22] Liebig, J. *Die organische Chemie in ihrer Anwendung auf Physiologie und Pathologie*, Vieweg, Braunschweig, Deutschland, **1842**; Reprint: Strothe, Frankfurt am Main, Deutschland, **1992**.

[23] Niedersen, U.; Pohlmann, L. (Hrsg.*), Selbstorganisation, Jahrbuch für Komplexität in den Natur-, Sozial- und Geisteswissenschaften*, Vol. 1: *Selbstorganisation und Determination*, **1990**, Vol. 2: *Der Mensch in Ordnung und Chaos*, **1991**, Duncker & Humblot, Berlin, Deutschland; Kratky, K. W.; Wallner, F. (Hrsg.), *Grundprinzipien der Selbstorganisation*, Wiss. Buchgesellschaft, Darmstadt, Deutschland, **1990**; vgl. auch [58].

[24] Mach, E. *Erkenntnis und Irrtum*, 5. Aufl., Leipzig, Deutschland, **1926**; Reprint: Wiss. Buchgesellschaft, Darmstadt, Deutschland, **1991**.

[25] Mach, E. *Prinzipien der Wärmelehre*, 2. Aufl., Barth, Leipzig, Deutschland, **1900**, 354.

[26] Pauling, L. *The Nature of the Chemical Bond and the Structure of Molecules and Crystals*, 2. Aufl., Cornell, Ithaca, New York, USA, **1948**.

[27] (a) Primas, H. *Ein Ganzes, das nicht aus Teilen besteht: Komplementarität in den exakten Naturwissenschaften* in: Fischer, E. P. (Hrsg.), *Neue Horizonte 92/93: Ein Forum der Naturwissenschaften*, Piper, München, Deutschland, **1993**, 91–93; (b) *Chemie i. u. Zeit*, **1985**, *19*, 109, 160.

[28] Hund, F. *Geschichte der physikalischen Begriffe*, Teil 2: *Die Wege zum heutigen Naturbild*, 2. Aufl., Bibliographisches Institut, Mannheim, Deutschland, **1987**, 134; Brush, S. G. *Kinetische Theorie*, Band 2: *Irreversible Prozesse: Einführung in die Originaltexte*, Akademie Verlag, Berlin, Deutschland, **1970**.

[29] Müller, A.; Römer, M.; Krickemeyer, E.; Bögge, H. *Naturwiss.* **1984**, *71*, 43; Müller, A.; Diemann E.; Jostes, R. *Naturwiss.* **1984**, *71*, 420; Müller, A.; Diemann E. *Adv. Inorg. Chem.* **1987**, *31*, 89.

[30] Erfkamp, J.; Müller, A. *Chemie i. u. Zeit* **1990**, *24*, 267; Renger, G. *Chemie i. u. Zeit*, **1994**, *28*, 118.

[31] Hohl, R. (Hrsg.), *Die Entwicklungsgeschichte der Erde*, 7. Aufl., Brockhaus, Leipzig, Deutschland, **1981**.

[32] Krumbein, W. E. (Hrsg.), *Microbial Geochemistry*, Blackwell, Oxford, Großbritannien, **1983**; Ehrlich, H. L. *Geomicrobiology*, Dekker, New York, USA, **1981**; Trudinger, P. A.; Walter, M. R.; Ralph, B. J. (Hrsg.), *Biogeochemistry of Ancient and Modern Environments*, Springer, Berlin, Deutschland, **1980**.

[33] Schlegel, H. G. *Allgemeine Mikrobiologie*, 6. Aufl., Thieme, Stuttgart, Deutschland, **1985**.

[34] Cairns-Smith, A. G.; Hartman, H. (Hrsg.), *Clay Minerals and the Origin of Life*, Cambridge University Press, Cambridge, Großbritannien, **1988**; Lahav, N. *Heterogeneous Chem. Rev.* **1994**, *1*, 159; Cairns-Smith, A. G. *Genetic takeover and the mineral origins of life*, Cambridge University Press, Cambridge, Großbritannien, **1982**.

[35] Vögtle, F. *Supramolekulare Chemie*, 2. Aufl., Teubner, Stuttgart, Deutschland, **1992**, 45; *Supramolecular Chemistry*, 2. Aufl., Wiley, Chichester, Großbritannien, **1993**, 27;

[36] Müller, A.; Reuter, H.; Dillinger, S. *Angew. Chem.* **1995**, *107*, im Druck; *Angew. Chem. Int. Ed. Engl.* **1995**, *34*, im Druck.

[37] Eschenmoser, A. *Nachr. Chem. Tech. Lab.* **1991**, *39*, 795; Eschenmoser, A.; Dobler, M. *Helv. Chim. Acta* **1992**, *75*, 218.

[38] Hörz, H. *Materiestruktur*, Deutscher Verlag der Wissenschaften, Berlin, Deutschland, **1971**, 89f.

[39] vgl. [35], 45.

[40] (a) *Aristoteles Werke*, Band 1, Teil 1, *Kategorien*, übersetzt und erläutert von Oehler, K., Akademie-Verlag, Berlin, Deutschland, **1984**, 22 (9a 32-33); (b) *Aristoteles' Metaphysik*, 1. Halbband, Neubearbeitung der Übersetzung von Bonitz, H., Meiner, Hamburg, Deutschland, **1989**, 223 (1020b 8-12); (c) Hegel, G. W. F. *Wissenschaft der Logik*, Meiner, Leipzig, Deutschland, **1951**.

[41] Mutschler, E. *Arzneimittelwirkungen*, Wiss. Verlagsgesellschaft, Stuttgart, Deutschland, **1970**; Vögtle, F. *Reizvolle Moleküle der Organischen Chemie*, Teubner, Stuttgart, Deutschland, **1989**; *Fascinating Molecules in Organic Chemistry*, wiley, Chichester, Großbritannien, **1992**.

[42] Ebeling, W.; Engel, A.; Feistel, R. *Physik der Evolutionsprozesse*, Akademie-Verlag, Berlin, Deutschland, **1990**, 285.

[43] Wächtershäuser, G. *Microbiol. Rev.* **1988**, *52*, 452.

[44] Reichholf, J. H. *Der schöpferische Impuls: Eine neue Sicht der Evolution*, Erster Teil, Kapitel 1, dtv, München, Deutschland, **1994**, 25; Russell, M. J.; Daniel, R. M.: Hall, A. J.; Sherringham, J. A *J. Mol. Evol.* **1994**, *39*, 231; Müller, A.; Schladerbeck, N. *Naturwiss.* **1986**, *73*, 669; *Chimia* **1985**, *39*, 23.

[45] (a) Hörz, H. *Philosophie der Zeit*, Deutscher Verlag der Wissenschaften, Berlin, Deutschland, **1990**; (b) Hörz, H. *Zufall: Eine philosophische Untersuchung*, Akademie-Verlag, Berlin, Deutschland, **1980**.

[46] Köhler, G. *Die Erkennungsmoleküle des Immunsystems* (Einleitung) und Burnet, M. *Die Mechanismen der Immunität* in: Köhler, G. (Hrsg.), *Immunsystem: Abwehr und Selbsterkennung auf molekularem Niveau*, Spektrum, Heidelberg, Deutschland, **1987**, 8, 12; Betz, A. G.; Neuberger, M. S.; Milstein, C. *Immunology Today* **1993**, *14*, 405.

[47] Tarassow, L. *Wie der Zufall will? Vom Wesen der Wahrscheinlichkeit*, Spektrum, Heidelberg, Deutschland, **1993**.

[48] Stiller, W. *Ludwig Boltzmann: Altmeister der klassischen Physik – Wegbereiter der Quantenphysik und Evolutionstheorie* (Kapitel: *Mitbegründer der (theoretischen) Biophysik und Evolutionsbiologie* sowie *Wegbereiter einer hierarchischen Kosmologie*), Deutsch, Thun, Deutschland, **1989**, 196, 199.

[49] Hörz, H. *Physik und Weltanschauung*, URANIA, Leipzig, Deutschland, **1975**, 57.

[50] Cramer, F.; Kaempfer, W. *Die Natur der Schönheit: Zur Dynamik der schönen Formen*, Insel, Frankfurt am Main, Deutschland, **1992**, 147.

[51] vgl. [9] [10].

[52] Kekulé, A. *Lehrbuch der Organischen Chemie*, 1. Band, Enke, Erlangen, Deutschland, **1861**, 3.

[53] vgl. [14].

[54]   Monod, J. *Zufall und Notwendigkeit: Philosophische Fragen der modernen Biologie*, 8. Aufl., dtv, München, Deutschland, **1988**, 87.

[55]   von Weizsäcker, C. F. *Aufbau der Physik*, 3. Aufl., dtv, München, Deutschland, **1994**, 167.

[56]   vgl. [8] [36] sowie Müller, A; Plass, W.; Krickemeyer, E.; Dillinger, S.; Bögge, H.; Armatage, A.; Proust, A.; Beugholt, C.; Bergmann, U. *Angew. Chem.* **1994**, *106*, 897; *Angew. Chem. Int. Ed. Engl.* **1994**, *33*, 849.

[57]   Voet, D.; Voet, J. G. *Biochemie*, VCH, Weinheim, Deutschland, **1992**.

[58]   Hörz, H. *Selbstorganisation sozialer Systeme*, LIT, Münster, Deutschland, **1994**.

[59]   Waldrop, M. M. *Inseln im Chaos: Die Erforschung komplexer Systeme*, Rowohlt, Reinbek/Hamburg, Deutschland, **1993**.

[60]   Gell-Mann, M. *Das Quark und der Jaguar: Vom Einfachen zum Komplexen – Die Suche nach einer neuen Erklärung der Welt*, Piper, München, Deutschland, **1994**.

[61]   Lederman, L.; Teresi, D. *Das schöpferische Teilchen: Der Grundbaustein des Universums*, Bertelsmann, München, Deutschland, **1993**.

[62]   Müller, A.; Hörz, H. *System u. Struktur*, **1995**, *3*, 171; Zurek, W. H. (Hrsg.), *Complexity, Entropy and the Physics of Information*, Vol. VIII (Santa Fe Institute Studies in the Sciences of Complexity), Addison-Wesley, Redwood City, California, USA, **1990**.

[63]   Eigen, M. *Stufen zum Leben: Die frühe Evolution im Visier der Molekularbiologie*, Kapitel: *Komplexität als physikalisches Problem*, Piper, München, Deutschland **1987**, 31.

[64]   Futuyma, D. J. *Evolutionsbiologie*, Birkhäuser, Basel, Schweiz, **1990**, 67.

[65]   vgl. [24], 3.

16   Über systemtheoretische Grundlagen der Gesellschaftstheorie
Niklas Luhmann

[1]   So etwa in der amerikanischen Diskussion über "Neofunktionalismus", angeregt vor allem durch J. Alexander; aber auch im immer noch zunehmenden Klassikerkult und seinen exegetischen Bemühungen, seien sie nun an M. Weber, an E. Durkheim, an G. Simmel orientiert oder mit der Entdeckung und Aufwertung weiterer Figuren jener Zeit befaßt.

[2]   Für diesen Stand der Diskussion siehe die (nach wie vor unveränderte) Position von J. Habermas, jetzt etwa in: *Der philosophische Diskurs der Moderne*, Suhrkamp, Frankfurt am Main, Deutschland, **1985**, 390; oder von P. H. Lilienfeld in: *The Rise of Systems Theory, An Ideological Analysis*, (Dissertation an der humanistisch engagierten New School for Social Research, Wiley, New York, USA, **1975**). Der etwa gleichzeitige Ansehensverlust der Parsonsschen Theorie des allgemeinen Handlungssystems hat mit dieser Debatte im übrigen nichts zu tun, und das gegenwärtig erneute Interesse an Parsons hat denn auch nichts zur Klärung von Grundfragen der Systemtheorie beigetragen.

[3]   Siehe: Brown, G. S. *Laws of Form*, Allen and Unwin, London, Großbritannien, **1969**. Bei einer gründlichen Darstellung müßten auch andere radikal differentialistische Ansätze behandelt werden (etwa die französische Linie, die von der Linguistik Saussures bis zu J. Derrida führt).

[4]   Die Veröffentlichung von G. S. Brown hat zwar in Kreisen, die an Kybernetik und Systemtheorie interessiert sind und fast nur hier, sofort Beachtung gefunden,

während Logiker daran nichts Neues sehen und Philosophen sie nicht zur Kenntnis nehmen. Eine explizite Anwendung auf die System-Umwelt-Unterscheidung findet man jedoch erst in den letzten Jahren. Siehe vor allem: Simon, F. B. *Unterschiede, die Unterschiede machen. Klinische Epistemologie. Grundlage einer systemischen Psychiatrie und Psychosomatik*, Springer, Berlin, Deutschland, **1988**, 27.

[5] Siehe: von Foerster, H. *Erkenntnistheorien und Selbstorganisation* in: Schmidt, S. J. (Hrsg.), *Der Diskurs des radikalen Konstruktivismus*, Suhrkamp, Frankfurt am Main, Deutschland, **1987**, 131.

[6] An Pädagogen richtet H. von Foerster daher die Frage, ob es wirklich ihr Ziel sein sollte, Menschen zu Trivialmaschinen zu erziehen, die auf bestimmte Fragen die richtigen Antworten geben können. Siehe ders.: von Foerster, H. *Sicht und Einsicht. Versuche zu einer operativen Erkenntnistheorie*, Vieweg, Braunschweig, Deutschland, **1985**, 12.

[7] Siehe: Maturana, H. R. *Erkennen. Die Organisation und Verkörperung von Wirklichkeit. Ausgewählte Arbeiten zur biologischen Epistemologie*, Vieweg, Braunschweig, Deutschland, **1982**; und als eher populär geschriebene Darstellung der entsprechenden Biologie: Maturana, H. R.; Varela, F. J. *Der Baum der Erkenntnis*, Goldmann, München, Deutschland, **1987**. Auch auf dieser biologischen Grundlage gibt es eine (dann aber durch eine Theorie der Sprache vermittelte) Epistemologie der Beobachtung zweiter Ordnung als Nachfolgekonzept für die europäische Ontologie.

[8] Maturana selbst schließt von der Begrifflichkeit her die Erweiterung auf kommunikative Systeme nicht aus; aber er fügt (mündlich) hinzu: man müsse es "zeigen" können. Eben das wäre dann die Aufgabe des Soziologen. Im übrigen ist der Begriff des sozialen Systems bei Maturana anders besetzt, nämlich im Sinne einer aus lebenden Menschen bestehenden (also niemals autopoietischen) Gruppe.

[9] Zu dieser Parallele von Repräsentationalismus und Adaptionismus und zur Ablehnung beider Theorietraditionen (mit freilich noch unklaren Nachfolgekonzepten) vgl. Varela, F. J. *Living Ways of Sense-Making. A Middle Path for New Science* in: Livingston, P. (Hrsg.), *Disorder and Order. Proceedings of the Stanford International Symposium (Sept. 14. – 16. 1981)*, Stanford, USA, **1984**, 208.

[10] Die Konsequenzen für eine Sprachtheorie kann ich an dieser Stelle nicht behandeln. Sie sprengen jedenfalls den Rahmen der Saussureschen Linguistik – allein deshalb schon, weil sie Sprache gerade nicht als System voraussetzen, sondern nur als Mechanismus der strukturellen Kopplung von Bewußtseinssystemen und sozialen Systemen. Das Gesamtkonzept bedürfte weiterer Ausarbeitung. Für einige Hinweise siehe auch: Luhmann, N. *Wie ist Bewußtsein an Kommunikation beteiligt?* in: Gumbrecht, H. U.; Pfeiffer, K. L. (Hrsg.) *Materialität der Kommunikation*, Suhrkamp, Frankfurt am Main, Deutschland, **1988**, 884.

[11] Wie immer, kann ein Beobachter dies natürlich so beschreiben, wenn er sich entschließt, entsprechende Kausalzusammenhänge für ausschlaggebend zu halten und anderes zu vernachlässigen.

[12] Siehe: Bachelard, G. *La formation de l'esprit scientifique, contribution a une Psychanalyse de la connaissance objektive*, Vrin, Paris, Frankreich, **1938**. Zit. nach der Ausgabe Paris, Frankreich **1947**, 13.

[13] Für Einzelheiten siehe: Luhmann, N. *Soziale Systeme. Grundriß einer allgemeinen Theorie*, Suhrkamp, Frankfurt am Main, Deutschland, **1984**.

[14]   Als erwägenswerte Alternative wird gegenwärtig eigentlich nur Handlungstheorie diskutiert. Aber der Begriff der Handlung verweist primär auf das handelnde Individuum und seine körperliche und mentale Ausstattung. Er hat keine notwendig soziale Referenz. Und wenn M. Weber von sozialem Handeln spricht, sieht er die Sozialität in der individuellen Intention, im "gemeinten" Sinn begründet. In der heutigen Diskussion wirkt deshalb die Empfehlung "Handlungstheorie" so, als ob sie trotz zahlloser Einwände, die bis ins 18. Jh. zurückreichen, zur Rettung des Subjekts notwendig wäre.

[15]   Die Verwirrung der üblichen Diskussion über Moderne und Postmoderne ist nicht zuletzt durch die These eines "Referenzverlustes" entstanden. Siehe für viele: Bürger, P. *Prosa der Moderne*, Suhrkamp, Frankfurt am Main, Deutschland, **1988**. Das liegt vor allem daran, daß man die Unterscheidung wahr/unwahr (den Code der Erkenntnis) nicht deutlich vom Referenzproblem unterscheidet. Wenn etwas für die moderne Gesellschaft charakteristisch ist, dann dies: daß sie zwischen den Unterscheidungen wahr/unwahr und Selbstreferenz/Fremdreferenz differenzieren und damit die Unterscheidung Subjekt/Objekt, in der diese Probleme zusammengezogen waren, aufgeben muß.

[16]   Dieser Begriff stammt aus der Linguistik und formuliert die Einsicht, daß eine Theorie der Sprache sprachlich formuliert werden, also Teil ihres Gegenstandes sein muß. Siehe z.B.: Löfgren, L. *Towards System. From Computation to the Phenomenon of Language*, in: Carvallo, I. E. (Hrsg.), *Nature, Cognition and System I. Current Systems-Scientific Research on Natural and Cognitive Systems*, Kluwer, Dordrecht, Niederlande, **1988**, 129.

Authors

Christof K. Biebricher studied chemistry at the University of Basel and Heidelberg; his graduate work was in molecular biology at the Max-Planck-Institute for Medical Research in Heidelberg. His postdoctoral work was at the Salk Institute for Biological Studies in La Jolla. 1973 he moved to the Max-Planck-Institute for Biophysical Chemistry in Göttingen. He teaches Biophysical Chemistry at the University of Braunschweig.
MPI für Biophysikalische Chemie, Am Faßberg 11, 37077 Göttingen, Germany.

Andrea Caneschi, born in 1958, graduated in Chemistry at the University of Florence in 1983 where he got his PhD in 1989 presenting a thesis on the chemistry of nitroxide radicals. Since 1991, he is a researcher at the same university. He is the author of more than 70 publications in international journals.

Frank Cordes, born 1961, studied physics at the Universität Hamburg and worked for his Diploma thesis with Dr. H. D. Bartunik, at DESY. Since 1990, he has been working on his PhD thesis in the group of Prof. Saenger at the Freie Universität Berlin.

Steffen Denzinger, born 1968, studied chemistry and received his diploma in 1993 at the University of Mainz. Since 1994, he is working on his PhD in the groups of Helmut Ringsdorf at the University of Mainz and of Hans-Werner Schmidt at the University of Bayreuth. His main interest is molecular recognition in ordered systems such as interfaces, with emphasis on hydrogen bonding interaction.

Ekkehard Diemann, born 1944, commenced his chemistry studies at the University of Göttingen. In 1971, under the supervision of A. Müller, he continued his studies at the University of Dortmund with a PhD thesis on preparation and spectroscopical investigations of thio- and selenometallates of the transition metals. He has been employed at the University of Bielefeld, under the Chair of Inorganic Chemistry I, since 1977. To-date, he holds the title of Academic Director and teaches on the subject of Analytical Chemistry.

Achim Dittrich, born 1964, studied chemistry and received his PhD in 1994 at the University of Mainz. After that, he went to the Center for Self-organizing Molecular Systems at Leeds (UK) where he got a postdoctoral fellowship. His main interests are reversible processes in self-organized assemblies, in particular those based on photochromism.

Andreas W. M. Dress, born 1938 in Berlin, studied mathematics, philosophy, and economics in Berlin, Tübingen, and Kiel where he received his doctorate (1962) and his "Habilitation" (1965) in mathematics. After visiting the Institute of Advanced Studies in Princeton from 1967 to 1969, he followed a call to the new University of Bielefeld where he is presently heading the interdisciplinary *Research Center for Studies of Structure*

Formation. For the last ten to fifteen years, he has concentrated mainly on developing new methods in discrete mathematics related to structure formation as studied in the biosciences, chemistry, and crystallography.
Universität Bielefeld, Fakultät für Mathematik, Postfach 100131, 33501 Bielefeld, Germany.

Werner Ebeling, born 1936, studied Physics at Rostock University and at Moscow State University, received his Diploma in Physics, his Dr. rer. nat. and Dr. habil. from Rostock University under Professor Hans Falkenhagen. He has served as Professor at the Universities of Rostock, Riga, Torun, Vera Cruz, Bruxelles and Paris and is now a Professor of Theoretical Physics at the Humboldt University. He is author and coauthor of several monographs on statistical physics and the theory of self-organization.

Dante Gatteschi, born in 1945, graduated at the University of Florence in Chemistry in 1969. At the same university, he became Assistant and from 1980 Professor of General and Inorganic Chemistry. He is the author of more than 250 publications in international journals and of one book. His research interests are in the area of molecular magnetism.
Università di Firenze, Dipartimento di Chimica, Via Moragliano 75/77, 50144 Firenze, Itali.

Joachim Granzin, born 1957, studied crystallography at the Universität Hamburg, where he received his PhD in 1987. After a postdoctoral year at the Institut für Mineralogie und Petrographie (Section physical crystallography), he joined the Institut für Kristallographie at the Freie Universität Berlin and moved to the Institut für Biologische Strukturforschung at the Forschungszentrum Jülich in 1994. Main interests are elucidation of structure and function of biological macromolecules by crystallographic methods.

Hanspeter Herzel, born 1957, graduated at Humboldt University Berlin in Theoretical Physics. Lectures on statistical physics, dynamical systems and nonlinear dynamics in biology at Berlin universities. Since 1993, he is a Heisenberg fellow at the Technical University Berlin and University of Iowa. About 50 publications and a monograph on stochastic processes, nonlinear dynamics, time-series analysis, complexity measures. Current fields of interest: nonlinear dynamics of the voice, modelling physiological rhythms, statistical analysis of biosequences.
Humbold Universität Berlin, Fachbereich Physik, Institut für Theoretische Physik, Invalidenstraße 42, 10115 Berlin, Germany.

Winfried Hinrichs, born 1950, studied chemical engineering and chemistry and received his PhD in 1983 at the Universität Hamburg. After holding a postdoctoral position at the Rijksuniversiteit te Leyden, Netherlands, he joined the Institut für Kristallographie, Freie Universität Berlin. One of his main interests are structure-function relationships of biological macromolecules.

Herbert Hörz, born 1933, studied philosophy and physics and received the degree Dr. Phil in 1960 with theoretical examination in philosophy and physics as well as the Habilitation 1962 at Humboldt-University, Berlin. He was Professor of Philosophical Problems of the

History of Science at the Academy of Sciences of the GDR, Vicepresident of this academy and is presently scientific collaborator at the Berlin-Brandenburgische Academy of Sciences (BBAW). He was full member of the Academy of Sciences of the GDR. He is full member of the European Academy of Science, Arts and Humanities (Paris) and received 1989 the honorary degree Dr. h.c. in philosophy (Erfurt-Mühlhausen). He has worked on fundamental problems of philosophy of science (evolution, law and chance, time, information), on relations between philosophy and natural sciences (heuristic function of philosophy, dialectical materialism and natural sciences, philosophical interpretation of quantum theory and of biological evolution theories) and on history of science. He has published numerous papers in the history and philosophy of science and 19 books. He is editor of materials of Hermann von Helmholtz from the archive at the BBAW.

Christopher Hunter began his research career with Dr J. K. M. Sanders at the University of Cambridge and received a PhD in 1989. He then took up a lectureship in the University of Otago, New Zealand where he stayed for two years. In 1991, he moved to the University of Sheffield where he is currently a lecturer in organic chemistry. He was awarded the 1992 Meldola Medal for the most promising chemist under 30 years of age by the Royal Society of Chemistry and the Society of Maccabeans, and in 1994, he became a Lister Institute Research Fellow.
University of Sheffield, Department of Chemistry, Sheffield S3 7HF, United Kingdom.

L. Jos de Jongh was born in 1942. He studied Physics at the University of Amsterdam, where he graduated in 1968, and received his PhD in 1973 on the thesis "Experiments on simple magnetic model systems" (with Prof. Dr. A. R. Miedema). Afterwards he became a member of the permanent scientific staff of the Kamerlingh Onnes Laboratory of the University of Leiden, since 1986 as a Professor. His scientific interests include: low-dimensional magnetic or electronic systems, phase transitions, superconductivity, strongly correlated electron systems, metal clusters, intermetallic compounds, and nanostructured materials or systems.
Faculty of Mathematics and Natural Sciences, Kamerlingh Onnes Laboratory, PO Box 9506, 2300 RA Leiden, The Netherlands.

Miguel Angel Jiménez-Montaño, born 1950, studied physics at the University of Mexico City. He received his Doctoral degree 1975 from the University of Torun (Poland) for a dissertation on Information-Thermodynamics of Chemical Systems. He is now a Professor of Physics at the Universities of Vera Cruz and Puebla (Mexico).

Caroline Kisker, born 1964, studied biochemistry at the Freie Universität Berlin and received her PhD in 1994 whilst working within the group of Prof. Saenger at the Institut für Kristallographie. At present, she is postdoctoral fellow with Prof. D. C. Rees, Caltech, USA. Her main interests are crystallographic and biochemical studies on the structure and function of medically relevant macromolecules.

Dirk Kostrewa, born 1961, studied biochemistry at the Freie Universität Berlin and received his PhD in 1991 at the Institut für Kristallographie. He joined the Department of Protein crystallography at Hoffmann-La Roche, Switzerland. Since 1994, he has been a permanent member in that department. His major interests are structure and function of enzymes.

Meir Lahav (born in 1936) studied chemistry and received his PhD in 1967 at the Weizmann Institute of Science. After two postdoctoral years at Harvard University, he joined in 1971 the Weizmann Institute of Science where he is full professor since 1982. His main interest is stereochemistry, solid-state chemistry, chirality, with emphasis on crystal engineering, resolution of enantiomers and crystal polymorphism. He received together with L. Leiserowitz the Prelog Medal from the ETH Switzerland.
Weizmann Institute of Science, Department of Materials and Interfaces, Rehovot 76100, Israel.

Leslie Leiserowitz (born in 1934) studied chemical crystallography and received his PhD in 1964 at the Hebrew University. After a postdoctoral year in Heidelberg, he joined in 1969 the Weizmann Institute of Science where he is full professor since 1983. His main interest is structural chemistry, chirality, crystal engineering and packing modes of molecules in two- and three-dimensional crystals, struture of molecular films at air/liquid and solid/liquid interfaces. He received together with M. Lahav the Prelog Medal from the ETH Switzerland.

Niklas Luhmann, born 1927, studied jurisprudence and worked as a jurist from 1954 to 1962. In 1966 he received his PhD and his Habilitation as sociologist at the Universität Münster. Since 1968 until his emeritation in 1993 he has been full professor for sociology at the Universität Bielefeld.
Universität Bielefeld, Fakultät für Soziologie, Postfach 100131, 33501 Bielefeld, Germany.

Klaus Mainzer, born 1947, studied mathematics, physics, and philosophy, and received his PhD degree (1973) and the Habilitation in philosophy (1979) at the University of Münster. Presently, he is Full Professor of Philosophy of Science at the University of Augsburg. He is a member of several national and international interdisciplinary academies and institutions. His scientific research programme includes philosophy and history of mathematical, natural, and social sciences. In particular, he is working on problems of symmetries, symmetry breaking, computer-assisted modeling, and nonlinear complex systems in the natural and social sciences. He has written various books, entitled, for example, *"History of Geometry"* (1980), *"Symmetries in Nature"* (1988, 1995), *"Computer: New Wings of Mind?"* (1994, 1995), *"Time"* (1995), *"Thinking in Complexity: The Complex Dynamics of Matter, Mind, and Mankind"* (1994, 1995). Besides these monographs and numerous original papers, he is editor and coeditor of nine books.

Charles N. Moorefield was born in 1954. he studied chemistry and received both his BS (1982) and MS (1984) at the University of Central Florida. He subsequently moved to Baton Rouge, Louisiana to attend Louisiana State University. During 1986, he transferred to the University of South Florida at Tampa, Florida to continue working with Professor George R. Newkome who relocated from LSU. He received his PhD in 1991 for work related to the preparation, characterization, and potential applications of cascade macromolecules at the Center for Molecular Design and Recognition at the University of South Florida.

Achim Müller, born 1938, studied chemistry and physics and received his PhD degree (1965) and the Habilitation (1967) at the University of Göttingen under the supervision of O. Glemser. He is now Full Professor of Inorganic Chemistry at the University of Bielefeld and a member of several scientific academies including also the Deutsche Akademie für Naturforscher Leopoldina and the Polish Academy of Sciences. He is working on problems of molecular physics (theory of mass influence on molecular constants), vibrational spectroscopy (matrix isolation spectroscopy, band contour analysis, resonance Raman effect, metal isotope effects), bioinorganic chemistry (model compounds and biological nitrogen fixation by free-living microorganisms) and molecular metal oxide and sulfide complexes and clusters (synthesis, geochemical relevance, electronic and molecular structure, topological aspects, their use in heterogeneous catalysis, and their model character for the understanding of self-organization processes) as well as aspects of natural philosophy. In these areas, he has published, besides numerous original papers, more than 30 reviews and edited eight books.
Universität Bielefeld, Fakultät für Chemie, Postfach 100131, 33501 Bielefeld, Germany.

George R. Newkome graduated (BS, 1961; PhD, 1966) from Kent State University, then did a postdoctorate at Princeton University. From 1968 to 1986, he was a faculty member at LSU, where he advanced through the ranks to Professor, and in 1986, he moved to the USF, where he was appointed Vice Provost for Research and Graduate Studies. In 1987, Dr. Newkome was promoted to Vice President for Research and oversees the Divisions of Sponsored Research and Patents and Licensing, as well as in the president of USF Research Foundation. In May 1991, he was honored by the American Chemical Society as the recipient of the 1991 Florida Section Award for the most outstanding chemist of the year. In 1992, he was appointed as a Distinguished Research Professor. Dr. Newkome has published over 250 papers in international journals and numerous books in the areas of organic and inorganic chemistry, architectural design from nature, energy utilization, natural resource conservation, and environmental.
University of South Florida, Center for Molecular Design & Recognition, 4202 E. Fowler Avenue, FAO 126, Tampa, FL 33620, USA.

Luca Pardi, born in 1957, graduated at the University of Florence in 1986. He is now at the Laboratoire Leon Brillouin, Saclay, France, where he is working on polarized neutron single-crystal diffraction. He is the author of about 50 publications in international journals.

Wolfgang Paulus, born in 1962, studied chemistry and received his PhD in 1993 at the University of Mainz. Since 1994, he is working in the Kunststofflaboratorium of BASF in Ludwigshafen. His main interest is synthetic polymer chemistry, with emphasis on self-organizing polymers and polymer networks induced by amphiphilic, mesogenic, or hydrogen bonding interactions. Other interests are polymers with special electro-optical properties such as photoconductivity and electroluminescence.

Michael T. Pope was born in Exeter, England in 1933. He received the D.Phil. in Chemistry from Oxford University (with R. J. P. Williams) in 1957. Following two years of postdoctoral study at Boston University (with L. C. W. Baker), he joined Laporte Chemicals Ltd., Luton, England. He went to Georgetown University as an Assistant Professor in 1962, has been full Professor since 1973, and is currently serving as Department Chairman. His research interests lie in the synthesis, reactivity, and applications of polyoxometallate complexes of the early transition metals. He has been a Petroleum Research Fund International and a Senior U.S. Scientist Awardee of the Humboldt Foundation.
Georgetown University, Department of Chemistry, Washington, DC 20057-0001, USA.

Helmut Ringsdorf, born 1929, studied chemistry in Frankfurt, Darmstadt and Freiburg where he also received his PhD in 1958. After two years as scientific assistant at the Polytechnic Institute of Brooklyn, New York, working with H. F. Mark, he joined the University of Marburg, where he received his habilitation in 1967. From 1967 to 1971, he was professor at the University of Marburg. Since 1971, he has been full professor at the University of Mainz. His main interests are molecular architecture and functionalization of polymeric liquid crystals, synthesis, structure and properties of functionalized supramolecular systems, and attempts to mimic biomembrane processes. He received numerous awards, among them the H. F. Mark Award of the Austrian Chemical Society, the H. Staudinger Award of the German Chemical Society, the A. v. Humboldt Award of the French ministery of Science, the Polymer Award of the Society of Polymer Science, Japan, the ACS Award in Polymer Chemistry, the Order "Chevalier dans l'ordre des Palmes académiques" of the French Republic, and Doctores honoris causa of the University Paris-Sud, France and the Trinity College, University of Dublin, Ireland.
Universität Mainz, Institut für Organische Chemie, Johann-Joachim-Becher-Weg 18-22, 55128 Mainz, Germany.

Natalie M. Rowley has been a Postdoctoral Research Fellow at the University of Birmingham since 1991. She was awarded her BSc in Chemistry in 1988 and her PhD in 1991, both from the University of Birmingham. She was elected to the Membership of the Royal Society of Chemistry in 1994. Dr. Rowley has research interests in the field of the photochemistry of metal-organic compounds, and it was in this area that she gained her PhD. She has also spent one year working in the area of the photochemistry of gas phase molecules. She is currently examining the incorporation of photochemically-active components into supramolecular moieties, with a view to developing compounds which may behave as photochemically-switchable molecular devices.

Wolfram Saenger, born 1939, studied chemistry and received his PhD in 1965 at the Technische Hochschule Darmstadt. After two postdoctoral years at Harvard University, he joined the Max-Planck-Institut für Experimentelle Medizin. In 1971, he received his Habilitation at Universität Göttingen. Since 1981, he has been full professor at the Institut für Kristallographie, Freie Universität Berlin. His main interest is structural biology with emphasis on the crystal structures of proteins involved in DNA binding, of enzymes, and of photosystem I. Other interests are cyclodextrin inclusion complexes and hydrogen bonding interactions. He received the Leibniz and Humbold awards.
FU Berlin, Fachbereich Chemie, Institut für Kristallographie, Takustraße 6, 14195 Berlin, Germany.

Claus Sandmann, born 1962, studied chemistry in Erlangen and biochemistry in Berlin. Since 1991, he has been working on his PhD thesis at the Freie Universität Berlin. One of his major interests is the computer based modeling of protein-ligand complexes.

Günter Schmid studied Chemistry at the University of Munich and took his doctor's degree in 1965 in Inorganic Chemistry, followed by the Habilitation in 1969 at the University of Marburg. From 1971–1977 he held a professorship at the University of Marburg. Since 1977 he has the chair for Inorganic Chemistry at the University of Essen. His main research field is, besides the investigation of boron-nitrogen compounds as complex ligands, the study of the development of the metallic state.
Universität/GHS Essen, Fachbereich 8 - Chemie, Universitätsstraße 5-7, 45117 Essen, Germany.

Armin Schmitt, born 1964, studied physics at the University of Heidelberg and received his Diploma for a Thesis in Theoretical Chemistry. He wrote his Ph. D. thesis about structural analysis of DNA sequences at the Humboldt University Berlin and is now dealing with protein research at the Weizmann-Institute of Science, Rehovot, Israel.

Roberta Sessoli, born in 1963, graduated in Chemistry at the University of Florence in 1987, where she got her PhD in 1992 presenting a thesis on molecular magnetism. She is the author of about 50 publications in international journals.

J. Fraser Stoddart has been Professor of Organic Chemistry at the University of Birmingham since 1990. He was appointed Head of the School of Chemistry there in 1993. Previously, he was a Reader in Chemistry at the University of Sheffield for eight years, where he was also Lecturer in Chemistry from 1970 to 1982. From 1978 to 1981 he was seconded from the University of Sheffield to the ICI Corporate Laboratory in Runcorn. He gained his BSc in 1964, his PhD in 1966, and his DSc in 1980, all from the University of Edinburgh. He was elected to the Fellowship of the Royal Society of London in 1994. He has received many awards, including the International Izatt-Christensen Award in Macrocyclic Chemistry in 1993, and has been a distinguished lecturer in many universities all around the world. Prof. Stoddart has published more than 320 communications, papers, reviews, and monographs and has wide ranging interests in supramolecular science. He is presently developing the transfer of concepts such as self-assembly between the life sciences and materials science. The template-directed synthesis of unnatural products with prescribed functions is being pursued within the context of gaining fundamental understanding about the nature of the non-covalent bond.
University of Birmingham, School of Chemistry, Edgaston, Birmingham, B15 2TT, United Kingdom.

Eörs Szathmáry was born in 1959 in Budapest. He obtained his PhD in ecology in 1987. He has always kept his position as a research fellow of the Hungarian Academy of Sciences in the Ecological Modelling Research Group, Department of Plant Taxonomy and Ecology, Eötvös University, Budapest. In 1987-88 he was on a Soros Foundation scholarship in the School of Biological Sciences at the University of Sussex, Brighton as a post-doctoral fellow working with J. Maynard Smith, FRS. He was a research fellow in the Laboratory of

Mathematical Biology at the MRC National Institute for Medical Research in London in 1991-92. Then he spent a year as invited fellow in the Institute for Advanced Study Berlin (Wissenschaftskolleg zu Berlin). In 1994 he was a guest professor at the Dept. of Zoology, University of Zürich. Currently, he is leader of an international group in theoretical biology at the Colleqium Budapest (Institute for Advanced Study).
Eötvös University, Department of Plant Taxonomy and Ecology, Ludovika ter 2, 1083 Budapest, Hungary.

Fritz Vögtle, born 1939 in Ehingen/Donau, studied chemistry in Freiburg as well as chemistry and medicine in Heidelberg. There he received his PhD for research with H. A. Staab on "Valence Isomerization of Double Schiff Bases". After his habilitation on steric interactions inside cyclic compounds, he was H2/H3 professor in Würzburg from 1969 to 1975. He then accepted a position as full professor and director of the Institut für Organische Chemie und Biochemie in Bonn. He received the literature prize for the book "Supramolecular Chemistry" and the Lise Meitner-Alexander von Humboldt prize. His research interests are in the fields of supramolecular chemistry and molecular recognition, and after working on topics concerning crown ethers, podands, and siderophores, he has concentrated on macrobicyclic compounds with large intramolecular cavities, ligands for supramolecular photochemistry, concave dyes, molecular tweezers, and spherical and tubular and threaded molecules. His further areas of research are strained and helical molecules, cyclophanes, and compounds with appealing architecture.
Universität Bonn, Institut für Organische Chemie und Biochemie, Gerhard-Domagk-Straße 1, 53121 Bonn, Germany.

Index